The End of Bias: A Beginning

The End of Bias

A Beginning

THE SCIENCE AND PRACTICE
OF OVERCOMING
UNCONSCIOUS BIAS

Jessica Nordell

METROPOLITAN BOOKS   HENRY HOLT AND COMPANY   NEW YORK

Metropolitan Books
Henry Holt and Company
Publishers since 1866
120 Broadway
New York, New York 10271
www.henryholt.com

Metropolitan Books® and ▥® are registered trademarks of
Macmillan Publishing Group, LLC.

Portions of chapters 4 and 7 appeared in different form in the *Atlantic* and the *New York Times*.

Every effort has been made to trace copyright holders and to obtain their permission
for the use of copyright material.

"Beaumont to Detroit: 1943" from *The Collected Poems of Langston Hughes* by Langston
Hughes, edited by Arnold Rampersad with David Roessel, associate editor, copyright © 1994 by the
Estate of Langston Hughes. Used by permission of Alfred A. Knopf, an imprint of the Knopf
Doubleday Publishing Group, a division of Penguin Random House LLC. All rights reserved.

Library of Congress Cataloging-in-Publication Data

Names: Nordell, Jessica, author.
Title: The end of bias : a beginning : the science and practice of overcoming unconscious
   bias / Jessica Nordell.
Description: First edition. | New York : Metropolitan Books, Henry Holt and Company,
   [2021] | Includes bibliographical references and index.
Identifiers: LCCN 2021022445 (print) | LCCN 2021022446 (ebook) | ISBN 9781250186188
   (hardcover) | ISBN 9781250186171 (ebook)
Subjects: LCSH: Prejudices. | Discrimination.
Classification: LCC HM1091 .N67 2021 (print) | LCC HM1091 (ebook) |
   DDC 303.3/85—dc23
LC record available at https://lccn.loc.gov/2021022445
LC ebook record available at https://lccn.loc.gov/2021022446

Our books may be purchased in bulk for promotional, educational, or business use. Please
contact your local bookseller or the Macmillan Corporate and Premium Sales Department at
(800) 221-7945, extension 5442, or by e-mail at MacmillanSpecialMarkets@macmillan.com.

First Edition 2021

Designed by Kelly S. Too

Printed in the United States of America

1 3 5 7 9 10 8 6 4 2

"... now waking
making

making
with their

rhythms some-
thing torn

and new"

—Kamau Brathwaite, from *The Arrivants*

CONTENTS

The End of Bias: A Beginning

Introduction

Years after his cancer treatment, Ben Barres recalled how he'd phrased the request to his oncologist. While you're removing my breast, he'd asked, could you please take off the other one? Cancer ran in the family, so the doctor agreed, but the truth is Barres just wanted the breasts gone. Christened with a girl's name and raised as a girl, he'd never been at ease with that identity—not as a four-year-old, feeling he was a boy; not as a teenager, uncomfortable with the changes of puberty; not as an adult, squeezed into heels and a bridesmaid dress. This was 1995. It was before Laverne Cox and Caitlyn Jenner were household names, before a Google search for "transgender" provided legal advice, before Google. Barres didn't understand what being trans was. But the double mastectomy was an enormous relief. A year later, he read an article about a trans man, and the lights came on.[1]

Barres was eager to begin hormone treatment, but he had a major concern: his career. At forty-three, he was working as a neurobiologist at Stanford and had recently made a groundbreaking discovery about the significance of the glia, brain cells whose role had been previously underestimated. Others in the scientific community had always perceived him as a woman. He had no idea how they would respond to this change. Would students stop wanting to join the Barres lab? Would invitations to conferences disappear?[2]

The scientific community did react, but not in the ways Barres had feared. After his transition, people who did not know he was transgender started listening to Ben more carefully. They stopped questioning Ben's authority. Ben, middle-aged, White, and male, was no longer interrupted in meetings. He was, again and again, given the benefit of the doubt. He even received better service while shopping. At one conference, a scientist who didn't know Ben was transgender was overheard saying, "Ben gave a great seminar today—but then his work is so much better than his sister's."[3]

Barres was astonished. Before his transition, he had rarely detected sexism—even overt examples hadn't registered. Once, when Barres was an undergraduate at MIT, and the only person in a math class to solve a hard problem, the professor said, "Your boyfriend must have solved it for you." Barres was offended. He had solved it himself, of course. He didn't even have a boyfriend. But he didn't think of the comment as discriminatory because he thought sexism had ended. And even if it hadn't, he didn't identify enough as a woman to think sexism could apply to him—he was just furious to have been accused of cheating. Pre-transition, Barres assumed he had been treated like everyone else.[4]

Now he had stunning evidence to the contrary. It was almost a scientific experiment: he had the same education, same skills, same achievements, same capacity. All the variables had been held constant except one. Barres saw, with searing clarity, that his daily encounters, his scientific career, his life had all been shaped by the gender others saw, in ways that had been invisible even to him. Before transitioning, his ideas, contributions, and authority had all been devalued—not overtly, generally, but in a way that was noticeable when that devaluing suddenly vanished. Now, the differences in the ways men and women are treated were discernible, the way new patterns appear on flower petals under ultraviolet light.

So when, in 2005, the president of Harvard University, Larry Summers, famously opined that the dearth of women in science might be due to innate differences between their capabilities and those of men,

Barres couldn't stay quiet. He penned a cri de coeur in *Nature* that demanded the scientific community pay attention to bias.[5]

"This is why women are not breaking into academic jobs at any appreciable rate," he said. "Not childcare. Not family responsibilities." After working in science as Ben, he added, "I have had the thought a million times: I am taken more seriously."[6]

It's not that Barres never encountered barriers and bias, he told me of his career before his transition. "It was just that I didn't see it."[7]

Many of us have experiences with others that lead us to wonder whether bias is playing a role. But those of us who have not lived through a dramatic discontinuity in how we appear to the outside world may not have the opportunity to confirm these hunches. We may be able to verify them to ourselves if we lose or gain significant weight or acquire a visible disability. We may see them if we travel to countries where our skin color carries a different meaning, like the Black student who told me about the strange sensation he felt while traveling in Italy, which he realized was the feeling of *not* being followed in stores by suspicious salespeople. People in heterosexual marriages whose spouses undergo a gender transition often come to recognize how much validation they'd previously received for being part of a heterosexual couple.[8] Eventually, many of us will feel the discrimination and disrespect that await the elderly. But often the bias we encounter remains difficult to pinpoint.

While it is challenging to know how much of a role bias is playing in any given interaction, a growing body of studies confirms that there are differences in treatment across nearly every realm of human experience and a dizzying variety of social groups. In the best of these studies, only one marker of identity is changed and all other variables remain constant. Studies have found that if you're a prospective graduate student with a name that sounds Indian, Chinese, Latino, Black, or female, you're less likely to hear back from faculty members than if your name is Brad Anderson. If you're a same-sex couple, you're more likely to be denied a home loan than a heterosexual couple; you may also be charged higher fees. If you're a White job applicant with a criminal record, one study found, you're more likely to get a callback than if you are a Black job applicant with a criminal record—and without one.[9]

The list goes on. If you're Latino or Black, you're less likely to receive opioids for pain than a White patient; if you're Black, this is true even if you've sustained trauma or had surgery. If you are an obese child, your teacher is more likely to doubt your academic ability than if you are slim. If your hobbies and activities suggest you grew up rich, you're more likely to be called back by a law firm than if they imply a poor childhood, unless you're a woman, in which case you'll be seen as less committed than a wealthy man. If you are a Black student, you are more likely to be seen as a troublemaker than a White student behaving the same way. If you are a light-skinned basketball player, announcers will more likely comment on your mind; if you are dark-skinned, your body. If you're a woman, your medical symptoms will be taken less seriously; if you're a woman seeking a job in a lab, you will be seen as less competent and deserving of a lower salary than a man with an identical résumé. Pursuing an academic fellowship, one classic study found, you must be 2.5 times as productive as a man to be rated equally competent.[10]

Across communities of color, bias turns to horror. An analysis of more than six hundred shooting deaths by police found that, compared to White people, Black people posing little to no threat to officers are three times as likely to be killed. On July 17, 2014, a forty-three-year-old former horticulturist on Staten Island named Eric Garner was approached by police officers who suspected him of selling untaxed cigarettes. One of them put him in a choke hold, a maneuver the New York City Police Department prohibits. Garner died an hour later. According to the medical examiner, his death at the hands of an officer was a homicide.[11] While police officers in many cases argue that they acted appropriately, a pattern of disparate use of force bespeaks the fact that Garner, along with Michael Brown in Ferguson, Philando Castile in Falcon Heights, Atatiana Jefferson in Fort Worth, Tamir Rice in Cleveland, and many, many others died because a police officer reacted to these individuals—a father, an unarmed teenager, a Montessori nutrition supervisor, a pre-med major, a twelve-year-old child—differently than they would have had these individuals been White.

At this moment in history, if I'm a woman and you're a man, the words I write (these words, even) will be regarded differently than if you had written the same words. If I'm White and you're Black, we will be treated differently by others for no other reason than that our bodies have a meaning in this culture, and that meaning clings to us like a film that cannot be peeled away.

Of course, some people intend to demean or devalue other people because they belong to a particular group, a fact to which violent White nationalism attests. Some people harbor overt prejudices, and mean to cause harm. Any advantages transgender men enjoy often depend on others not knowing they are trans, and they can disappear in an instant: trans people today face abysmal rates of physical and sexual violence, harassing experiences in health care, and rejection from workplaces, family, and faith communities. Transgender women of color in particular are often subject to a pernicious combination of anti-trans bigotry, misogyny, and racism. Unvarnished cruelty is real. The slow-motion murder of George Floyd by a Minneapolis police officer in the summer of 2020 revealed a casual savagery so dehumanizing and horrific it shook the world.

But most people do not go into their professions with the goal of hurting others or providing disparate treatment. And for those who intend and value fairness, it is still possible to act in discriminatory ways. That contradiction between values of fairness and the reality of real-world discrimination has come to be called "unconscious bias," "implicit bias," or sometimes "unintentional" or "unexamined bias." It describes the behavior of people who want to act one way but in fact act another. How we work to end it is the focus of this book.

GROWING UP IN THE 1980s and '90s, I'd been in many ways protected from understanding or even perceiving bias. Racialized as White in a majority-White town, and so undetectably Jewish that I was invited onstage at a Christmas pageant to share "What Jesus means to me," I moved through the racial landscape as most White people do: like a coddled, swaddled baby, never having to seriously contend with the problem of racism, always able to opt out of its consideration. I was also protected somewhat from gender bias by the structure of academia. If

I aced a calculus test at my small Catholic high school, that was a hard, indisputable fact. It didn't matter that I ditched pep rallies to lounge with the stoners across the street, and it didn't appear to matter that I was a girl. Grades seemed to overshadow the specifics of my body, shielding me from gender-based discrimination. In college, I majored in physics. When at times my serious questions in classes in various fields were rebuffed or ignored, I, like Barres, did not routinely link these dismissals to sexism. I had been internalizing messages about women and about myself since childhood, but bias felt more like a background hum than a siren.

That changed. A handful of years out of college, I was struggling to break into journalism, pitching ideas to editors at national magazines and hearing only stony silence. Discouraged, I decided to try sending a story out under a man's name, conducting an experiment of my own. I created a new email address and pitched the same outlets again, this time as J.D. Within hours, a response showed up in my in-box—the piece was accepted. I had spent months trying to place this very same essay as Jessica. J.D. succeeded within a day.

That essay started my career. As J.D., not only was I more successful, but I also felt freer in my self-expression. I was more direct, less apologetic. I wrote one-line emails without caveats or justifications. I saw up close how bias, and its flip side—advantage—are dynamic and penetrating forces, transforming their recipients from the inside just as they strike from the outside. As others changed how they treated me, I changed, too. But I'm a bad and anxious liar, and managing these dual identities became exhausting. After a few years, I said good-bye to my swaggering alter ego, and I started to write about bias. Along the way, I worked for many organizations, racking up a tidy collection of gendered workplace experiences, from having my ideas credited to others to being told my successes were due to luck.

People often enter into justice-related issues through a door swung open by their own experience. Gender bias cracked the door for me, before I understood its place within a massive, multidimensional phenomenon. It can be tempting to overlook ties between diverse forms of bias because the contexts and levels of severity are so different. As the Barbadian author George Lamming explained at the first International Conference of Negro Writers and Artists in 1956, when a person's own life has been deeply shaped by one kind of oppression, it is easy to lose sight of the connection between "the disaster which threatens to reduce him, and the wider context and condition of which that disaster is but the clearest example."[12] The differences between the expressions and virulence of unconscious bias experienced by people of various religions, races, ethnicities, abilities, sexual orientations, and genders are vast, ranging from lost job opportunities to lethal bodily harm. But in each instance, the brute mechanics are the same. The individual who acts with bias engages with an expectation instead of reality. That expectation is assembled from the artifacts of culture: headlines and history books, myths and statistics, encounters real and imagined, and selective interpretations of reality that confirm prior beliefs. Biased individuals do not see a person. They see a person-shaped daydream.

Over time, I came to see bias as a kind of soul violence, an attack not just on the material conditions of one's life—on one's choices and possibilities—but an assault on one's sense of self. This soul violence was there for all to see in what became known as the Clark Doll Study, which was used as evidence in the 1954 *Brown v. Board of Education* ruling to desegregate schools. In the study, psychologists Mamie and Kenneth Clark had shown Black children dolls that appeared to be Black or White. When asked to point to the nice or pretty doll, most children chose the White doll. When asked which doll "looks bad," they selected the Black doll. Then, asked which doll looked like them, the children again chose the Black doll. Some became so upset, they cried or ran out of the room. Decades later, Kenneth Clark told an interviewer that their findings had been so disturbing that they sat on the data for two years before publishing it.[13]

While there has been progress, Clark added, contemporary racism is more insidious. The racial bias of today, whether stealthy or overt, continues to alter one's inner experience. Repression becomes, as poet Dawn Lundy Martin writes, "so much a part of you that you hardly feel it. . . . Your heart rate increases when you see the police drive by, but you feel relief the second the car turns the corner."[14] On cold currents, bias travels from the outside world into a person's deepest interior.

The more I studied the problem, the more I wondered what could be done about it. Advice has long abounded for people who encounter bias. (Women in the workplace: act less threatening and wear feminine silhouettes! Black men: keep your driver's license visible!) But these commands do not solve the problem, they simply trade responsibility for it. One series of studies in fact found that "Lean In"–type messages lead people to think workplace gender inequality is women's fault—and women's responsibility to solve.[15] These commands are insufficient: there will never be a smile wide enough, a sweater soft enough, a tone unassuming enough, or a license and registration visible enough to outmaneuver another person's misjudgments.

Yet, if those on the receiving end of bias can't stop it, who can? Can anything be done to lessen discrimination itself?

Journalism is usually concerned with discovering and probing problems, not solutions; optimism is left to public relations firms and self-help books. But this problem *has* been probed and proven. I wanted to discover how it can be overcome. In a quest to unearth remedies, I set off in search of people and places that have successfully reduced everyday bias and discrimination on the basis of race, gender, religion, ability, and beyond. I sought out settings as diverse as hospitals, preschools, and police precincts and drew on more than a thousand laboratory, field, and case studies. I conducted hundreds of interviews with researchers, practitioners, and everyday citizens, casting a wide net in terms of geography and approach. I looked for interventions that have transformed not just people's biased thinking but their real behavior, and that have reduced bias not in pristine experimental lab settings but in the messy, imperfect workplaces and schools and cities where we actually live.

As I shadowed trauma surgeons, attended police trainings, and met with social psychologists and neuroscientists, I found a hidden topography of interventions, a patchwork of scrappy, inventive organizations, researchers, and lay people rooting out discrimination through curiosity, creativity, and brute trial and error. Sometimes the approaches worked exactly as intended. Sometimes problem solvers backed into solutions by accident—intending to generally improve a process and inadvertently making it less biased as well.

I ran into obstacles, too. In science, we come to accept that something is true, like the existence of gravity and the efficacy of penicillin, through a preponderance of evidence. Since researchers have not been trying to change unconscious racism, sexism, or other forms of discrimination for very long, many of the interventions I include here have not yet been replicated many times. It's important to view them as promising but not yet absolutely definitive.

Additionally, prejudice research is geared largely toward gender and racial bias, so this book focuses on these categories. There is less research about class and disability-related unconscious bias, and very little about age-related bias. Moreover, gender bias research assumes a male/female binary, and racial bias research in the United States mostly addresses bias against Black Americans. There is less rigorous data about the growing number of people with multiracial identities, or the way identities combine to generate new forms of discrimination.

Gender bias studies, for their part, have focused largely on White women's experiences, and racial bias research on that of Black men, inhibiting a full understanding. Black women, for instance, experience more workplace harassment, more penalties for error, and greater obstacles to promotion than White women or men of any race. They may also endure less backlash than White women for displaying dominant behavior, while Black men are seen more negatively than White men for acting dominant. As a tenured White professor with a masculine appearance who was able to choose whether to disclose his status as trans, Barres gained advantages after his transition that are far from universal among trans men. Black trans men after transition, for instance, are newly subject to the racism specific to Black men, including police harassment. One Black man reported being asked, at his blue-collar job, to now play the suspect in training exercises. He went from being seen as an "obnoxious Black woman" to a "scary Black man." Another was repeatedly told he was "threatening."[16] Biases aren't simply additive; they are unique to their intersection, the way blue and yellow glass overlap to create an entirely new color.

These gaps in our knowledge matter. The dearth of studies on bias faced by Indigenous people, people of Asian origin, and other groups echoes the ways these groups have been blotted from public consciousness more broadly. As psychologist Stephanie Fryberg of the Tulalip Tribes points out, any true understanding of prejudice must take into account not only actions but omissions. The discrimination Native Americans face, for instance, often takes the form of not being considered at all. This, too, is a form of bias. What isn't counted or even perceived remains outside the circle of attention and care. These omissions are even written into the history of prejudice research. I found, more than once, that observations and discoveries made by White social scientists had been articulated in the writings of Black women outside the academy decades prior. "Discoveries" are made by those with access to tools and institutions. All silence, wrote the poet Adrienne Rich, has meaning.[17]

Over the course of writing this book, I ran into my own silences, too—the way my own very specific knowledge determines the nuances I do and do not see, the questions I do and do not know to ask. It's a challenge that mirrors the larger challenges of addressing bias at all: people in the majority, for instance, often see an entirely different reality from those in the minority. Social psychologist Evelyn Carter points out that members of the cultural majority may only see intentional acts of bias, while those in the minority may register unintended discrimination, too. White people might only notice a racist remark, while people of color might be aware of more subtle actions, such as someone scooting away slightly on a bus—behaviors White people may not even be aware they're doing.[18] Bias is woven through culture like a silver thread woven through cloth. In some lights, it's brightly visible; in others, it's hard to distinguish. And your position relative to that flashing thread determines whether you see it at all.

Of course, discrimination is more than the moment-by-moment distortions of individuals: it is also institutional and structural, and the past bleeds into the present—legalized oppression and prejudice against some groups, the compounding advantages of wealth and resources for others. Individual acts of bias are concentrations of a vast, diffuse legacy, like light rays focused through a lens into a single burning point. Any effort to reduce injustice and inequality requires foundational legal and policy solutions, as well. But laws and policies are not supernatural inventions: people support them, write them, pass them, and enforce them. As psychologist Jennifer Eberhardt's lab has shown, the biases in people's minds predict the policies they support—in one study, the "Blacker" a prison population was depicted as, the more punitive the policies White voters accepted. Moreover, laws and policies create guardrails, but they don't dictate what happens within those boundaries. As civil rights lawyer Connie Rice says, laws merely put a limit on how bad discrimination can get.[19] They don't change the more subtle, fleeting human interactions. Laws create a floor; people determine the ceiling.

In the space between floor and ceiling, the interpersonal moments matter. Their cumulative effect endangers individuals and societies. Bias in education can constrict student achievement; bias in medical providers can diminish health outcomes; bias in police officers can be lethal. Taken together, these encounters can drive people out of jobs and careers and undermine the health and safety of families and neighborhoods. In this way, bias not only robs individuals of their futures, it robs fields of talent, companies of ideas, and culture of progress. It robs science of breakthroughs, art and literature of wisdom, and politics of insight. By constricting the makeup of who asks questions, it shapes what questions are asked, compressing the scope of human knowledge. It is a habit that reduces the potential of individuals and undermines the gifts and resources of an entire society.

AFTER HE TRANSITIONED, BARRES FELT angry, really angry—not just about his own treatment, but about all the others who face unnecessary obstacles, like the Black faculty he saw hired by his university only to leave a few years later. "We destroy them. These are the best of the best people, and we just destroy them."[20]

"These young scientists kill themselves for years to develop as a scientist," Barres said. "Just when they're most ready to contribute to society, they're facing barriers. . . . It's insane to put barriers in the way of half of the very best talent." While Barres's whiplash of privilege is not universal, when sociologist Kristen Schilt interviewed trans men about their work lives, many expressed disbelief and anger at the ways men and women are treated differently. "Do you know how smart I am?" said one interviewee about his life post-transition. "I'm right a lot more now." Others reported being asked for their input more frequently and given more support; one transgender man noted that when he opines in a meeting, everyone writes it down. Personality traits that had been viewed negatively before are now seen as positives. "I used to be considered aggressive," said one man. "Now people say, 'I love your take-charge attitude.'"[21]

Transgender women, by contrast, may run into a looking-glass version of what Barres encountered. Joan Roughgarden, a White biologist who transitioned in her early fifties, has said that any challenge she now presents to a mathematical idea is met with the assumption that she doesn't understand it. That never happened before. Likewise, Paula Stone Williams, a pastoral counselor who began her transition in her sixties, was stunned to have her expertise newly doubted. Her confidence wavered. "The more you're treated like you don't know what you're talking about," she says, "the more you begin to question whether you do in fact know what you're talking about."[22]

It can be alarming to face evidence of others' biases. It can also be deeply uncomfortable to see confirmation of one's own. Over the course of writing this book, my own flawed assumptions and reactions became increasingly visible to me, as though they'd been written in invisible ink and were now held over a decoding fire. Like many people, I initially rejected what I saw. When others pointed out paternalistic assumptions I'd made in an article I'd written, I reacted with denial. Then I felt angry. I justified, too: *If I had just been granted that one interview, I wouldn't have had to make assumptions.* Denial, anger, bargaining—the reactions were familiar. If I was grieving anything, perhaps it was the loss of my own innocence. When Elisabeth Kübler-Ross first developed the stages of grief, they were meant to describe the reactions not of the bereaved but of those who learned they were ill. Here, my illness was a cultural pathology so saturating it took me years to recognize. The writer Claudia Rankine distinguishes between *understanding* how contemporary imaginations are polluted by the bigotry of the past and *grasping* it. Before undertaking this project, I may have understood this, but I did not grasp it.[23]

The emotions that accompanied me on the journey mutated over the years from anger to curiosity to deep humility, and finally to hope pierced with urgency. For these habits can change. I saw it, in the people I profile here who revised the way they act toward others, and in the places that transformed their operations in order to be more fair. I saw it in data that measure the degree to which biased behavior can diminish. I saw it in myself, in the way I learned to pause, notice my own reactions, and hold them up to the light. I also witnessed how gaining a deep understanding of bias motivates people to fight against it. Before he died in 2017, Ben Barres worked as a vocal advocate, lobbying the National Institutes of Health and Howard Hughes Medical Institute to create less discriminatory processes for recognizing and funding scientists and pushing academia and science to evolve.

In the field of ecology, there's a notion of an "edge," a place in the landscape where two different ecosystems meet, like the salt marshes where land meets sea or the riparian zone where a stream cuts a hillside. This edge is often the most fertile and generative area in an entire landscape, providing nurseries for fish and stopover points for migrating birds.[24] Where one human meets another is also an edge. It's the place where bias appears, a space thick with potential for harm. But it's also the place where we can interrupt bias and replace it with different ways of seeing, responding, and relating to one another. In the ferment of that edge, something new can grow—insight, respect, a mutuality that has evaded us for too long. The stakes are high, the repercussions are serious, and the problem is solvable. There is so much we can do. This book is one beginning.

PART I

How Bias Works

The Chase

It didn't make sense.

Patricia Devine sat hunched over the desk in her cramped office, staring at a piece of paper. Her elbows were splayed, her chin propped up on the heel of each hand. She was twenty-five years old. On the paper were two graphs. She squinted. *Nope, still nothing.* "This is driving me crazy," she said to her officemate. She'd been sitting in the same position for weeks, trying to make sense of the graphs. She'd blink, stare, trek to the nearby Wendy's for food, and trek back to stare some more. Her life had shrunk to a blur of graphs and chicken sandwiches, with an occasional visit to Buck-i-robics, the official Ohio State University aerobics class. She was starting to feel desperate.

"How could the data be so wrong?" she asked herself. "How could *I* be so wrong?" It was March 1985. She was supposed to defend her dissertation by August, then immediately start her first academic job. But this experiment—one that she'd meticulously designed and carried out, and on which she'd staked her entire dissertation—was falling apart. Worse, it was her first independent project. Her advisor had even tried to steer her away from it. It was too risky, he'd said; the approach required new tools. Besides, the subject was too far outside his area of expertise. But she had persuaded him that it was a good idea. "Maybe he was right," she now thought miserably. "Maybe I'm not cut out for research."[1] In fact, Devine's experiment was about to provide a new window into the way we understand prejudice. It would, shortly, alter the social science landscape.

Devine had set out to test the sincerity of White people who said they opposed racism. At this moment, in the mid-1980s, psychologists were flummoxed by a phenomenon we might call the "prejudice paradox." On the one hand, White Americans overwhelmingly opposed racial prejudice: when asked, they denied holding racist beliefs. On the other, many still acted in racially discriminatory ways, both in lab settings and in the real world. Prominent psychologists of the era, faced with this contradiction, concluded that these people were hiding their true beliefs in order to protect their image. White people who said they weren't racist were lying.[2]

Devine wasn't so sure. This verdict didn't ring true to her—it didn't match her experience of people and her knowledge of the world. What about White people who actively fought against racism? Were they faking it, too? She was White. She knew she sincerely opposed racism. The notion that all these White people were engaging in a mass game of make-believe was hard to accept. There must be something else happening inside their minds.

OUR DATA ABOUT RACIAL ATTITUDES don't go back very far, because the study of racial prejudice is not very old. In the nineteenth and early twentieth centuries, American and European scientists accepted the notion of White superiority prima facie. Researchers in anthropology and medicine—mainly White, Anglo-Saxon men—were in the business of trying to prove racial hierarchies, sometimes resorting to baroque methods like filling human skulls with mercury and pepper seeds to assess relative brain volume. By the turn of the century, psychologists had joined the quest, publishing and promoting manufactured "evidence" of White greatness. A paper in the *Psychology Review* in 1895, for instance, reported that a handful of Black and Native American subjects had faster reflexes than White subjects and took this as "proof" of the former's "primitive constitution."[3] The same paper argued that men had faster reflexes than women because of their greater "brain development." Reconciling these two conclusions was left, apparently, as an exercise for the reader.

Black scholars long denounced this project (Frederick Douglass had, in 1854, neatly summed up the arguments as "partial, superficial, utterly subversive of the happiness of man, and insulting to the wisdom of God"), and Black and White social scientists like W. E. B. DuBois, Franz Boas, and W. I. Thomas forcefully rejected what came to be known as scientific racism. But the financial resources, authority, and imprimatur of science at the time were largely lassoed to the cause of White supremacy: proving that groups of people White scientists deemed "inferior" possessed immutable, inherited differences that placed them lower in a natural hierarchy. In the meantime, the meaning of the invented category "White"—and who exactly this "superior" group included—was constantly changing, expanding and contracting over centuries. (One study concluded that Nordic Europeans were more advanced than Mediterranean Europeans, declaring "the mental superiority of the white race."[4]) Nonetheless, well into the twentieth century, social scientists largely considered what we now think of as prejudice as simply the truth.

Then, in the 1920s and '30s, the psychology community began an about-face. What had been taken as "evidence" was crumbling under scrutiny. Analyses of "intelligence tests" of World War I army conscripts, for instance, showed that Black conscripts from northern states in fact outscored White conscripts from southern states.* In 1930, Carl Brigham, a psychologist who had analyzed the army tests and concluded Whites were superior, publicly retracted his verdict as "without foundation" (though not before it was used to promote immigration restriction and eugenics). Black civil rights efforts in the United States and anti-colonial movements around the world further propelled psychologists to

---

* Questions from the army tests included whether a Percheron was a goat, horse, cow, or sheep, and whether Crisco was a medicine, disinfectant, food, or toothpaste.

begin viewing beliefs about White supremacy as prejudiced and worthy of study. This evolution may also have been hastened by the arrival of ethnic minority immigrants into the profession, including Jewish and Asian newcomers; alarming news from Europe about Hitler's uses of "race science" provided additional fuel. Eventually, even the psychologist who had crowned Nordic Europeans mental monarchs proposed that psychologists were "practically ready" for "a hypothesis of racial equality." The task now shifted to understanding the origins of this irrational, unethical way of thinking.[5]

It was as if astronomers suddenly decided to investigate why so many people believed the moon was made of cheese after spending decades trying to separate its curds and whey. Throughout this radical transformation, as psychologist and historian Franz Samelson wryly notes, the researchers did not question their own "superior rationality."[6]

It wasn't until World War II, however, that the government began collecting information about people's racial attitudes—not out of ethical concern, but because racism threatened the war effort. In Detroit in 1942, the KKK and other White protesters rioted to protest housing built for Black defense workers who had moved north to the factories turning out bullets, ball bearings, and B-24s. The next year, twenty-five thousand White assembly line workers walked off the job to protest laboring next to their Black peers. Detroit's production, as historian Herbert Shapiro notes, was seen as essential to winning the war: now racism was interfering with victory.[7]

Racism caused another problem for the government: it undermined the legitimacy of the fight. Black Americans were being asked to crush the Nazi ideology of racial supremacy on behalf of a country whose racism enforced their own second-class citizenship. As an editorial in the NAACP's *Crisis* proclaimed, "*The Crisis* is sorry for brutality, blood and death among the peoples of Europe. . . . But the hysterical cries of the preachers of democracy for Europe leave us cold. We want democracy in Alabama and Arkansas, in Mississippi and Michigan. . . ." Langston Hughes pointed out the symmetry in his poem "Beaumont to Detroit: 1943":

You tell me that hitler
Is a mighty bad man.
I guess he took lessons
From the ku klux klan.[8]

In fact, the parallels were more than coincidental: Nazi lawyers closely studied American race laws as they institutionalized anti-Semitism. Transcripts of a 1934 meeting meant to work out the details of eliminating "racially foreign elements from the body of the *Volk*" reveal Hitler's minister of justice and others debating the merits of Jim Crow. If only these laws included Jews, said the state secretary in the Ministry of Justice, American jurisprudence *"würde für uns vollkommen passen"* [would suit us perfectly].[9]

As young Black men tore their draft cards in half and tossed them at police in Detroit, alarm bells went off in Washington. Anxious officials in the new Office of War Information commissioned surveys of White and Black people's beliefs about race. This was the first wide-ranging effort to collect such data, and it confirmed that Black Americans "are deeply devoted to American ideals, asking only that these ideals be realized in relation to themselves." It also showed, quantitatively, that the racism enshrined in laws and institutions (segregated schools wouldn't be ruled unconstitutional for another ten years) flourished in the minds of individual White Americans. Most of the thousands surveyed in 1942 and 1944 didn't think Black people should have the same job opportunities. They approved of separate housing and disapproved of interracial marriage. They thought it was best for schools to be segregated.[10]

These surveys, carried out by the National Opinion Research Council, Gallup, and others, continued over the next five decades. By the late 1980s—after desegregation, after civil rights reforms—the numbers had flipped: most White people disapproved of housing discrimination and segregation and responded that Black people should have the same job opportunities as Whites. So few of them supported school segregation that the question was dropped from surveys altogether. As sociologist Lawrence Bobo writes, among White Americans, commitment to legalized discrimination collapsed and was, by the early twenty-first century, replaced at least publicly by "broad support for equal treatment, integration, and a large measure of tolerance."[11]

But, contrary to expressed opinions, racial discrimination in the 1980s had not gone away. It was, in fact, pervasive. To wit: Black renters and would-be homeowners were disproportionately rejected, Black job seekers were less likely to be granted interviews or be hired than equally qualified White people, Black employees were steered to less desirable positions, and Black borrowers were denied loans. These cases found their way to the courts. In 1985, a federal judge determined that the city of Yonkers in New York State had purposely restricted its Black residents to one square mile. In 1993, the American chain restaurant Shoney's settled for almost $135 million for charges of pushing Black workers into low-paying jobs. In 1999, the Department of Agriculture paid more than $1 billion in a settlement for more than a decade of discrimination against thousands of Black farmers applying for loans.[12]

Psychologists found that this gap between word and deed played out on a personal level, too. White individuals denied being prejudiced, but they were observed, unobtrusively, displaying all sorts of discriminatory behavior: in experiments, they acted more hostile to Black people, and, given the opportunity, moved away from them physically. In one set of studies that would not be considered ethical today, White men were provided with fake controls for delivering electric shocks. They were told the study they had joined was examining how punishment affects people's learning. They delivered more aggressive shocks when they were led to believe the recipients of the shocks were Black.[13]

Witnessing this disconnect between White people's responses on questionnaires and their actual behavior, social scientists concluded the rosy surveys couldn't be trusted. People were lying—there was no other way to make sense of the chasm between word and deed. It was all a façade.

Even the studies themselves were suffused with bias. As psychologist Nicole Shelton notes, even in prejudice research, White people have traditionally occupied a higher status: studies like these were designed to learn from the behavior of White subjects, with Black people cast in passive roles and treated as homogenous. When the internal experience of Black people was studied, the focus was often narrowly trained on how they meet with oppression. The same has often been true of other groups that face discrimination. Studies purporting to study prejudice were and continue to be plagued by racist assumptions as well.[14]

DEVINE HAD GROWN UP IN fairly homogeneous communities in New York State. She was the third youngest of eight children in a Catholic family that uprooted and resettled every few years as her father quit or got fired from a series of jobs. Devine kept her head down and made school her job, though she was a terrible test-taker and a worse speller. Once, her mother saw Devine boxing with her older brother, hauled her inside, and made her write, "I am a girl" five hundred times. Devine wrote, five hundred times, "I am a gril."[15]

College was a near disaster. She couldn't find like-minded peers; her philosophy professor told her she'd asked the dumbest question he'd ever heard. She was lost. She had resolved to drop out entirely when a psychology professor named Roy Malpass invited her to assist in his lab—he'd seen something in her, a seriousness. Malpass studied criminal eyewitnesses, and together they staged crimes. Now *this* was interesting. In one experiment, they orchestrated a crime that took place during a college lecture in front of 350 people. The "criminal" (in actuality, a high school wrestler they'd recruited to help) shattered a rack of electronic equipment and screamed obscenities at a professor. Then he ran out the door and jumped into a getaway car. Devine was the driver.[16]

The goal of the experiment was to test whether changing the instructions given to eyewitnesses would alter their responses to a police lineup. At the time, real eyewitnesses were often asked, simply, to choose the suspect from a lineup. But these instructions were biased in that they implied that the suspect was present. Malpass wanted to see whether pointing out that the suspect may or may *not* be in the lineup would change the number of false accusations. When students viewed the lineup of suspects, some were given biased instructions; others were given the unbiased version, stating that the disruptive student might not be present.

After they gathered the data, Devine scribbled it on the chalkboard in their lab. Then Malpass walked in, and Devine watched as his eyes lit up. The numbers revealed that when people were given unbiased instructions, they made fewer errors: they were less likely to mistakenly blame a suspect, but equally likely to correctly identify him.[17] People's perceptions of others didn't always line up with reality, but when prompted to think more carefully, those perceptions could change for the better.

In psychology, Devine discovered, you could make a prediction about human behavior and then set up a piece of theater to test it. And you could learn something new. Not just new to you. New to the world. Devine was hooked. She hustled to graduate school at Ohio State, where she began hunting for a meaty dissertation topic.

At the time, racial prejudice was not widely covered in university psychology courses. Psychology professor James Jones had recently written a pathbreaking book called *Prejudice and Racism*, which traced the way different levels of racism—individual, institutional, and cultural—shape one another, and argued that an institution or culture could be racist through its customs and policies even if its members didn't intend to be.[18] At Ohio State, there were courses on "race relations," but none on the topic of prejudice.

The prejudice paradox bewildered her. The conclusion that all White people were lying to conceal their racist attitudes didn't account for those who were troubled by racism. Devine began requesting reprints of studies from researchers all over the country.

At the same time, she was reading about a new discovery in the psychology research community called "priming": planting a thought in a person's mind in ways that could influence how they then perceived the world. For instance, if you presented someone with words like "care-less," then gave them a story about a whitewater kayaker, they'd be more likely to see the kayaker as reckless. If you primed them with words like "independence" and "self-confidence," they'd see the kayaker as adven-turous. It was as though once one concept entered through the mind's stage door, it would lurk in the wings and nudge others onto center stage.

And priming could affect people's reactions even when it was done subliminally. If you flashed the word "hostile" at someone for mere microseconds, they would judge another person's ambiguous behavior as more hostile, even though they hadn't registered having seen the word. The word would hit the retina, flow through the visual system to the brain, activate the concept of hostile, and then affect people's evaluations—without their awareness.[19]

In addition to nudging people's reactions, priming also seemed to open up a new way of understanding how knowledge was organized inside the mind. When a person was primed with the word "bread," for instance, and then asked to pick out words from a list, they were faster to recognize the word "butter" than the word "chair." This suggested that "bread" and "butter" were closely connected in the mind. Knowl-edge, it seemed, was organized in networks, each concept connected to myriad other concepts, like a web. Tapping one seemed to tap the others in the network as well, the way plucking a single string in a web sets the whole web aflutter.[20]

Taken together, these discoveries suggested that there were now surreptitious ways to investigate a person's mind. As she read about priming, Devine began to imagine it might provide a means of assess-ing White people's true racial attitudes. Perhaps you could prime not just a specific object, like bread, but a social category, like "White" or "Black." If White people were truly racist, she reasoned, the category of "Black" in their minds would be connected to a whole network of rac-ist beliefs and stereotypes. And if you primed them with the category alone, their network of racist notions would cause them to interpret some other scenario in a racist way. Because you could prime people subliminally, their interpretation would be a genuine reflection of their network of beliefs. They wouldn't know their racial attitudes were being tested; they wouldn't have an opportunity to lie.[21]

By contrast, Devine reasoned, if people were truly *not* prejudiced, they would not have a network of racist beliefs connected to the category "Black." There would be no web of stereotypes to pluck, no assumptions to activate. Priming them with the category "Black" would not influence their interpretation of another scenario. Trying to elicit stereotyped responses from an unprejudiced person would be like trying to strike a match against a vacuum. If White people said they were not prejudiced, and priming them with the category "Black" had no effect on their behavior, that would mean they were in fact telling the truth. Priming, she imagined, would be a way to illuminate people's hidden beliefs—a truth serum she could administer.

Devine's advisor said no. Subliminal priming was still so new, he warned. It was risky. There was no way she'd be able to develop the expertise she needed in time. But Devine persisted, and finally he relented.[22]

In the spring of 1985, Devine got to work. She gave 129 White students a questionnaire. Embedded among questions related to politics and gender were questions from the Modern Racism Scale, a tool developed a few years earlier and designed to unearth, indirectly, signs of racial prejudice. Students were asked whether they agreed or disagreed with statements like "It is easy to understand the anger of Black people in America" and "Discrimination against Blacks is no longer a problem in the United States."[23] Based on their answers, she marked them as high or low in prejudice.

Weeks later, she brought students into the lab, telling them they were there to participate in a project related to visual perception. When students came into the lab, she had them put their chin on the chin rest of a device called the tachistoscope, a box that could flash words or images for fractions of a second. They placed their forehead against a strap and looked through goggles, staring at the center of the screen. Devine told them they'd see flashes of light. The "flashes" were in fact words. Some were meant to bring to mind the concept "Black," like "Black" or "afro" or "Harlem." Others, like "something" and "water," were not meant to evoke anything in particular. One group of students saw mostly words relating to Black people; the other group saw mostly neutral words. Because the words appeared for microseconds, the students could tell they'd seen a flash but weren't aware of the specific words they'd seen.[24]

After a short break, the subjects were told there was a second, unrelated experiment about how people form impressions of others. They were asked to read a short story about a person named Donald, whose race was intentionally not specified. As Donald went about his day, he refused to pay rent until his apartment was painted; he also made a purchase and then asked for his money back.

Devine asked them to share their impressions of Donald. How dependable was he? How unfriendly, how boring? How interesting, kind, conceited, or hostile? Prior studies had established hostility as a racist stereotype White Americans held about Black Americans. Devine predicted that for White subjects who were high in prejudice—who had scored higher on the Modern Racism scale—that stereotype would spill onto their view of Donald if they had been heavily primed with words suggesting "Black." They would see him specifically as hostile.

She also predicted, by contrast, that people who had scored low in prejudice would not see Donald as more hostile, even if they had been primed with the same group of words associated with Black people. Those people didn't hold racist beliefs, she reasoned, so priming them with the category "Black" would not activate a network of negative stereotypes. If these people didn't judge Donald as hostile, Devine predicted, it would prove that they had been sincere all along. It would prove that their minds were free of racial prejudice.

BUT THE DATA SHE COLLECTED blew her hypothesis apart.

This is what she expected to see.

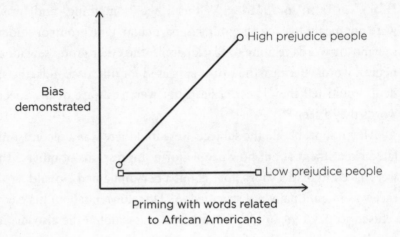

This is what she saw.

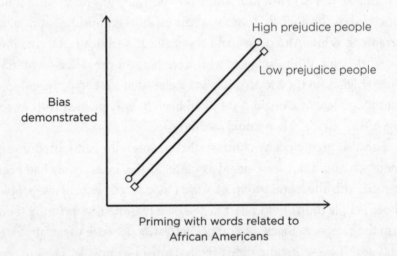

What Devine found—what the bewildering graph in front of her showed—was that the subliminal messages affected *everyone* who was heavily primed with words relating to Black people. Both subjects who held prejudiced beliefs and those who did not judged Donald as hostile. They didn't see him as generally negative—boring or unfriendly. They specifically saw him through a racist lens, as hostile.

It didn't make sense. Why would those who scored low on the prejudice scale show bias? The primes were meant to pluck a web of beliefs people already had. For unprejudiced people, there shouldn't be any racist stereotypes to trigger.

"How could I be so stupid?" Devine asked herself. "Why can't I design an experiment?" Or was it that everyone *was* truly racist, just as other psychologists had concluded?[25]

As DEVINE DESPAIRED, OTHER IDEAS spun in her head. One was from the field of cognitive psychology. Researchers had begun to see the human mind as having two distinct modes of operating: effortful, deliberate thinking and rapid, automatic thinking. The first kind comes into play when we are engaged in something that requires a lot of attention, like riding a bicycle for the first time or following a challenging conversation. The second arises when we are doing something familiar, like riding a bike for the hundredth time or touch-typing a keyboard. When we engage in the same actions or thoughts repeatedly, they become effortless habits of mind. These two modes, it appeared, could operate independently. They could even contradict each other: studies showed that people could act one way automatically and then, upon reflection, decide they'd been wrong and make an effort to correct it.[26] Automatic and deliberate reactions could oppose each other in the same human brain.

Slowly, then all at once, Devine understood the graphs.

"It all came together in my mind," she told me when we met in Madison, Wisconsin, where she now teaches. "I could understand how automatic processes could set people up for failure. I could understand the predicament of egalitarian people. That's when I realized that prejudice could be a habit."[27]

Devine deduced that people could consciously reject prejudice on one hand and behave in biased ways out of habit on the other. They might be aware of decisions they made based on their conscious beliefs but not alert to reactions that were influenced by their deep associations. These were habits, she concluded, and people can engage in habits without thinking, just as someone might chew their fingernails down to the quick before realizing what they were doing.

Devine was far from the first to propose that people's behavior could be influenced by stereotypes that arise surreptitiously from hidden chambers in the mind. Lena Olive Smith, the first Black woman to become a lawyer in the state of Minnesota, presented a precise analysis of the phenomenon nearly a century ago. In 1928, she wrote, "It is common knowledge a feeling can be so dormant and subjected to one's subconsciousness, that one is wholly ignorant of its existence. But if the proper stimulus is applied, it comes to the front, and more often than not one is deceived in believing that it is justice speaking to him; when in fact it is prejudice, blinding him to all justice and fairness." In the years preceding Devine's discovery, other researchers had also proposed that people might have prejudices they hadn't acknowledged or faced, and that embedded stereotypes might influence reactions. Psychologists John Dovidio, Sam Gaertner, and others observed that people might believe themselves to be egalitarian but feel discomfort or have other negative reactions to people of different races.[28]

Devine's argument was that the prejudice paradox could be explained without the requirement that White people were lying: the human mind could contain beliefs people consciously endorse alongside stereotypes or associations that they do not. A belief, she claimed, is something people actively choose, while an association is something that they absorb from their surroundings—cultural knowledge gained without their consent or even awareness. In this view, a belief is like a newsletter you intentionally subscribe to; an association is like spam from a company that somehow obtained your address. You did not choose the spam, and you don't want it, but there it is, clogging your in-box, and you can't seem to get off the mailing list. People who are explicitly prejudiced, on the other hand, do not experience a conflict between chosen beliefs and hidden stereotypes. They subscribed to the spam.

This distinction laid the groundwork for the concept of implicit bias. It was a new way to think about discriminatory conduct: a habitual reaction rooted in deep associations. According to this perspective, beliefs and associations both exist in the mind, and both can steer our reactions to others. When people's words and deeds oppose one another, the contradiction might stem from an internal struggle between the values a person holds dear and the stereotypes they do not. And these stereotypes could be pressed into service without one's bidding, just as a person might drive all the way home from work with no awareness of having turned left or right. Discrimination could be unintentional, even unconscious—stains from our culture, found all over our behavior.

Devine published her findings in 1989. In people who report being unprejudiced, she wrote, the "activation of stereotypes can have automatic effects that if not consciously monitored produce effects that resemble prejudiced responses." Devine's officemate Mahzarin Banaji told Devine that she had revealed "the dark side of the mind." Devine disagreed. Having biased associations didn't mean you were a bad person. It meant you existed in a culture.[29]

The idea of implicit bias suggests that bias functions like a circuit. The circuit begins when we absorb "cultural knowledge" from the world around us, as our families, the media, our classrooms, our neighborhoods shower us with information about different groups of people. Some of this knowledge is true—there are, for instance, statistical differences between men's and women's average height. Some is not true: boys are not on average better at math than girls. Over time, this information becomes deeply embedded as associations and stereotypes. When we encounter someone or something that triggers those associations, our cultural knowledge affects how we react to the situation at hand, including the way we act, what we say, and how we feel. The discriminatory behavior that emerges as a result can contribute to disparities, which further feed the cultural knowledge that sets the whole process in motion. And we do not see a single dimension of identity at a time, but multiple categories, including race, gender, age, and more, each of which carries associations that may be integrated into a perceiver's mind.[30]

This notion of implicit bias as a circuit can help explain a range of encounters. Take a little boy named JJ Powell, for instance. At age four, JJ was bright and gregarious. He could expertly write his name and his younger brother's name. He loved playing school and was usually pretty well behaved. But in the spring of his first year of preschool in his hometown of Omaha, Nebraska, his mother, Tunette Powell, began receiving calls for her to come pick JJ up. He'd been suspended for dribbling spittle on a classmate. Another time for throwing a chair; another time for not listening during nap time. Powell was bewildered. Her shining, upbeat boy? "Wow," she recalled during an interview. "I failed as a parent."[31]

But then she attended a birthday party for a child in JJ's class and began speaking with other parents at the school. One mother shared that her son had hit a boy so hard the child had been sent to the hospital. Her son hadn't been suspended; she'd only received a phone call. More parents told stories of their kids' behavior problems. There had been no other suspensions. In fact, the other parents didn't even know the school used that punishment. JJ had been suspended three times. The only difference, as far as Powell could tell, was that her son was Black and the other children were White.

Powell's son was not alone. One study in Texas looked at literally millions of school and disciplinary records, among them the school records for every student who started seventh grade between 2000 and 2002, all the way through twelfth grade. These included all the disciplinary actions for codes of conduct violations—things like tardiness or inappropriate dress. Responding to these behaviors is discretionary—the school can dispense whatever punishment it sees fit. The study found that Black students were more than twice as likely to be suspended the first time they violated the rules.

Studies confirm that same pattern. In one set conducted by psychologist Phillip Atiba Goff and colleagues, subjects were presented with a story about a boy who had behaved in an antisocial way, ranging from a misdemeanor to a felony. They were then asked questions about the child's responsibility for his actions, and his intentions behind them. Evaluating the same behavior, the subjects saw the boy as more culpable for his actions when he was Black; they also overestimated his age by more than four years, such that a thirteen-and-a-half-year-old was perceived as a legal adult. While the Black child was seen as more culpable for a felony than a misdemeanor, the opposite was true for the White child: as the severity of his behavior increased, the White child was seen as less responsible for it. In another study, researchers showed teachers a school record of a student who misbehaved. When the student was Black, the teachers were more likely to label the student a "troublemaker" and were more likely to see a second infraction as part of a larger pattern of bad behavior.[32]

According to the notion of implicit bias, JJ's teacher absorbed racist stereotypes about Black children. When JJ refused to take a nap, the stereotypes in her mind shaped her interpretation of him and his behavior, leading her to see the behavior as worse than if he had been a White child—and JJ as more deserving of punishment.

In the case of a young computer science student named Philip Guo, biased treatment worked in his favor. Guo had grown up in a Chinese American family of humanities majors. He had tried to teach himself BASIC in sixth grade but had given up because it was too hard. Later, he took one high school computer science class, taught by a teacher who'd learned the material weeks before the class started. It sparked his interest in programming, but when he arrived at college in 2001, he was essentially a beginner, especially in contrast to his classmates, many of whom, by freshman year, had already had ten years of programming experience. He took the entry-level computer science class and started doing summer internships where he noticed something unusual. "It always seemed like during meetings," he recalls, "people assumed that I knew what I was doing."[33] He didn't. When he was silent because he was lost, his coworkers thought he was silent because he understood.

During the school year, Guo cold-emailed professors and obtained a research position. Over the next few years, he landed job after job that required skills he knew he didn't yet have. In these jobs, he had the opportunity to acquire knowledge while getting paid to do so, all the while benefiting from people's presumption of his competence. He recalls, "Other people of my similar sort of technical skill level just did not get the encouragement I did."

Guo's supervisors had likely absorbed a stereotype linking men of his background to technical competence. When they saw him puzzling over a problem, these stereotypes sprang to mind and colored their interpretation of what he said and did, of his questions and even his silence. Guo's experience highlights something important about bias: it not only creates disadvantages in some cases, but generates advantages in others. The same group of people can be subject to stereotypes that help and those that harm, sometimes even in the same stereotype. The "model minority" stereotype associated with Asian Americans, for instance, obscures challenges such as harassment, racism, poverty, violence, and discrimination; it suggests homogeneity where there is none. In school settings, it can obscure students' need for resources and support. And it does not protect people from dehumanization: an analysis of student admissions records at Harvard even alleged that Asian applicants routinely received lower scores on the "personality" part of their evaluations.[34]

For Ben Barres, the biologist, the bias circuit would have functioned differently before and after his gender transition. Before his transition, he had been seen through the lens of stereotypes about women—that women were less scientifically capable. These stereotypes influenced colleagues' perceptions of his work, words, and behavior. They saw him as less authoritative, talented, and valuable and responded by interrupting him, questioning him, and denying his expertise. After his transition, Barres had the opposite experience: he was seen as more competent, more knowledgeable and authoritative, and less appropriate to interrupt. Had Barres been a scientist of another race or ethnicity, or had had a disability, for instance, his experience both before and after his transition would have been different.[35]

A video clip from 2015 revealed the bias circuit in real time when the chairman of one of Silicon Valley's biggest venture funds was interviewed on Bloomberg TV. Asked to explain why there were no women among the partners at his firm, he responded quickly. "We think about that a lot," he said. "I like to think—and genuinely believe—that we are blind to somebody's sex." He went on to cite a recent hire, a young woman from Stanford "who is every bit as good as her peers. And if there are more like her, we'll hire them." But they were "not prepared," he added, "to lower our standards."

The tech community derided the comments as a gaffe, a slip of the tongue. In truth, they were an illumination. The interview showed the bias circuit in action: the chairman hadn't been asked whether the recent hire was "as good as her peers," or whether he'd be willing to lower his standards. He was asked about women. But the mere mention of hiring women instantly triggered a defense of his company's standards, revealing an association between women and lack of ability that was spontaneous and automatic.

As the interview continued, the chairman's responses demonstrated how difficult it can be for a person to perceive bias flowing either from or to them. He attributed the hiring difficulty to the fact that women weren't studying science, and that his firm would be happy to hire women who were "really interested in technology." But he had just explained that he'd gotten started in venture capital knowing nothing about technology, and without any experience in the field. He had been a history major; after working as a journalist, he was hired because the firm's founder "took a risk" on him. No technical track record, no technical background.[36]

In fact, an analysis by tech executive Sukhinder Singh Cassidy of the top venture capitalists showed that while 80 percent of the women on the list had STEM degrees, the same was true of only 61 percent of the men. The chairman asserted that 100 percent of women must have backgrounds that nearly 40 percent of men lack. But the problem, he concluded, lay with the women.[37]

THE NOTION OF IMPLICIT BIAS suggests that discrimination need not necessarily emerge from malice or strongly held prejudices. Some people who act in discriminatory ways are unapologetically racist or sexist. But many others hold egalitarian convictions yet behave in detrimental ways. Psychologists Mahzarin Banaji (Devine's former officemate) and Anthony Greenwald labeled the phenomenon "implicit racism," and debates abounded about the merits of this seemingly exculpatory explanation for bad behavior.

In the meantime, other ideas arose, like social dominance theory, which suggests that biases persist because all societies are organized so that some groups dominate others and keep disproportionate resources to themselves. While individuals vary in the degree to which they prefer hierarchies, inequality between groups is part of a larger social order organized around maintaining power. From this perspective, implicit bias and the stereotypes that underpin it are just some of the many instruments that maintain group-based inequality, a human pattern that is a long-lived and resilient around the globe, as evidenced by the thousands-year-old caste system in India. According to social dominance theory, what matters most is not how people stereotype one another but which group has the ability to dominate others, and how motivated people are to maintain their group's status. Oppression can never be eliminated, only attenuated. The cocreator of this theory, James Sidanius, told Devine she was the "queen of optimism" while he was the "prince of darkness."[38]

In the late 1990s, Anthony Greenwald and colleagues developed the Implicit Association Test (or IAT), a tool that promised to ferret out implicit bias by gauging how strongly social identities are linked with certain associations or stereotypes in a person's mind. In an IAT designed to assess anti-gay bias, you might be presented with a list of words like "smiling" or "rotten" or "homosexual." You are asked, word by word, to decide whether to put the word into the category "gay or bad" or the category "straight or good." (Because "smiling" is good, it fits "straight or good," while "rotten" is bad, and therefore in the category "gay or bad.") Then, you are shown another list and are asked to sort each word again, but this time the categories are "gay or good" and "straight or bad." If you are faster at sorting the words into "gay or bad" than "gay or good," this suggests that the connection in your mind between "gay" and "bad" is stronger than the connection between "gay" and "good," revealing an implicit negative association with homosexuality.[39]

A review of more than 2.5 million such tests revealed that most test-takers (85 percent of whom were from the United States) show a bias in favor of straight people over gay, able-bodied over disabled, and young over old. In many cases, people in stigmatized groups themselves also show an implicit preference for the culturally dominant group. Overweight people exhibit anti-fat bias. White people, Native Americans, Asians, Latinos, and multiracial people all display an implicit bias in favor of White people. Black people are the only racial group that does not express an implicit pro-White bias; some research finds that Black students at historically Black colleges show a pro-Black implicit preference.[40]

These tests also revealed that most people associate men more with work and women more with family, and men more with science and women more with humanities. All racial groups, Black people included, associate Black more than White people with weapons. Research by psychologists Phillip Atiba Goff, Jennifer Eberhardt, and others has found that White subjects implicitly associate Black people with apes. This specific dehumanization, describing people of African origin as not fully human, was explicit in European writings of the eighteenth century, and further accelerated through mainstream medicine and academia throughout the nineteenth century. The fact that this lie persists in the White consciousness centuries after its invention is a testament to how thoroughly and aggressively it was promoted, though some may have trouble facing this reality. In her book *Biased*, Eberhardt recounts how even her scientific colleagues disbelieved these findings as evidence of racist stereotypes, grasping for alternative explanations based on "color-matching."[41]

The Implicit Association Test seemed, at first, to be the holy grail of implicit bias: a laser pointing at the source of biased behavior. But this view, too, has come into question. The test has a number of weaknesses, two of which are particularly problematic. First, the IAT has, in scientific parlance, low "test-retest reliability": the same person might end up with different scores at different times. (If a bathroom scale says you weigh 210 pounds today and 190 tomorrow, you might feel skeptical about the scale.) Second, there's only a modest relationship between a person's IAT score and their actual behavior. A score indicating bias does not necessarily mean a person will act in discriminatory ways toward others—and an unbiased score does not necessarily predict fairness.*[42]

---

* It's also unclear that the IAT measures deep associations alone: it may also measure people's ability to control their responses to the test. It's well known, for instance, that one's level of self-control can diminish with age. Older adults have IAT scores that suggest more bias, but this could also come from the fact that they also have less ability to control their reactions to the test.[43]

But these weaknesses of the IAT may in fact point to a more complex, nuanced way to understand implicit associations. Researchers have posited that having different IAT scores at different times might indicate that the associations themselves are not stable quantities but wobbly, malleable connections that are subject to a person's state of mind. People in one experiment, for instance, revealed positive associations with rich foods when they were prompted to focus on taste and negative associations when they were prompted to focus on health. Associations may also vary according to context.[44] If I see a hulking man holding a knife, I'm going to have one kind of reaction if he is in a dark alley, another if he's onstage, another if I'm on an operating table. In this view, we wouldn't expect implicit associations to be static and re-testable.

And while they do not perfectly predict whether individual people will discriminate, neither do explicit beliefs. There is, in fact, no single mental construct that precisely dictates how a person will act, because people are also guided by social norms, personal goals, others' expectations, and more. Implicit associations as measured by tests like the IAT may be most usefully seen as a portrait of a culture. They do reveal social trends: the same preferences for certain social groups over others reappear across millions of tests. Looked at as a group, these associations may uncover the extent to which people in a culture have been exposed to specific knowledge; they show the contours of a society's stereotypes. They also reveal how cultures can change over time. A recent review of millions of test results found that implicit bias about race and sexual orientation has decreased markedly over a recent decade, while negative associations about the elderly and people of heavier body weight persist.[45]

But implicit bias is not the only plausible explanation for the prejudice paradox. Another perspective holds that beliefs and associations are *not* separate things. According to this view, a person's true belief about, for instance, a marginalized group may be buried but become apparent given the right conditions. A psychologist named Russ Fazio proposed that this true belief will emerge or stay hidden depending on the extent to which a person possesses the motivation and opportunity to reveal or conceal it. If someone shows on an implicit test that they link women with incompetence, this means they have not had the motivation or opportunity to conceal their actual attitude.[46] Indeed, "implicit" stereotypes influence behavior especially when people are tired or stressed, under time pressure, or otherwise mentally taxed, while more "explicit" beliefs dominate when people have the motivation or mental resources to think carefully about their actions.

Does this mean people are lying? Not necessarily. It's possible that many of us have simply not fully investigated our beliefs, especially if those beliefs clash with our values. Indeed, Devine's original study relied on a survey to divide people into "low" or "high" prejudice. But it's possible that people answered those survey questions while still harboring unexamined racist beliefs, and that it was those latent beliefs that were later triggered and appeared on the graphs. The prejudice paradox may not be a sign that people are lying, but simply that they haven't fully scrutinized the interior of their own minds. In this case, the conflict may not be between people's true, egalitarian beliefs on one hand and their habitual associations on the other, but between people's unexamined beliefs and their moral values.

THE IDEA OF IMPLICIT BIAS suggests a sharp distinction between the prejudiced and the unprejudiced, but the distinction is perhaps not so clearcut. Even the idea that there are two distinct processes in the mind—an automatic one and a deliberate one—is still debated; some see the idea of two processes as overly simplistic. There are, psychologists propose, *many* processes that unfold in a person's mind between any stimulus and any response—between, say, seeing a pair of words and pressing a button, or seeing a female candidate's résumé and making a judgment about her ability to do the job. Our behavior is likely governed by processes that are automatic, deliberate, and combinations of the two.[47]

People's behavior may also be shaped by the person with whom they are interacting. Indeed, psychologist Nicole Shelton has argued that examining prejudice and discrimination in individuals is itself flawed and limited because bias happens dynamically, between people. People don't just project stereotypes onto passive others; both parties respond to each other's actions; misunderstandings and different perceptions can shift behavior in real time. Each person in an interaction exerts pressure on the other's behavior.[48]

Some researchers have begun avoiding the term "implicit bias" altogether. Instead, they refer to bias "measured implicitly," distinguishing the tool, not the attitude. Devine prefers to call it "unintentional bias." A more straightforward term for bias that opposes a person's values is simply "unexamined bias." Practically speaking, the difference between this and other more overt kinds of prejudice is the large gap between what is consciously intended by one person and what is experienced by another.

The exact sequence of mental events that causes well-intentioned people to engage in bias is still a roiling question. The truth is that for those of us who hold the fundamental equality of all people as a value, our behavior toward people of different genders, races and ethnicities, religions, ages, abilities, sexual orientations, and beyond, may stem from an unknowable combination of associations we do not endorse and beliefs we have not fully examined. What is important is the fact that acting in prejudicial ways conflicts with our values. When, faced with our actions, we feel unease or guilt, that may be our conscience speaking, creating the crucial entry point for changing our own biased behavior.[49] We'll explore the many routes this can take in detail in the coming chapters. We can start by exploring the inner workings of the human mind.

Inside the Biased Brain

If you happened to be scrolling through Facebook in April 2015, you might have seen an ad for the movie *Straight Outta Compton*, a biopic about the iconic West Coast music group N.W.A and an ode to pioneers of gangster rap. If you clicked on the ad, you would have seen documentary footage of real-life members of N.W.A, Dr. Dre and Ice Cube, driving through their old neighborhood. You'd have heard Dre talk about N.W.A as a form of nonviolent protest and about the group's efforts to inspire the next generation: "We put it all in our music— our frustration, our anger." You'd have seen the men hugging young people from the neighborhood. "I'm so appreciative," says a man in a barbershop. "I'm just trying to keep the flame lit." Then, when this mini-doc cut to film footage, you would have seen musicians at work and a White manager speaking about raw talent. You'd have seen these words flash across the screen: "In the most dangerous place in America, their voices changed the world."[1]

At least, that's what you would have seen if you were Black.

If you were White, you'd have been shown something different: within the first twenty seconds, a Black woman brandishing a shotgun.

Lights, and sirens, and young Black men drinking dark liquor in a bar. Police handcuffing Ice Cube and then slamming him onto a car. Eazy-E asking, "Where the money at?" to which someone in an undershirt replies, "Why you gotta be so ruthless?" Then: Eazy-E pulling a shotgun from a duffel bag, a row of Black men facedown on the pavement, the White manager demanding that his clients be freed.[2]

Straight Outta Compton was the first time Universal Studios used Facebook's "ethnic affinity" labels to target market a film. Instead of creating one trailer for everyone, Universal created race-specific trailers. What you saw depended on how Facebook identified the color of your skin. And as author and journalist Annalee Newitz points out, "They look like they're advertising two completely different films."[3]

The film grossed $200 million at the box office, and Universal's EVP of marketing, Doug Neil, said this race-targeted advertising was partly responsible for the film's triumph. Its success, Neil said, was a surprise. Then he corrected himself. "I shouldn't say a surprise. A breakout hit."[4]

WHY DID IT WORK? AND why did White audiences thrill to this kaleidoscope of Black stereotypes? One possibility is that holding and confirming stereotypes make people feel good. Holding them provides an illusion of certainty in uncertain situations; finding evidence that they are right is also affirming. Like listening to music, or fitting a jigsaw puzzle piece into place, having one's stereotypes confirmed may even be physiologically pleasing.[5]

What these activities have in common is that they involve a sort of prediction. When we listen to music, our minds predict each note that comes next, and we derive pleasure from hearing the pattern we've been expecting. Solving a jigsaw puzzle, too, is a prediction: we guess a piece's location and delight when it snaps into place. In each case, the outcome is uncertain, and research shows that correctly predicting an uncertain outcome feels, in our brains, like pleasure.

Stereotyping, too, is an act of predicting an uncertain outcome—say, seeing a woman and assuming she is bad at math. If she does indeed turn out to be bad at math, then the prediction was correct. Even correctly predicting something negative can feel good, like the smug satisfaction we feel when a habitually tardy friend shows up an hour late. "Aha!" we think. "Just as I predicted." It's irritating, but strangely gratifying: we *knew* what would happen, and we were right. It's as if our brain is constantly running a movie of what we expect will happen milliseconds before it actually happens, and then comparing that movie to reality. Our brains light up when our predicted reality and actual reality match. Our brains love to be right.

We also don't like to be wrong, and we feel irked and threatened when our stereotyped predictions don't come true. A series of studies by psychologist Wendy Berry Mendes and others revealed this phenomenon. Mendes asked White and Asian college students to interact with Latino students who had been hired as actors by the researchers. Some of the Latino students portrayed themselves as socioeconomically "high status," with lawyer fathers, philanthropist or professor mothers, and summers spent volunteering or traipsing around Europe. Others portrayed themselves as "low status," with unemployed fathers and summer waitressing jobs. Participants were then asked to solve puzzles together. The researchers found that when participants interacted with the Latino students who appeared to come from wealth and thus defied American stereotypes, they responded physiologically as if to a threat: their blood vessels constricted and their heart activity changed. In these interactions, participants also saw the students who violated stereotypes as less likable. Similar responses were seen when participants encountered others who defied expectations, like Asian American women who spoke with strong U.S. southern accents—hired actors who were given vocal training.[6]

Being disliked for violating a stereotype is a particularly common experience for women. When women behave in ways that differ from female stereotypes, for instance by not being warm and helpful, they are seen as unpleasant and unlikable. In this way, stereotypes that are *descriptive* can easily become *prescriptive*. Even a so-called positive stereotype can have negative consequences because failing to adhere to it becomes grounds for condemnation. The phenomenon known as backlash, it turns out, may have a neuroscientific explanation: it's an angry protest from the brain's reward system.

In the case of the "White" N.W.A trailer, images of Black men were embellished with guns, drugs, and sirens, reflecting ubiquitous stereotypes of Black male criminality. There was even, in the form of the manager, a White savior, a figure long present in the American imagination. For the White audiences driven into the theaters, the stereotypes on display in the trailer fulfilled expectations and registered in the brain as pleasing.

The reward of having one's expected stereotypes confirmed is enhanced by another feature of prediction. When predictions are only intermittently correct, the act of predicting becomes difficult to stop. This is why refreshing one's in-box or glancing at one's phone is so hard to quit: they offer what's called an intermittent reward. Stereotyping is in the same category, psychologist Will Cox suggests. Sometimes our predictions are right and sometimes they're wrong. This inconsistency makes stereotyping others a textbook intermittent reward cycle. At a neural level, our tendency to stereotype may be a kind of dependency. As media scholar Travis Dixon writes, "Thinking in a stereotyped way becomes almost addictive."[7]

But where do these expectations come from in the first place? How do we become people who find stereotypes neurally rewarding? The best way to understand the process is to watch it unfold in real time, to track people who do not harbor stereotypes as they transform into people who do. To understand how we change from being unbiased to biased, in other words, we have to begin with kids.

ONE SPRING DAY IN 2010, Rebecca Bigler walked through the main door at a private elementary school in Austin, Texas, to check on the status of an experiment. At the time, Bigler was a developmental psychologist at the University of Texas, Austin, where she'd been conducting research on prejudice for nearly two decades. Her aim was to understand how bias begins, and she frequently collaborated with schools to do her research. At this particular school the children were young—three and four. And her experiment had gone haywire.

That day, Bigler remembers, the teacher she'd been working with rushed up to her in the hallway. "We are canceling the experiment," the teacher said. "You've made monsters of my children."[8]

Bigler's experiment had been an attempt to artificially increase gender bias in the classroom. The project was part of her larger effort to pinpoint the conditions that cause prejudice to thrive and those that cause it to wither. To do this, she would alter facets of children's classrooms and then watch what happened to their attitudes and behavior. As Bigler systematically introduced changes into children's environments, she acted much like a botanist who varies soil composition, light levels, and nutrients to observe their effects on vegetation.

Schools are a useful setting to study prejudice because bias starts early. Children as young as three or four can show gender bias; by age six or seven, girls' belief in female brilliance drops, and they gravitate away from games that they are told require being "really, really smart." Discrimination on the basis of skin color arrives around age five or six, though recent research suggests that White children may form biases at the intersection of race and gender as early as four, reacting more negatively to Black boys than to Black girls, White girls, or White boys.[9]

Crucially, children do not come into the world with any of these prejudices. But they are raring to arrange what they see into categories. And they're talented at it. By infancy, they've figured out enough about the category "dog" to connect a furry creature in a photo with a cartoon illustration. And they can recognize that even though dogs and cats might both be four-legged and furry, with fuzzy paws and twitching tails, they belong to different groups.

It's good they have this particular genius. Categorizing—turning raw sensory data into meaningful information by grouping things that belong together—allows humans to perceive the world, make predictions about it, and survive as a species. If a lion appears on the veldt, you have to know to run, but first you have to correctly identify it as a lion and not your grandmother. Categories are essential.

They're also reinforced constantly, something Bigler noticed when visiting schools. Elementary school teachers labeled children throughout the day: "Girls line up," "Sit boy-girl-boy-girl," "Good morning, boys and girls." At Bigler's own 1982 high school graduation, in St. Cloud,

Minnesota, the school administrators wanted to put all the young women in white robes and the young men in blue. "They wanted to mark gender," Bigler told me, "for absolutely no reason."[10]

This ubiquitous labeling, she believed, might be having serious consequences. The idea had been first proposed by pioneering psychologist Sandra Bem, and Bigler wanted to explore it. In the early 1990s, Bigler persuaded parents and teachers at a midwestern summer school to participate in an experiment. Teachers in one set of classrooms were instructed to make the boy and girl categories the main method of organizing the children. They put all the girls' desks on one side of the room and all the boys' desks on the other. They gave boys and girls different colored name tags. They told boys to sit down first, or the girls to line up first. When the children made self-portraits, they displayed the boys' portraits on one bulletin board and the girls' on another. Teachers didn't favor one sex over the other. They simply emphasized the categories to which each child belonged. In another set of classrooms, by contrast, teachers were instructed to address children by name.

At the start of the experiment, Bigler tested all the children to see how much gender stereotyping they did—specifically, to what extent they thought certain occupations, like housecleaning or plumbing, were appropriate for "only men" or "only women." The two groups stereotyped men and women equally at the beginning of the study. She then tracked how this changed after the daily labeling. After four weeks, she found, the gender-labeled children as a group now did far more gender stereotyping than the control group. They described more jobs as being appropriate for men or women alone; they also were much more likely to describe a majority of girls as gentle, neat, and prone to crying, and a majority of boys as adventurous and math-loving and athletic.

Crucially, the teachers had been instructed not to give the children any extra information, either fact or fiction, about what girls or boys are like or what jobs they have. They had only made children pay attention to gender, relentlessly. Simply insisting that children notice gender seemed to inflame any nascent stereotypes they held.[11]

Over the next two decades, Bigler continued to explore the idea that emphasizing categories is a precursor to bias. She realized, for instance, that the studies she had conducted had a shortcoming, which was that the children had arrived in the classroom having already been exposed to gender stereotypes. They were, in scientific terms, contaminated samples. While the experiments showed that stereotyping could be dialed up or dialed down, they didn't reveal much about its emergence. To catch the birth of a stereotype, Bigler had to first create categories of people that didn't exist before. So, armed with boxes of child-size shirts, she began a series of studies in which she created bias from scratch.

In one test, she outfitted all the children in a summer program in either yellow or blue T-shirts she called "work shirts." The kids would don the same shirt at the start of each day, and Bigler and her assistants would wash the shirts every night. In one set of classrooms, she asked the teachers never to mention the shirt colors. In another, they were told to organize the children based on their shirt colors: the blues should line up first, or the yellows should get their art supplies. The blues' desks were clustered together, and so were the yellows'. The blues' and yellows' projects were posted on different bulletin boards. In the mornings, teachers noted children's color with greetings like, "Good morning, blues and yellows!"

Over time, Bigler found, the children did in fact begin to develop beliefs about blues and yellows, but only in rooms where the teachers emphasized shirt color: yellows thought yellows were smarter, while blues thought blues were smarter. In the rooms where teachers ignored shirt color, the children ignored it, too. In psychology terms, Bigler had created "in-groups." An in-group is the group to which a person feels they belong, and people tend to favor people in their in-groups. But these color in-groups, and the favoritism that followed, only emerged if the children lived in a world where the color categories mattered.[12]

In the real world, of course, groups are never neutral: status is everywhere. Some groups have more prestigious jobs or bigger houses, and children readily absorb that information, too. Some have more power. As James Baldwin wrote, "The world has innumerable ways of making this difference known and felt and feared."[13] In a later study, Bigler replicated the blue and yellow T-shirt setup, but she made it more realistic: this time she gave one group higher status. She decorated all the classrooms with images of successful yellows: spelling bee winners, student leaders, and athletic champions all wore yellow shirts. The experiment followed the usual pattern. In some of the rooms, the teachers emphasized shirt colors when they spoke with children; in others, teachers never mentioned yellows and blues, although the evidence of yellows' superiority remained on display.

In classrooms where teachers reinforced the importance of shirt color, the children not only developed stereotypes, but the yellows became enormously certain of their own group's greatness. Seeing yellow superiority reflected all around them, while being told that shirt color mattered, the high-status group developed a juvenile form of prejudice. Children in the other classrooms, who had been surrounded by the same information but in which shirt color was not mentioned, did not. Thus, the children only developed stereotypes if they were taught to focus on group distinctions. Ambient information about different groups needed an armature in order to cohere into prejudice, and teachers' insistence that these groups were important offered the required structure.[14]

What lays the foundation for prejudice, it seems, is not the perceptible differences between people, but how much the culture tells us these differences matter. By directing a spotlight on gender, Bigler was able to ratchet up gender stereotyping. By playing up shirt colors, she was able to conjure stereotyping that had never before existed. By then situating children in a culture where yellow shirts had higher status, she was able to create yellow shirt supremacy. Crucially, stereotypes still emerged when there were no real innate differences between the groups in the classrooms at all: the T-shirt assignments had been random.

For ethical reasons, Bigler was unable to replicate a crucial element of the real world. In everyday life, children learn not only that some groups are wealthy and powerful but that others are more frequently poor or powerless or incarcerated. Bigler didn't plaster the walls with posters showing low status blues. But, of course, children absorb all that information, too.[15]

As for the private school experiment, which was the first time she had worked with preschoolers, it had succeeded, and dangerously so:

the teachers who had sorted children by sex now had classrooms of children who were mean and unmanageable. They divided themselves along gender lines and refused to play with the other sex. The investigation was called off, Bigler recalls, and she hustled into the classroom the next day to try to reverse its effects and bring the children back to their less-tampered-with state of grace. The study was supposed to have lasted for weeks. It was shut down after only three days.

But why would simply insisting that children pay attention to who's a girl and who's a boy, or who's a yellow and who's a blue, lead them to stereotype? It seems to have something to do with what the brain does with its categories. The act of categorizing, it turns out, makes way for a bevy of phenomena in the brain that lead directly to discrimination.[16]

When we see beings as belonging to a particular group, for instance, we start to believe there's something fundamental and biological that unites all the creatures in that group, that there's some invisible essence that makes a dog a dog or a cat a cat. We do the same thing with humans: if we are told that a category is important, we infer that the people in that category share a fundamental essence. We essentialize them. And the more a category is emphasized, the more we think its members have a unifying thread. One form of emphasis is segregation. In Northern Ireland, for instance, children who attend religiously separate schools are quicker to believe that Catholics and Protestants are fundamentally different from one another than are children at integrated schools.[17]

From essentializing it's a straight shot to stereotyping. If all members of a group share fundamental similarities, it's easy to make assumptions about them based only on their membership in that group and to let those assumptions guide our behavior. This sequence—categorize, essentialize, stereotype—has been observed in studies all over the world.[18]

We can see the effects of essentializing in the medical field's current tendency to emphasize the biological basis of mental health challenges. One analysis of twenty-five studies found that the more people believe mental illness has a neurobiological basis—that there are essential differences between those with and without mental health challenges—the more dangerous they find those who suffer and the more they want to avoid them. They also see those with mental health difficulties as less likely to recover. Emphasizing the idea that mental illness is a disease with a physiological origin, like diabetes, has been presented as a way to combat discrimination. In fact, it makes stigma worse.[19]

Essentializing is one trick of the categorizing brain, but the brain performs other stunts with categories, too. Given two categories, we overestimate the differences between them. We underestimate the variation among members of each group, imagining it as monolithic. We also tend to see our own group as beautifully diverse and people outside it as homogeneous. The technical term for this is "outgroup homogeneity," and it helps explain, for example, the different ways American media cover violence. When White, Christian people commit hate crimes, they are largely portrayed in mainstream media as disturbed individuals whose actions issue from their own particular psychological state. Crime committed by a Muslim person is likely to be ascribed not to individual pathology but to group identity: violent acts by people from marginalized groups are seen as a reflection of that group to which they belong. The term "Black on Black crime" exists, while the term "White on White crime" does not, even though more than 80 percent of White murder victims in the United States are killed by other White people. In the cultural imagination, the White perpetrator and the White victim are not seen as part of a meaningful group. White people are, as philosopher George Yancy puts it, "simply human."[20]

THE SIMPLE ACT OF TEACHING children how to categorize others is something we do constantly, unthinkingly. Every time children see one group of people occupying one sort of job or living in one part of town, it teaches them that these categories matter. Every time we separate by gender, race, ethnicity, religion, age—whether by language, attention, the location of people in space—we are trumpeting the importance of these different categories and pointing to the fundamental essence that underlies them. Every time teachers say, "Good morning, boys and girls," children learn that the distinction between the two is important.

This instruction generally happens imperceptibly, but in my own life I can trace one memorable instance of developing a category from scratch. On a high school foreign exchange trip to France, I stayed with a family in the Lorraine region, near the German border. I learned, from them, how to be French: how to eat cheese after dinner, drink tiny cups of coffee, and go home from school for lunch. During the week, I followed my host student from class to class at her lycée. One day when I was on my own during a break, a group of French students I didn't know motioned for me to join them. They were curious about me and my impressions of France, about the music I liked and the movies I watched. We stretched out in the grassy courtyard under a cloud of cigarette smoke and laughed, each venturing words in the other's language. When the break ended, I waved good-bye to my new friends and headed back to class.

Later, one of the host students sidled up to let me know that I had been hanging out with Arabs. The kids in the courtyard and their families, it turned out, had come to France with the wave of immigration from Morocco and Algeria. Unaware of the French political or social context, I had noticed no difference between these French-speaking students and any others. Where I was from, a largely White, Catholic town in northeastern Wisconsin, the category "Arab" carried no immediate significance at the time, and thus I held no stereotypes, harbored no prejudice, and, crucially, didn't even notice anything about my new friends that might have singled them out. Any characteristics that might have indicated their membership in a group known as "Arab" escaped my attention.

It may be that I did not even see these characteristics. Research suggests that our vision itself is partly a product of our culture: the categories and associations we learn affect how we process visual information. Studies by psychologist Amy Krosch and others showed, for instance, that when White Americans feel threatened, they perceive Black people as darker in skin tone. They are also quicker to categorize mixed-race faces as "Black." Likewise, when White Americans feel threatened, those who associate Arabs with danger perceive their faces to be more angry. Psychologist Jennifer Eberhardt has found that subliminally priming the category "Black" also changes how people see. In her studies, when people were subliminally shown images of Black faces, they were faster to visually discern a gun from a low-resolution image. Likewise, when people were subliminally exposed to images suggesting crime (say, a picture of a gun or handcuffs), their eyes focused more on Black faces than White faces.[21]

As a teenage American in a foreign country, once I learned that Arab was a category, new details came into focus. I saw hair and skin color, and I noticed that these students tended to cluster together. As my vision became tuned to the categories I learned, I stepped through the sequence that a child goes through. I was taught a category and told it was important, and then I began to pay attention to attributes I had not previously known to register.

BUT WHILE THE URGE TO categorize is universal, the perimeters of our categories are not. How any community defines group boundaries and who belongs within them is changeable, and wholly specific to its time and place. A two-gender system, for instance, is nowhere near universal. Historically, myriad Native American cultures have had gender systems that included men, women, and what have come to be called "two-spirit" people, a distinct third gender. These individuals could take on roles that diverged from those dictated by their sex. "Two-spirit" men might become skilled weavers, while "two-spirit" women could be warriors. In the Bugis culture in South Sulawesi, in Indonesia, there have historically been five genders: biological males who occupy masculine roles, biological females who occupy feminine roles, biological males who occupy feminine roles, biological females who occupy masculine roles, and a fifth gender, the *bissu*, whose sex is considered immaterial and who are thought to transcend gender. They act as priests, shamans, and healers. Indeed, in the Indigenous cosmologies of Southeast Asia, in which God is the unification of male and female, two-spirit people, embodying both elements, were seen as more God-like. Even in the contemporary American context, the gender binary has in many places become less rigid.[22]

The boundaries we draw around racial and ethnic categories, too, are malleable. While today the term "ethnicity" may refer to people with a common ancestry, archaeological research suggests that in ancient Egypt, for instance, no matter where you came from, if you spoke, prayed, dressed, and acted like an Egyptian, you could be considered "ethnically Egyptian." As an "ethnic Egyptian," you could enjoy the advantages of Egyptian society. This was true whether you were a lighter-skinned Canaanite from what is now Lebanon or a darker-skinned Nubian from the region that is now Sudan.[23]

Ethnic distinctions have long been recognized, but the concept of racial groups—the idea that people can be broadly classified into types with inherited, biological differences both physical and mental—is a recent invention and can be traced to the enslavement of Africans in the sixteenth and seventeenth centuries. Earlier slaves across Europe were often Georgian, Armenian, and Circassian, the word "slave" originating from *sclavus*, or "of the Slavic people." With the Ottoman conquest of Constantinople in 1453, this regional source of slaves was blocked, and slave trading was largely redirected to sub-Saharan Africa. The idea that Black people were a "race" arose as a result. Historian Ibram X. Kendi traces the notion that Black people constitute a unified group to a Portuguese writer named Gomes de Zurara, whose 1453 account of the African slave trade chronicled a slave auction in Portugal that included individuals of many ethnicities, including those "white enough, fair to look upon" and those "black as Ethiops." Together, they were "that miserable race." Indeed, slavery created anti-Black racism directly, writes historian David Brion Davis, in that "negative stereotypes that had been applied to slaves and serfs since antiquity, regardless of ethnicity, were ultimately transferred to black slaves and then to most people of African descent after bondage became almost exclusively confined to blacks."[24]

European thinkers pursued the idea of races with gusto, devising and revising various arrangements. In 1684, French physician François Bernier proposed a classification system consisting of four groups: Indigenous Americans, people from North Africa, parts of India and Asia, and all of Europe "except a part of Muscovy"; those from sub-Saharan Africa; those from other parts of Asia and the Middle East, including those "who live along the Euphrates towards Aleppo"; and finally Laplanders from Finland, whose faces were "much partaking of the bear." Meanwhile, in the American colonies, the idea of a "White race" was taking shape as a consequence of racialized slavery. Writes anthropologist Daniel Segal, "the rendering of 'Africans' as a singular race-class meant that settlers from Europe . . . became a singular race." Toward the end of the 1600s, the Europeans scattered across the American colonies were commonly described as "White."[25]

The use of the term "Caucasian" to mean a human group was introduced in 1785 by German philosopher Christoph Meiners, who divided humanity into two types: Caucasians and Mongolians. Caucasians were either Celtic or Slavic. (Meiners thought the former more spiritual and virtuous.) European Jews, to Meiners, were not Caucasian but an "Asian" people. Ten years later, German physician J. F. Blumenbach introduced his system of five races: Caucasian, Mongolian, Ethiopian, Malay, and American. He based these groupings on his extensive collection of human skulls and agreed with Meiners about the Jews, professing that their eyes "breathe of the east."[26]

The boundaries of the "White race" have shifted wildly over centuries to meet the needs of a changing social order. The institution of racialized slavery meant that all Europeans were joined together as "White." After 1880, when a cascade of immigration from southern and eastern Europe engendered hostility and fear of "undesirable" White ethnics, the "White race" was subdivided. An 1897 book published by the U.S. Bureau of Immigration, the *List of Races and Peoples*, catalogs forty-six "races or peoples," among them Romanians, Poles, South Italians, and North Italians. After immigration restrictions shredded the number of visas available to southern and eastern Europeans including Italians and Jews, European ethnicities again merged into a single "White race."[27]

BEYOND LEARNING TO IDENTIFY CATEGORIES our culture deems important, of course, children also learn what each category means. What does it mean to be a boy or a girl, a race, a sexual orientation? Children are motivated to find meaning in those designations, says Bigler. "Children hear 'Girls, line up' and 'Hello, girls,' and they think, 'I better figure out what a girl is.'"[28]

And the meanings we associate with particular categories are just as culturally specific as the boundaries of those groups. While history is often conceptualized as a linear trajectory toward ever-greater enlightenment and equality, its geometry is closer to a spiral. The idea that women have steadily made inroads in economic activity, entering the working world in the twentieth century, is ahistorical; the domestic feminine ideal is an elite White eighteenth- and nineteenth-century invention, and women have always had essential economic roles. Women milled grain in the Neolithic era and did investigative reporting in the 1800s. Written records from ancient Mesopotamia show that women were running textile businesses in 1900 BC; they operated linen and wool workshops and negotiated the sale of their fabrics with their husbands and sons, who traveled to other cities to sell them. In one letter, an entrepreneur demanded of her son, "Why haven't you sent me the profits from my textiles?"[29]

Likewise, tests of implicit associations today show that men are associated with leadership and women with being supporters. But in other cultures at other times, femininity was associated with authority and power. Indeed, one of the oldest images of worship we have is the Warka Vase of 3000 BC, which shows a procession of men offering gifts to a female deity. The ancient Sumerian cities of Uruk and Babylon were thought to be protected by the deity Inanna, the goddess of sexuality, fertility, and war. Elsewhere in the world, women held positions of authority to a far greater extent than they do today. In the Haudenosaunee (or Iroquois) Confederacy, a coalition founded as early as 1142 that included the Onondaga, Mohawk, Seneca, Oneida, Cayuga, and later Tuscarora Nations, women participated in all major decision-making and commanded essential leadership positions. Women had the power to veto any act of war, selected the chiefs, and could disqualify or remove a chief from office. They also meted out justice; to men who committed sexual assault, they might prescribe banishment, scarring, or capital punishment.[30]

In fact, Haudenosaunee women's roles perplexed early European Americans, who were irritated that women attended treaty signings and other political meetings. As Oneida chief Conoghquieson explained in 1762, "It was always Custom for [them] to be present on

Such Occasions, being of Much Estimation Among Us." At a meeting with the Haudenosaunee that same year, Sir William Johnson, the British superintendent of Indian Affairs, expressed his bafflement:[31]

> Brethren,
> When I Called you to this Meeting I really could not Discover any
> Necessity there was for the presence of Women. . . . And altho' I
> am Obliged to your Women for their Zeal and Desire to promote
> good work, And I know it is their Custom to Come down on Such
> Occasions, I could heartily wish that no more persons would Attend
> any meeting that were necessary for the Discharge of the business on
> Which they were Summoned.[32]

This parity may be why many European women captured by the Native Americans opted to stay with their captors even after they had the opportunity to return.[33]

JUST AS THE ATTRIBUTES ASSOCIATED with gender are culturally contingent, so is the meaning of skin color, and the idea of White supremacy is a modern fiction. There is no evidence, for instance, that people of Nubian origin in ancient Egypt faced discrimination based on skin color. In fact, they rose to the highest ranks throughout the Egyptian military and political establishment. One tomb excavation revealed a Nubian man named Maiherpri (which meant "Lion on the Battle Field") buried alongside the most powerful Egyptian rulers of the day, surrounded by enormous riches, including a quiver of stone-chipped arrows and gilded leather collars for his dogs. Archaeological evidence of mingled Nubian and Egyptian-style cookware suggest that Nubians and Egyptians married one another.*[34]

Skin color carried different values and meanings in antiquity,

---

* This is not to say group-based animus or desire for domination did not exist in ancient Egypt. The soles of King Tut's shoes were studded with pictures of Libyans and Nubians so he could perform the act of stomping his adversaries. But the antipathy was political, not racial.

some of which would be unrecognizable to us today. The Greek physician Hippocrates and his disciples promoted the idea that a person's physical and mental makeup are determined by their humors (phlegm, blood, and yellow and black bile), which are in turn determined by the climate. According to a Hippocratic text designed to guide traveling doctors, the cold, wet climates of northern Europe blanched skin and produced moisture, which in turn made people dim-witted. The hot sun in Egypt and Ethiopia, by contrast, darkened the skin and dried out the humors, making Africans highly intelligent. This worldview can be seen in some Roman scientific manuals. "The southern nations," wrote Roman author, architect, and military engineer Vitruvius, "are quick in understanding and sagacious in council." Lamented the Roman author Vegetius, "We were always inferior to the Africans in wealth and unequal to them in deception and stratagem."[35]

In these ancient minds, the hot, dry African climate conferred shortcomings as well as benefits. If desiccated humors gave Ethiopians intelligence, it was also thought to make them cowardly because they had little blood to spare in battle. In contrast, northerners, flush with fluids, were brave and bellicose. Unfortunately, bravery didn't make up for doltishness. Because northern people charged their enemies unthinkingly, wrote Vitruvius, "their attacks are repulsed and their designs frustrated." Wrote Roman poet and orator Florus, they had "the spirit of wild beasts." If pale northern Europeans were brawny but thick, the "dumb jocks" of the Roman world, Africans were the nerds.[36]

In an unabashed display of in-group favoritism, the Romans saw in themselves a perfect blend of traits. Wrote Vitruvius, "The people of Italy excel in both qualities, strength of body and vigorous mind. . . . Italy enjoys a temperate and unequalled climate between the north on one side and the south on the other. Hence it is, that by stratagem she is able to repress the attacks of the barbarians, and by her strength to overcome the subtilty of the southern nations." This moderate climate, wrote Pliny, supported "gentle customs, clear thoughts, and temperaments open and capable of understanding all of nature."[37] In other words, Romans possessed a charming balance: enough cold-fueled brute strength to repel the more clever Africans, and enough brain-enhancing southern sun to out-smart the pale thugs from the north.*

The relationship between skin color and beauty is also ever chang-ing. The idea that lighter skin is more beautiful than darker skin is now ubiquitous around the world. Skin lightening creams like Fair and Lovely are sold from Mali to China to the United States. The prod-ucts, containing toxic ingredients including mercury and hydroqui-none, are a multibillion-dollar industry. (Amid backlash, Fair and Lovely changed its named to "Glow and Lovely" in 2020.)

This global obsession with the desirability of Whiteness matches contemporary racial hierarchies, of course, but it can also be traced at least in part to an archaeological mistake. Enlightenment thinkers, marveling over Greek and Roman marble statuary, concluded that the ancients' aesthetic preference was for skin that was milky white. Since their catalog of statuary was pale, the thinking went, Romans must have venerated pale skin. But Roman statuary wasn't white at all: new technical analysis, including ultraviolet photography, has revealed that the iconic snowy marble statues of our museums and imagina-tions were in fact vividly colored and riotously patterned. Romans didn't think Whiteness was beautiful—to them, northern Europeans looked bleached. When stressed, wrote Florus, they sweated "as snow is melted by the sun."[38]

ALTHOUGH ALL THESE HUMAN CLASSIFICATIONS—THEIR boundaries, their significance—are taught, the means by which we learn them is rarely noticed or documented. When Toni Morrison scoured the lit-erary canon, she did find the practice portrayed in fiction, in Flannery O'Connor's story "The Artificial N*****," set in Georgia in the 1950s. In the story, a White man and his grandson, Nelson, board a train. A portly, well-dressed man passes them, and the grandfather asks the boy

---

* Not every ancient agreed. Roman citizen Apuleius insisted climate was not des-tiny: "Was not the philosopher Anacharsis born among the idiotic Scythians? Was not Meletides the idiot born among the clever Athenians?"

to describe him. A man? No, says the grandfather. A fat man? Nelson offers. An old man? he tries again. "That was a n*****," the grandfather says. The lesson is complete. "Nelson must be educated," historian Nell Irvin Painter writes. "Nelson must undergo the process of unseeing a well-dressed man and reseeing a 'n*****.'"[39]

For children, these lessons rarely happen so directly. Instead, children observe the world in front of their eyes, seeing that some groups are rich and others are poor, some occupy jobs with high status and others low, some are jailed and some are free. They learn from their parents—even well-intentioned parents—who may not understand the impact of their actions, whose biases leak out through an averted gaze or a move to a bus seat far from someone in another group.

Of course, children also learn through the media, that simulation and recapitulation of the world that we now carry with us in our pockets. We can see how this process unfolds in the curious case of one small logging village in British Columbia. Until the early 1970s, the village was without television. Others in the area had television, but this one was TV-free as a fluke of geography—the community was tucked into a place in the Rocky Mountains where surrounding peaks blocked all the signals. Because the nearest decent reception was fifty miles away, the villagers just didn't watch TV.

In 1973, the townspeople heard they would finally get reception. Tannis Macbeth, a sociologist at the University of British Columbia at the time, thought its arrival would offer perhaps the first opportunity to witness the effect of TV. It would be, in scientific terms, a natural experiment, because the events were happening in the real world, without meddling from researchers. Macbeth scrambled to get to the village before the installation happened. She and her colleagues dubbed it "Notel"—"No Television"—to preserve residents' privacy and gave the residents a battery of tests and questionnaires, gathering data on everything from how they entertained themselves, to how violent they were, to which jobs and behaviors they found appropriate for men and women. How suitable, they asked children, is it for boys to do dishes, play rough sports, show off? For girls? Should boys or girls become doctors, librarians, or the prime minister of Canada? The researchers also canvassed neighboring communities in order to compare Notel to nearby populations who had been exposed to TV for up to fifteen years.

The surveys conducted before television came to Notel showed that its children generally did less gender stereotyping than those in neighboring villages—they were less likely to link roles or behaviors exclusively to boys or girls. But two years after Notel received television, Macbeth returned and found that much had changed. There were fewer sporting events than there had been, and elderly people participated less in public life than they had before. Children had become more aggressive. And their levels of stereotyping had risen: it was now at roughly the same level as that of children who had had television for years. Watching the depiction of the sexes on TV had narrowed the children's vision of which jobs were for men or women, and which activities were appropriate for boys or girls—being prime minister was only for men, they were more likely to report. Doing dishes was only for girls. No major changes had occurred in Notel other than the introduction of TV.[40]

CHILDREN LEARN AND ABSORB THE boundaries around each social category, and how the members of each category should be; like sand poured into buckets, the child's culture rushes in with information. And then children become adults. With adulthood come new, various, and sophisticated inputs: the wider world, friends and neighbors, newspapers and radio, television news and social media.* The categories erected in childhood fill to the brim with more and more cultural debris.† Every time a person sees scientists as men, for instance, that association is strengthened. As associations strengthen, they're more

---

* Social media may reinforce stereotypes even further: one analysis of hundreds of selfies on Instagram found them to be more gender stereotyped than magazine advertisements.
† While children's stereotypes and beliefs and prejudices are often more rigid and crude than those of the adults they become, there's evidence that the biases formed in childhood linger as traces, even if just below the surface. For instance, work by psychologist Laurie Rudman and others has shown that people who were nurtured as children by overweight mothers later hold more positive implicit associations about overweight people than those nurtured by slim mothers, even if they outwardly prefer thinness.[41]

likely to pop to mind. The more one associates "scientist" with "male," the more the word "scientist" will spontaneously conjure up a man.

Some stereotypes, of course, have a basis in reality. Dutch people, for instance, are stereotyped as tall, and they are, on average, tall. Corporate leaders are assumed to be men, and this is true of most CEOs. Economists call discriminating on the basis of perceived real-world disparities "statistical discrimination." The problems with this kind of discrimination are many: first, judging unique individuals on the basis of average group differences creates errors, like telling a woman who is six foot two that she cannot play basketball because women are on average only five foot four. Second, we tend not to stereotype different groups based on which features appear most frequently but rather on what features are perceived as most distinctive about that group, even if they are not very common at all. People stereotype Republicans in general as wealthy, for instance, because wealth is more present among Republicans than Democrats, but the median household income in Republican districts is less than the median in Democratic districts. If a feature is more common in one group than another, it becomes a stereotype and is blanketed over every member of the group. Further, statistical disparities often have complex causes, but these are frequently overlooked in favor of simple essentialist arguments.[42]

Critically, cultural knowledge people receive about groups often does not reflect reality. While some politicians routinely denigrate Mexican immigrants as criminals, a study by the Cato Institute that looked at convictions and arrests in Texas in 2017 found that undocumented immigrants were convicted of 47 percent less crime than Americans born in the United States. Legal immigrants were convicted 65 percent less than native-born Americans. El Paso, Texas, is a working-class city across the border from Juarez, Mexico, one of the most dangerous cities in Mexico. About 80 percent of El Paso is Hispanic, the majority of Mexican origin. As criminologist Aaron Chalfin points out, if Mexican immigrants bring crime, El Paso should be riven by it. However, El Paso has been until recently one of the safest cities in the country, with a homicide rate of 2.4 per 100,000 residents—comparable to safe global cities like London. That safety was riven in 2019 when a White man shot and killed twenty-three people in an anti-Latino attack.* Another analysis found disproportionate news coverage for terrorist acts committed by a Muslim. The researchers found that in the years from 2006 to 2015, controlling for the number of people who died in the attack, an act by a Muslim received 357 percent more news coverage than an act by a non-Muslim.[43]

These representations cement our associations, a fact made vivid to me one snowy day several winters ago. I had bundled up books and writing materials and hiked over to the university library near my house. I was hoping for sweet silence, and I was in luck. It was finals week, so while the giant reading room I entered was packed with students, it was hushed except for the tapping of keyboards. I slid into one of the few empty chairs and got to work. About an hour later, I heard rustling and looked up to see a young man walking down the aisle toward a crowded end of the room. Another man followed him. They unrolled small prayer rugs and knelt down. As I watched, I felt the palms of my hands grow sweaty. My heart rate quickened. My breath became shallow. To my own bewilderment, I displayed all the signs of a fear response because I was watching two Muslim men pray.

Up until that moment, I hadn't realized that there was a link in my mind between Islam and fear. The truth is I'm more likely to be killed by an armed toddler than by a Muslim terrorist. But in the years leading up to that winter day, I—like every news watcher and reader in our society—had seen Muslims linked to terrorism.

Studies have also shown that Whites and people of color involved in the criminal justice system are represented differently. One analysis of Los Angeles television news found White individuals overrepresented as police officers compared to their actual numbers, while Latinos were underrepresented as police officers; the same analysis found White individuals overrepresented as crime victims, while Latinos were underrepresented, compared to crime reports. Studies have found that Black

---

* Another study tracked crime in Arizona following legislation that led to a decline in the foreign-born population. Property crime decreased by 20 percent, but this was driven by a disproportionate exodus of young foreign-born men, who appeared to commit crimes at no higher rate than young native-born men.[44]

individuals have been historically overrepresented as perpetrators of crime. They have also been represented in more threatening ways, with news stories about Black men accused of violent crime more likely to show them being bodily restrained than White men accused of violent crime. Over time, associations strengthen: Whiteness and order, Blackness and danger and the need for restraint. It's not surprising then that people who regularly watch local TV news that overrepresents Black people as criminals are more likely to perceive African Americans as violent and believe that Black suspects are guilty.[45]

Women, as a group, are still represented in media differently than men. In 2019, women made up only 37 percent of major characters in Hollywood's biggest films; in 1942, they made up 33 percent. An analysis of two thousand Hollywood screenplays found that even in movies with female protagonists, like Disney's princess films, men speak more dialogue than women. When women are depicted, they are often shown as supporters rather than protagonists. Likewise, in video games, female characters usually aid others, often as assistants to a male lead. In this post-Weinstein era, it has become increasingly clear that the images of women issuing from Hollywood for the last several decades were, in large part, even quite literally controlled by someone who viewed them as disposable nonhumans. The effect of this sensibility on contemporary notions of womanhood is incalculable.[46]

These associations between "women" and "support" are now also embedded in the digital infrastructure of our lives. The voice-activated AIs we increasingly rely on to play music or send messages for us are by default presented as female. There's Siri (Norse for "beautiful woman who leads you to victory"), Amazon's Alexa, and Microsoft's Cortana, named after a voluptuous video game character who wears a holographic body stocking. We ask Siri, Alexa, and Cortana for help and they offer it cheerfully, seamlessly. Some analysts attribute these decisions to research that people prefer women's voices. Nevertheless, every time a female AI attends to the administrative tasks of our complex lives, associations between womanhood and servility are reinforced.[47]

Our culture pummels us with these images. Associations seep into our mental trenches, settling like silt. And once these associations are there, they're very hard to eradicate. Stereotypes don't just lounge in our minds; our brains put them to work at every opportunity.

One way they're deployed is as self-reinforcement. While the concept of confirmation bias suggests that we overvalue evidence that supports our beliefs, we often receive no evidence at all when we stereotype. If a person assumes a woman is less competent and therefore declines to assign her a challenging assignment, that person receives no feedback about her capacity. If a person assumes a Black man is dangerous and therefore crosses the street to avoid him, that person receives no information about whether their assumption was right or wrong. If a doctor's stereotypes about women or people of color cause them to take symptoms less seriously, and a patient seeks care elsewhere, the stereotype is not contradicted. It may thus be employed again and again.

ANOTHER REASON STEREOTYPES MAY BE so hard to eradicate is that they are functional: they help us make sense of the world. On a basic level, we need to believe that the world around us is rational, and we look for ways to understand our environment and explain to ourselves why things are as they are. If stereotypes are true, then the world is organized in a way that is just and fair. If a country bans Muslims, surely they are terrorists. If a nation is building a wall, undocumented immigrants must be driving up crime. Stereotypes persist in part because they are culturally useful: they legitimize the status quo.

Take women in leadership. A common argument for women's underrepresentation in positions of power is that there's something about them that makes them less adapted to these roles. These "inherent" reasons include the ideas that women are averse to risk and competition and that they're more "neurotic," helpful, and agreeable. The implication, of course, is that there's also something inherent in men that makes them more suited to leadership.[48]

These claims about essential female characteristics fall apart upon examination. Let's look at "agreeableness," which psychologists consider one of the "big five" personality traits. Being agreeable includes being compliant, altruistic, and modest. But each of these is also highly incentivized for women, who are routinely punished for *not* being agreeable. A study by psychologist Madeline Heilman found that when women opted not to help coworkers, they were seen less favorably, while men who didn't help suffered no penalty. Another study that looked at the real performance reviews of hundreds of tech professionals found that only 2.4 percent of men's negative evaluations included critiques of their personalities. By contrast, 76 percent of women's contained criticism about their personalities, admonitions that they should hang back and let others shine more, or use a milder tone when speaking, or act less judgmental. In short, they should be more agreeable.* Their jobs depended on it.[49]

Insisting that women are naturally more agreeable is like offering a rat a spoonful of cod liver oil and a spoonful of peanut butter, delivering electric shocks every time the rat touches the peanut butter, and then concluding that rats have a natural taste for cod. Testing women for "innate qualities" merely measures behaviors that have been conditioned and reinforced as the only socially acceptable choices. But such stereotypes provide an efficient way to explain the predominance of male leadership: women are just too naturally meek and agreeable to do what it takes to get ahead.

Another argument put forth for women's absence from powerful roles is that women are risk averse. Some evolutionary psychologists, for instance, argue that women evolved to be more fearful, while men evolved to take risks, competing with other men for mates. But research suggesting that women take fewer risks often focuses on a few domains like gambling and investing. A study by psychologists Selwyn Becker and Alice Eagly found that in some contexts women assume *more* risk than men: women, for instance, were more likely than men to risk their lives to shelter Jews during the Holocaust. They are also more likely to donate a kidney, an act that poses significant risks to one's own health.[50]

Moreover, studies of risk typically do not account for the risks that

---

* Being disagreeable doesn't seem to pose any similar problem for men. In fact, a set of four studies found that disagreeable men earn an income premium.[51]

women routinely take, like having a child, which can imperil health and financial security, or engaging in intimate partnerships, which poses the risk of violence. Globally, 30 percent of women who have been in a relationship have experienced violence by a partner. Having a child also decreases salaries for women, so for women, having children also risks future earnings. Furthermore, if one faces more difficulty gaining traction in a career, change such as taking a new role may be more risky because it may be harder to recover from failure.

The truth may be that women are not more risk averse but that all people have a "risk budget," and women's lives include added risk that is not accurately accounted for in many studies of risk. Women's apparent risk aversion may simply be their accurate risk assessment.[52]

Another stereotype ubiquitous in American culture is that of the "model minority." Indeed, though Asian Americans are 5.9 percent of the U.S. population, they make up a fifth of students at Ivy League universities. Amy Chua and Jed Rubenfeld's *The Triple Package* claims that qualities inherent to Chinese culture, like "impulse control," explain why Chinese American children succeed academically.[53]

But sociologists Jennifer Lee and Min Zhou have traced Asian American academic success to other causes. In 1965, when the Immigration and Nationality Act opened up migration from Asia, it encouraged the arrival of skilled workers like scientists and doctors. This fostered a new profile of Chinese immigrants; today, more than 50 percent of Chinese immigrants have a college education, compared to only 4 percent of the population in China and 28 percent of native-born Americans overall. In China, entrance into top universities is based on a single test, and there's copious extracurricular tutoring; students sometimes study seven hours a day. Skilled Chinese families brought this emphasis on tutoring to the United States, setting up classes in community centers and churches and offering SAT prep for seventh graders—four years before students typically take the test. While in American culture, test prep is generally geared toward affluent students, this tutoring is often free and accessible to working-class Asian students who, Lee and Zhou argue, absorb academic habits and values.[54]

Further, because immigrants from many diverse countries of origin are racialized as "Asian," the children from these countries benefit not only from the practical academic help but from the image of Asian accomplishment. Some Chinese and Vietnamese students Lee and Zhou interviewed even described having been put in Advanced Placement courses despite not having met the required criteria. What are frequently thought of in America as "essential" parts of Asian culture, Lee and Zhou argue, were actually introduced structurally. Consequently, these patterns are not necessarily present in other countries. Children of Chinese immigrants in Spain, for instance, leave school at high rates and have the lowest educational expectations of all second-generation immigrant groups.*

When we see these patterns—women failing to ascend to power, or Asian students succeeding in school—it is far easier to attribute them to inherent characteristics. Psychologists Andrei Cimpian and Erika Salomon call this tendency "the inherence heuristic." It's less mentally demanding to believe that groups occupy the positions they do—in a company, in a society, in the world—because there's something intrinsic to these groups that explains it. It is far more difficult to see social patterns as contingent, the result of human intervention, or to trace the accumulation of subtle forces that cause these patterns' emergence.[55]

BECAUSE WE WANT TO BELIEVE that inherent properties explain the patterns of the world, stereotypes can also be deployed, strategically, to maintain the status quo. In the antebellum period in the United States, for instance, stereotypes served as rationalizations for slavery. As historian George Fredrickson writes, "If slaves could be regarded as contented and naturally servile, their enslavement could be more easily justified." Before the Civil War, enslaved people were often portrayed as docile, lazy, and

---

* In the United States, second-generation Chinese immigrants graduate college at about the same rate as their parents did—60 percent. Second-generation Mexican immigrants, on the other hand, graduate from college at double the rate of their fathers and more than triple the rate of their mothers—from 5 percent to over 17 percent.

even happy—representations that helped validate for religious Christians a practice that was religiously, ethically, and morally indefensible.[56]

A set of Civil War–era cartoons titled "Quashee's Dream of Emancipation" depicts this "laziness." In one panel, an enslaved Black man lounges with his feet up, reading a newspaper; in another, the same man appears in a crisp tuxedo, dreaming of "the light and easy employments at the North." These stereotypes provided ballast for the institution of slavery, which depended on the belief in and propagation of such myths. After all, if the enslaved person was lazy, then forced labor was necessary. If the enslaved person was content, then a brutal practice was acceptable. If the enslaved person was childlike, then a paternalistic system was justified.[57]

Abolitionists used stereotypes to support their cause, too. The symbol of their movement showed a half-naked enslaved person, kneeling in chains, pleading with an unseen White savior—creating yet another stereotype: the slave who cannot help himself. Black Americans consciously overrode these stereotypes when they represented themselves. Frederick Douglass, the most photographed man of the nineteenth century, intentionally projected an image of strength and dignity that directly opposed the stereotypes used by both slaveholders and abolitionists.[58]

After the Civil War, as Black citizens began to occupy public office and their economic and political autonomy and power became threatening to the White population, alternative stereotypes of Black Americans gained prominence. Suddenly, those who had been seen as "naturally lazy" were dangerous and required policing. As historian David Levering Lewis points out, the contrast between this image and the antebellum image of the "reliable, faithful" slave was stark. "Now, all of a sudden, the African American becomes lascivious, beastly, demonic, a threat." Indeed, as Fredrickson writes, "The key to understanding the larger history of white supremacist imagery in the U.S., both during slavery and afterwards, is this sharp and recurring contrast between 'the good Negro' *in his place* and the vicious black *out of it*." The apparent menace was made vivid in the 1915 film *Birth of a Nation*, in which a Black man (played by a White actor in blackface) threatens and pursues a White woman until she plunges to her death from a cliff. The film was screened for President

Woodrow Wilson, U.S. senators, and the chief justice of the Supreme Court. Newspapers around the country called it "the greatest of films."[59]

One hundred years after *Birth of a Nation*, these images of violent, hostile Black men, whose bodies must be restrained, appeared on the screen of millions of White Americans perusing Facebook. The shotgun in the duffel bag. The man facedown on the hood of a car. A White savior who is not an abolitionist but a manager who secures his clients' freedom. In the *Straight Outta Compton* trailer, the images multiply until the closing words flash across the screen. These men, White viewers are told, are "The World's Most Dangerous Group."

ALL THIS EVIDENCE THAT CATEGORIZATION paves the way for prejudice may suggest that the solution is simply to reduce our reliance on categories. Certainly if we had never made any distinctions among people, it would be difficult to become prejudiced. But the human tendency to form groups is universal, from families and kinship groups to religions, cities, and nations. Groups also arise and fracture spontaneously: children form sports teams and rivalries at a gym teacher's command, gossiping creates an instant in-group. Just as we do not live simply in a home in the universe, but in a neighborhood, city, province, or country, we do not simply exist among humans generally. We exist within communities.

Even if we could somehow achieve a sort of utopian grouplessness, it's not clear that this would be ideal. Groups also give us a sense of belonging. Tradition, culture, and identity can be sources of pride and meaning. Moreover, when one has been part of an oppressed group, others' efforts to ignore that aspect of one's identity (like "color blindness") can create feelings of erasure. When Armenian American writer Meline Toumani lived in Turkey, people would often react to news of her background by simply changing the subject: "I'm Armenian!" she'd say. "The weather's been beautiful lately, hasn't it?" they'd respond. "I felt, for the first time," she writes, "what it was like to be made invisible." When we fail to or pretend not to notice these aspects of a person's identity, it is a way of disregarding their suffering.[60]

Fortunately, there's an important exception to the link between observing categories and stereotyping. A study by psychologist Inas Deeb of Israeli children in schools that were entirely Jewish, entirely Arab, or integrated found that children in ethnically mixed schools became more sensitive to ethnic categories over time, but they also became less essentialist in their thinking. Likewise, a study found that babies who are familiar with different races and can therefore correctly categorize races at six months later display less racial bias than babies who can't. Meaningful connections with people from other groups can undermine the tendency to essentialize and stereotype. This is an idea we'll examine in depth in future chapters.[61]

INDEED, THE TENDENCY TO DISCRIMINATE based on unconscious, unintended, or unexamined bias is well established, but the question that remains is how much impact these repeated, momentary differences in treatment, behavior, and reactions have. Some forms of bias have obvious, life-threatening consequences: a patient whose symptoms are dismissed erroneously may become more gravely sick, an individual mistaken for an armed suspect may face a fatal threat. But what effects do daily, fleeting experiences have, even those that may at times be (as in Barres's case) subtle enough to be invisible to the person experiencing them? Do everyday biases have a substantial cumulative effect—over a career or a lifetime? How much might they shape the trajectory of a population, a community, or a culture?

I asked dozens of researchers in the field of prejudice whether they could quantify the impact of these everyday biases, whether there was a hard number that could begin to capture some dimensions of the problem. But no one had an answer. No one has done the long-term studies it would require, following individuals over decades, carefully tracking every biased interaction and measuring the disparities that emerged as a result.

Finding an answer, it turned out, would take many months, the help of a computer scientist, and a working knowledge of ant colonies.

How Much Does Everyday Bias Matter?

What impact does routine, ongoing bias have? The question was taken up by twelve jurors during the 2015 proceedings of the gender discrimination suit brought by venture capitalist Ellen Pao. Pao had been a junior partner at Kleiner Perkins Caufield & Byers, a powerful VC firm and early investor in Google and Amazon. When Kleiner axed Pao in 2012, she sued the company. Among her charges: gender discrimination had kept Pao and other women from being promoted and paid fairly, and blocked from opportunities to succeed. Not only had the firm ignored sexual harassment, she maintained, but it had overlooked myriad instances of less overt discrimination. She had, for instance, been asked to take notes at a meeting and excluded from a networking dinner with Al Gore because women "kill the buzz."[1]

Pao had also been criticized for being both "overly opinionated" and needing to "speak up more" and penalized for behaviors that went unchallenged in her male colleagues. Relative to them, she maintained, her contributions were undervalued; she'd made a case for investing in Twitter early on and had been dismissed, though the company invested two years later. And she didn't receive proper credit for her work: although she had been responsible for investment in one tech company, a male colleague received the credit—and a seat on the company's board. The reasons given for her termination? Among them, her personality, which was deemed, as recorded by a reporter covering the trial, "territorial, difficult, harsh, and demanding credit." But one of her male colleagues was described in much the same way, and he was promoted, despite having performed poorly.[2]

At the end of the trial, the jury concluded that gender discrimination had not played a large role in her case, that Pao's gender was not a "substantial motivating reason" for her not being promoted to senior partner. As law professor Deborah Rhode later told a reporter, the evidence in the Ellen Pao case is typical for Silicon Valley. "There are no smoking guns; much of it is what social scientists call micro-indignities." Even the prefix "micro" (as in microaggression) implies these biases are negligible: the prefix "micro" means one part per million. Indeed, in a statement to the press afterward, one juror referred doubtfully to a brief conversation Pao had cited as evidence of bias. The juror wondered aloud if an experience that unfolded over such a short time period could be truly consequential—a question rarely posed about, say, a rapidly executed bank robbery or a hastily accomplished scam.[3]

A few years earlier Supreme Court justice Antonin Scalia separately came to the same conclusion that daily biases do not add up to a large harm. In 2011, he wrote the majority opinion for a class action lawsuit filed on behalf of 1.6 million female Walmart employees. The lawsuit maintained that Walmart denied women promotions, paid them less than male workers, and steered them to low-wage positions. Betty Dukes, the lead plaintiff, maintained she'd been barred from opportunities, blocked from the training she needed to advance, and pushed to a low-wage position. As proof of discrimination, the plaintiffs cited company records that showed massive gender disparities in pay and promotions across the company. The lawyers for the plaintiffs also provided 120 statements from women stating they'd experienced discrimination, from being told to "doll up," to doing the work of a manager but without the title, to being punished more harshly than men for the same infractions.[4]

But the idea that the behavior of individual managers had led to the massive disparities shown in the company's records was too great a leap for Scalia. In his majority opinion, he maintained that it would be impossible for a company to reach the kind of disparities seen at Walmart without a coordinated master plan of prejudice. "It is quite unbelievable," he wrote, that managers would all discriminate in the same way unless they'd been instructed to do so. "Most managers in any corporation—and surely most managers in a corporation that forbids sex discrimination—would select sex-neutral, *performance-based criteria for hiring and promotion* that produce no actionable disparity at all," he wrote. (The emphasis is mine. Reading this opinion, one comes away with the distinct impression that Scalia had never had a job.) There were perhaps other reasons for the disparities, Scalia mused— perhaps male and female employees had different qualifications. Left to their own devices, managers would never in aggregate be able to produce the kind of pay and promotion gaps seen at Walmart. Thus spake the United States Supreme Court.*5

The trouble with these two examples is not only with a jury's confusion about the timescale of wrongdoing or with Scalia's faith in humans' perfect rationality. The problem is with the way bias is typically assessed. Academic studies about bias—whether gender, racial, or anti-LGBTQ bias, and whether at work, in education, or in health care—typically capture one instance of discrimination at one time and place. They document reality at a particular moment. What they do not do is account for how bias is actually experienced in the real world, where individuals are not the targets of prejudice once, or twice, or three times, but experience it continuously over weeks, months, and years. Just as one photograph (or even a series of photographs) cannot fully capture the trajectory of an object in motion, studies that take a snapshot of bias do not capture its actual effect on people over time.

These snapshot studies also do not capture the dynamic, interactive nature of bias. Bias exists between and among people, and being on

---

* Ultimately, the court disqualified the lawsuit, ruling that female employees didn't have enough in common to be a "class" for a class-action lawsuit.

the receiving end of discrimination may affect a person's subsequent decisions and behaviors, which may shape further interactions, which in turn affect more decisions, behaviors, and options. This cascade can lead to serious, life-altering consequences. Psychologist Jason Okonofua proposes that educational disparities, for instance, arise in part from the accumulation of such interactions. Reacting to deeply ingrained negative stereotypes about Black boys, a teacher may be more likely to label a Black student's actions as misbehavior. That student might already fear that his teacher will treat him unfairly, and when he is punished, he may feel unjustly targeted and act out, which will confirm the teacher's bias and lead to stronger punishment. This cycle, played out many times, writes Okonofua, is a feedback loop that ultimately results in greater disparities. "The problem arises not solely from either teacher or students, but from both acting together and perceiving and misperceiving one another."[6]

This feedback loop may also play an important role in the "school-to-prison pipeline," the phenomenon in which Black students experience higher rates of suspension and other punishments, later drop or fail out of school, and end up being arrested and entering the criminal justice system. A longitudinal study of over a thousand students found that when students were kicked out of school, they were more than twice as likely to be arrested later that same month than during months when they had not been kicked out.* Suspension and expulsion, at least partly driven by bias feedback loops, led to higher rates of arrest.[7] As such, there appears to be a direct link between the dynamic interactions Okonofua describes and serious, life-altering outcomes.

The court's ruling notwithstanding, Pao's case also illustrates the dynamic, interactive nature of bias. She was denied credit for her work, which may have caused her to become more vocal about being acknowledged, which resulted in her being seen as "difficult" and "demanding credit," which further limited her opportunities to succeed. What matters is not the single instance, or even the sum of experiences. It's

---

* This was particularly true for low-risk students who did not tend to disturb class, fight, or use drugs.

the compound effect of many interactions, one that only emerges over time.

Antonin Scalia and Pao's jury might have seen things differently had they seen the workplace as what's known as a complex system, an environment made up of many individual members who, through their interactions, bring about a set of conditions that may not be easy or intuitive to envision. A city can be thought of as a complex system, arising from the interactions of its residents. So can an ecosystem and its many animal, vegetable, mineral, and fungal parts. In a complex system, unexpected outcomes can materialize: extreme, sometimes surprising results can emerge from seemingly limited exchanges.

Take ants. Ants interact according to some simple rules. They react to chemical scents, such as those of other ants, larvae, and food, and they leave behind their own scents. They also react to sound. Over time, these behaviors compound and allow ant colonies to solve difficult problems, like finding the best foraging route to and from food and avoiding traffic jams. As they react to one another's chemical traces while foraging, for instance, they spontaneously form a highway system—a central inbound lane going from the food source to the nest, flanked by two outbound lanes from the nest to the food source. These ants are not directed by an ant overlord. They are merely engaging with one another according to basic rules. If we examine only the behavior of individual ants, however, we might never step back and witness the larger patterns that arise over time. And if we only look at the macroscopic patterns, and not the relationship between those patterns and individual behaviors, we might insist, like Scalia, that there must be a centralized ant authority sending out detailed policy directives.[8]

Scalia could not imagine that large disparities might emerge in the absence of a master plan. Pao's jurors could not imagine that fleeting biases compounded over time might result in a person not being promoted or even being fired. These are understandable conclusions if one sees the interactions as discrete events. But if we begin to perceive the workplace as a complex system, new insights become possible.

To assess any true impact of bias, we must look not at the single moments but at the result of many, many interactions. When I set out to explore this approach, I could not find real-world longitudinal studies, observing individuals and the biases they encounter over an extended period of time. To quantify the cumulative impact of one kind of bias, I came to realize, I'd need to carry out a virtual experiment of my own: to build a computer simulation that envisions a given environment as a complex system and allows us to observe changes over time.

A computer science professor named Kenny Joseph agreed to collaborate on the project, and we decided to design our simulation to quantify gender bias in the workplace, as these patterns of bias have been well studied and documented, giving us ample real-world data upon which to draw. We chose a hierarchical workplace as the setting so we could see the impact of bias at different levels of an organization. We drew upon the research of psychologist Richard Martell and colleagues who have used computer simulations to evaluate the impact of gender bias in the workplace. But we expanded on their research by introducing the very specific types of bias that women encounter on an everyday basis.[9]

The first step in building our simulation was to set up a virtual workplace that would serve as our complex environment. For this, Kenny and I used an approach pioneered by economist Thomas Schelling, who in the 1960s was looking for a way to demonstrate that people's interactions could generate unintended results. Schelling noticed that whenever Americans clustered in ballparks, clubs, and neighborhoods, they tended to organize themselves by age, income, and race. He was curious about the role of individuals' choices in creating these patterns. Personal preference is of course only one of many contributing factors; in the case of housing, structural discrimination and disadvantage, like redlining, contract buying, and unequal access to loans, play large and important roles. Schelling wanted to better understand whether people's individual choices might also have an impact.[10]

He began his exploration by borrowing his twelve-year-old son's coin collection. He created a checkerboard pattern on a coffee table and then laid out pennies on the board. Each coin on the board had at most eight neighbors. Some of the coins were regular copper pennies;

some were the gray zinc pennies used during World War II when copper was rationed. The coins represented two groups of people. Now Schelling began testing out what happened when the individual coins had "preferences." What if each coin "preferred" to have at least two of its neighbors be the same color? Coin by coin, Schelling moved the pennies to new squares until this condition was met. Then he started over. What if each preferred half its neighbors to be the same color?

What Schelling found, to his surprise, was that even if the coins preferred to have only some similar neighbors, the results were extreme. For instance, if each coin wished to have half its neighbors be the same color, the final result was almost always total segregation: the zincs were clustered entirely with zincs, and the coppers were clustered with coppers. Even if the preference was for balance, the result was homogeneity. Schelling's simple model powerfully showed how large divisions can unexpectedly emerge from moderate preferences alone. And they can arise even when no one actually wants that outcome at all.[11]

Schelling's experiment is considered one of the first simulations of a social environment to track the impact of the actions of individual members. What if, I wondered, we injected into our simulated workplace the specific biases that Ellen Pao and the Walmart employees experienced? Would we see, over time, the kinds of disparities that were evident at Walmart? Could subtle, fleeting, or everyday kinds of bias have a big effect?

Making a computer simulation is a little bit like playing God—creating a world and its rules, setting up the initial conditions, and then pressing Go and watching what happens. Over many months, Kenny and I, and later his graduate student Yuhao Du, designed a miniature world in lines of code.

We call the company NormCorp. There are eight levels of corporate hierarchy, from the cubicle dwellers in entry-level positions to the titans of senior management. There are five hundred people at the entry level; we might imagine them toiling in gray cubes under fluorescent lights. That number decreases as you go up the hierarchy ladder, until there are only ten people in mahogany-paneled rooms at the top. There's a range of skill and talent at NormCorp, and employees begin with a "promotability" score that dictates whether or not they will advance. Because we want to see what disparities emerge after the introduction of bias, we begin with equal numbers of men and women at every level of the hierarchy.

In NormCorp, men and women are equally able; there is a normal distribution of "promotability" scores. All the employees are regularly assigned projects, and men and women are also equally likely to succeed or fail. If an employee does a stellar job and their project is a success, they are rewarded with a boost in their promotability score. If they fail and the project is a flop, their score drops. Some projects are solo; others are team projects. Once a year, the top performers at every level are called into a (conceptual) glass-walled office and told they're being promoted to the next level. They bump up in rank and the cycle of projects commences once again.

Here's where we introduce a handful of the biases that creep into workplace interactions. To create a realistic simulation of a workplace, we chose five core biases that are known to affect women at work, biases like those Pao and the Walmart employees experienced and that show up repeatedly in research and case studies. We decided to make NormCorp not an aggressively sexist place, so the biases we introduce are small: there is on average a 3 percent difference in how men and women are treated.

It's important to note that studies of gender bias have typically focused on the experiences of White women, just as studies of anti-Black prejudice have often focused on men, as though women are by default White and Black people are by default male. As legal scholar Kimberlé Crenshaw has pointed out, focusing on a single category of identity at a time, such as race, gender, sexual orientation, disability or religion, excludes and even erases the experiences of many groups; analyses of race or gender discrimination by themselves cannot, for instance, accurately capture the experiences of women of color.[12]

In reality, gender bias takes on different dimensions depending on a person's race and ethnicity, not to mention sexual orientation and other characteristics. One American Bar Association study found that while White women and Black men in law experience a similar amount of discrimination, Black women experience more discrimination than either of them do. Black women also face greater harassment at work than any other group. At the same time, research suggests that Black women may be less bound by expectations of passivity and warmth than women of other races. One study found, for instance, that Black women were not perceived negatively for displaying dominant behavior, while White women were. Another found that they were seen positively, and as especially feminine, when smiling.[13]

Psychologist Robert Livingston and management scholar Ashleigh Shelby Rosette propose understanding these differences by looking at how threatening a group is seen as in relation to White men, and also how interdependent, or "necessary or important from a social, biological, or practical standpoint." White women at work are seen as unthreatening but interconnected; they may therefore be protected and praised but also seen as unworthy of power. Black women, by contrast, are seen as neither threatening nor interdependent. As a result, they are marginalized—made invisible and blocked from advancement, but perhaps less scrutinized for compliance with gender expectations.*[14]

Management scholar Erika Hall and colleagues suggest that what is happening here is that gender-biased evaluations are amplified or diluted depending on a person's race. For instance, she points out, research shows that Black people are often implicitly associated with men, while Asian people are implicitly associated with women. Thus, Hall proposes, Black women face diminished stereotypes about how women should behave, while for Asian women these stereotypes are exaggerated. One experimental study found, for instance, that dominant Asian women were more disliked as coworkers than dominant White women: they were violating pronounced stereotypes prescribing that they behave in passive and communal ways.[15] As a woman of Chinese origin, Ellen Pao may have faced particularly virulent backlash for her assertive behavior.

---

* Black men, the researchers point out, are seen as threatening and therefore evaluated negatively for behaving in a dominant, assertive way.

Of course, the ways different identities can intersect to elicit different forms of bias are myriad. While the specific biases we introduce
into our simulation affect women in general, in the real world, their
intensity depends on many other dimensions of identity.

1. **Devaluation of women's performance.** Relative to men's
   performance, women's contributions are routinely devalued. For instance, MIT sociologist Emilio Castilla found
   that in a large American firm of twenty thousand employees, women and minority employees had to achieve better performance ratings to get similar salary increases as
   nonminority men. There was no specific policy that caused
   this devaluation; the performance ratings of those who
   were not White men were in effect "discounted." Another
   study found that when science faculty were asked to rate
   the application of a student for a job as lab manager, they
   tended to rate the male contender as more competent and
   more hirable than the female, though the applicants were
   identical except for the name on the documents. And a
   study of over five hundred thousand physician referrals
   showed that after a good outcome, surgeons who are men
   receive double the referrals, while women receive 70 percent more—a difference of thirty percentage points.[16]

2. **Greater penalty for women's errors.** When men and women
   fail, women are punished more. The above study of physician referrals showed that after a bad outcome, women surgeons suffer a 34 percent drop in the number of referrals
   they receive, compared to a negligible drop for men. In the
   financial services profession, women are 20 percent more
   likely to lose their jobs when they commit misconduct than
   men—though men are three times more likely to engage in
   misconduct. Research shows that Black women in particular
   are more penalized for failures than White women (or Black
   men). Betty Dukes was punished for minor infractions like

returning late from lunch, while male colleagues were not disciplined for the same behavior.[17]

3. **Losing credit to a male colleague.** Just as Pao lost credit to a male colleague, a series of studies by psychologist Madeline Heilman showed that when men and women work together on a project, both groups believe that women played less of a role in its success. A study by economist Heather Sarsons found that the more women economists coauthor papers with men, the less likely they are to receive tenure. (Men do not suffer as a result of coauthoring with anyone, and women who coauthor with women aren't penalized as much.) Even Icelandic recording artist Björk noticed this pattern: she produces the majority of her records, but when she coproduced a recent album with a male DJ, he was frequently cited as the sole producer. "It wasn't just one journalist getting it wrong, everybody was getting it wrong," she told a reporter. "I've done music for, what, 30 years? I've been in the studio since I was 11." Before they worked together, the DJ had never produced an album.[18]

4. **Personality penalty.** Women are often penalized for defying societal expectations that they be agreeable and deferential. A study of real-world workplace performance evaluations revealed that men's negative performance evaluations almost never included critiques of their personalities, while more than three-quarters of women's critical evaluations contained negative comments about their personalities. A survey of women in Silicon Valley called "The Elephant in the Valley" found that 84 percent of those questioned had been told they were "too aggressive." Here, Asian women may face especially harsh consequences—a survey of science and engineering professors found that Asian women experienced more punishment for self-promotion and assertiveness than Black, Latina, or White women. Indeed, Ellen Pao was described by colleagues as having

"sharp elbows," and she was chastised for it. A male colleague was described as "highly aggressive" and was promoted.[19]

5. **Opportunity bias.** New challenges provide a chance to grow and develop, and they can have an outsize impact on one's career, but women frequently have less access to these opportunities. For instance, 44 percent of women attorneys of color report being denied desirable assignments, but only 2 percent of White male lawyers experience this obstacle. This pattern is due in part to the fact that men are frequently evaluated for opportunities based on their potential for future achievement, while women are evaluated based on proof of their past achievements. In the movie industry, for instance, male directors are often given the opportunity to direct a major film after directing only small independent films, while even women who have already directed a major film often struggle for the opportunity to direct another.[20]

IN OUR MODEL, THE BIASES are factored in every time an employee completes a project. When employees succeed with solo projects, they are rewarded with a score boost, but women receive on average 3 percent less of a boost than men, according to the devaluation bias. Women at NormCorp also receive a 3 percent bigger penalty than men do when a project fails. When they collaborate with men, NormCorp women receive 3 percent less of reward than those working alone or with other women. And when NormCorp women are denied credit and ask that their contributions be acknowledged, they, like Pao, are seen as difficult and demanding, which results in an additional 3 percent penalty in their scoring. Finally, "stretch" opportunities that have triple the rewards occasionally arise, but NormCorp women must have 20 percent more past successes to be given these projects.

We introduced these biases and then ran the simulation a hundred times to find out, on average, how biases affected promotions over twenty promotion cycles. This is what we found.

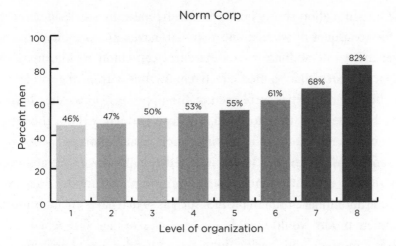

Even with just a 3 percent bias on average, over twenty promotion cycles, men came to represent 82 percent of the top positions.

Finally, we introduce one last bias. Research suggests that gender stereotyping increases as the number of women in an organization drops.[21] So we ran the simulation again, this time introducing another rule: as the proportion of women at a given level decreases, the bias they face increases, and keeps increasing the smaller their proportion. Once the ratio of women in a level of the hierarchy drops to 30 percent, the 3 percent bias increases to 4 percent. Once the ratio drops to 10 percent, the bias increases to 5 percent. This additional effect—called "downward causation"—made our graph look like this:

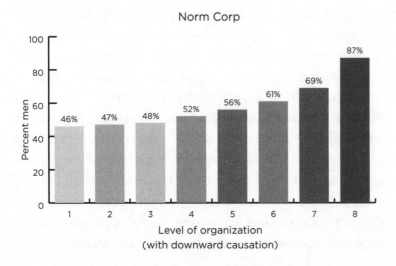

Our simulation shows how even mathematically small differences in the treatment of women and men can have an enormous impact. Over many interactions, these differences can cumulate into massive disparities such that women effectively disappear from the top level. The biggest disparities were seen at the upper echelons, after biases have had the most time to build. This can happen without the kind of coordinated, mandated discrimination Scalia thought necessary to generate such an imbalance. It can also happen without any one individual's explicit intent to harm: our simplified model does not include additional common hurdles such as sexual harassment. A more complete model would include the obstacles women face when they become mothers, the specific influence on gender bias of ageism, the precise amplification or dilution of biases experienced by women of different racial or ethnic backgrounds, and more. Here, even with a small amount of bias, NormCorp quickly became a workplace in which 87 percent of the leaders are men.

But bias creates disparities not only by blocking people from advancing: it can also lead people to quit. Reaching a plateau at work, for instance, may cause a person to decide, quite rationally, to leave their job. One study by the American Bar Association found that 86 percent of women lawyers of color in the United States quit large firms within eight years—not because they want to but because "they feel they have no choice." Some researchers have concluded that subtle bias can have *more* detrimental consequences than overt bias because its ambiguity demands more mental and emotional resources. A person can't be sure whether they were passed over for a promotion because of discrimination or some other reason, and they are left questioning their own perceptions, a sort of internal gaslighting.[22]

Moreover, bias can also erode individuals' capacity to succeed. In the moment, worries about being perceived through the lens of stereotypes, a phenomenon known as stereotype threat, can hijack working memory, distracting and derailing a person into underperforming. If the stereotype is of incompetence, stereotype threat creates a self-fulfilling prophecy. One study found that women treated with subtle discrimination suffered performance deficits, while those treated with overt prejudice did not. Another found that seeing instances of subtle bias had a greater negative impact on African Americans' performance than did viewing overt prejudice.[23]

Psychologist Robert Rosenthal found that the converse was true as well. In the 1960s, working with different classrooms at an elementary school, he randomly labeled some students "bloomers." Teachers were told they could expect great things from these students. At the end of the year, these students, particularly the youngest, made dramatic gains on IQ tests. In other words, if others perceive you as talented, you become more talented. These perceptions also help provide the grit necessary to push through adversity. One man who had been part of a longitudinal study of giftedness by psychologist Lewis Terman revealed how others' perceptions fuel internal drive. Being labeled as gifted, he explained, propelled him through the most challenging times of his career as a NASA researcher. "Sometimes, the problems got so complex," he recalled, "I would ask myself, Am I up to this? Then I would think, Dr. Terman thought I was." Our vision of what's possible for our lives can be partly fueled by others' recognition. If one group of people receives less recognition, they also experience less of the boost in confidence and resolve.[24]

What is apparent in NormCorp, however, is not only that bias makes women disappear but that afterward a vast homogeneity remains. Discussions of diversity in the workplace tend to focus on the benefits (or sometimes drawbacks) of including people from different genders, races, and other backgrounds in a team or organization. And while research has found that diverse teams can generate better and more creative ideas, they can also experience greater conflict and challenges in relationships. But just as diversity has pluses and minuses, so does homogeneity. Homogeneous teams may have smoother communication, but they may also have serious blind spots. Studies find that homogenous teams are less accurate at assessing their own performance, for instance. Homogenous juries tend to make more inaccurate statements; they also consider a more limited set of facts when deciding a case.[25]

Homogeneity can have even more far-reaching—and surprising—effects. For most of the twentieth century, for example, scientists underestimated the role of cooperation in the natural world, viewing interactions between organisms as mainly competitive, opposing factions vying for limited resources. In reality, cooperation is ubiquitous, from pollinator-flower interactions to forest-fungi interdependence. Mutually beneficial interactions are crucial to the survival of plants and animals and the key to the origin of DNA-based life-forms. Ecologists now believe almost every species on Earth is involved in some form of cooperation.

Why did scientists miss this? Ecologist Paul Keddy proposes that one reason was the homogeneity of the scientific community itself. Ecologists and biologists in the twentieth century were primarily male, working within a Western capitalist economy, and in intense scientific competition among themselves. This cultural background, suggests Keddy, may have caused them to overestimate the role of competition in ecology. "Scientists consciously and subconsciously draw upon their culture for models," he writes, and they "can only draw models from the possibilities of which they are aware."[26] The advancement of knowledge in the field of ecology was likely hindered by the fact that the scientists who published and debated among themselves were a homogenous group. Absent were people who could offer alternative, cooperative models with which to view nature.

Without understanding the workplace as a complex system—without a deep understanding of how interactions shape outcomes over time—people often revert to explaining disparities in simple terms. One favorite explanation is "inherent differences" between two groups, like the nineteenth-century arguments that women were simply intellectually incapable of contributing to the advancement of science. If we don't see women or other groups represented in a field, or at the top of a hierarchy, we may assume it is because they have less natural aptitude, flawed leadership skills, or, as Scalia suggested, inferior qualifications. This error in reasoning is akin to what's known in public health as an "ecological fallacy," a mistake that occurs when we observe an association happening at a population level and then use it to infer something about the individuals that make up that population.

The same reasoning happens in contexts far beyond the workplace.

Consider the way bias operates in health contexts. The accumulation of stressors of racism, research shows, leads to a process called "weathering," an accelerated aging process and decrease in health and well-being. Furthermore, when a person encounters racial discrimination in a health-care setting, this not only hinders the quality of care the person receives but may influence whether or not that patient seeks care in the future, including necessary checkups and screenings for deadly diseases. Add to this structural problems, like the fact that the pulse oximeter, reliant on the passage of light through skin, is calibrated to lighter-colored skin and can give incorrect readings for those with darker skin. While the cumulative impact of these injustices is significant health disparities between racial groups, medical education often oversimplifies the cause of disparities as "racial differences."[27]

Everyday biases, we can see, have broader consequences than may initially be apparent. At NormCorp, individuals' interactions spiraled over time and led to near homogeneity. The fact that even minute actions can influence others in such outsize ways makes ending unintended bias all the more important. But biased behavior need not be permanent. It is routine, yes, and it is unwanted and unintended. But it is not innate; it is something we learn. And if it's acquired, it can be cast off. If it's a habit, it can be broken.

PART II

Changing Minds

Breaking the Habit

On a gray day in February, psychologist Will Cox stood in a long, high-windowed room at the University of Wisconsin, Madison. A couple dozen graduate students sat in rows before him. On the screen behind him, he began projecting a series of words. The word "blue" was written in blue, "red" was written in red. Cox gave the students a task—to quickly name the color in which each word was written. Easy: the students called out the colors in unison. Now Cox showed another set of words, but this time red letters spelled out the word "blue," blue letters spelled out the word "brown." The group stumbled, laughing at how absurdly difficult it was to say the word "red" if red letters spelled out a different color.[1]

The reason for this difficulty, Cox explained, was the force of habit. Reading is such a strong habit that it can arise spontaneously and automatically. Interrupting this habit can be a challenge. Even when one intends to do one thing, one can inadvertently do another.

Cox then projected two newspaper images of New Orleans in the aftermath of Hurricane Katrina. In one of them, a young Black man clutches a carton of soda under his arm. Dark water swirls around him;

his yellow shirt is soaked. In the other, a White couple is in water up to their elbows. The woman is tattooed and frowning, gripping a bag of bread.

Cox read aloud the captions that ran alongside the images. Beneath the photo of the Black man: A YOUNG MAN WALKS THROUGH CHEST-DEEP WATER AFTER LOOTING A GROCERY STORE. Under the photo of the White couple: TWO RESIDENTS WADE THROUGH CHEST-DEEP WATER AFTER FINDING BREAD AND SODA.

"Looting." "Finding." A murmur spread through the rows of students.

Then Cox turned and volleyed the presentation to the woman standing next to him.

Patricia Devine is in her early sixties now. As she strode down the rows of students, her voice boomed. She could have been on the stage at a tech summit, not speaking in a narrow classroom in the university's education building.

"There are a lot of people who are very sincere in their renunciation of prejudice," she said. "Yet they are vulnerable to habits of mind. Intentions aren't good enough." Prejudice could be as habitual and automatic, she said, as reading the word "blue" instead of saying "red." "What we're not going to do today is point fingers," Devine continued. "This is not about blaming. The goal is to work together."

About ten years ago, one of Devine's undergraduate students proposed creating a workshop to try to reduce people's biases. Devine agreed, and she and her colleagues have been refining and testing this workshop ever since. Their approach is based on principles from cognitive behavior therapy. For someone to change their behavior, the thinking goes, it's not enough for them to be aware of their problem. They also have to be sufficiently motivated to make an effort. And they need concrete strategies to use so they can replace their old responses with something new.

Thus, the Madison workshop has three parts. The first is designed to increase people's awareness that they may be acting in biased ways without realizing it. The second builds their motivation to stop this behavior. And the third provides strategies to help people start to change. Devine and her colleagues have tested this approach with students and faculty, and in companies and organizations. The goal is to interrupt people's automatic responses. And their studies' results suggest that teaching people about bias this way can begin to shift their behavior.

DIVERSITY TRAINING IS NOW A multibillion-dollar-a-year industry. Nearly every Fortune 500 company uses some form of it. In the last ten years, trainings have expanded to include "unconscious bias trainings," which are now de rigueur at organizations across business, law, government, and beyond, and have given rise to a cottage industry of trainers, speakers, and consultants.* The goals of diversity trainings are generally to reduce discrimination, to increase people's skills in engaging with people from different groups, and to create more positive interactions in the organization. There's only one problem: trainings are rarely tested and evaluated in the field. This is a problem because they could be having literally *any* effect.[2] As a candid acupuncturist once told me before needling began, "After this treatment, you might get better, you might get worse, or you might stay the same."

When diversity trainings are evaluated, the results are often mixed. One meta-analysis of 260 studies of diversity training found the biggest effect of diversity training was on people's emotional reactions to the training itself—their positive feelings about the training, rather than their attitudes about other groups. It also found that while people did retain concepts they had learned in these trainings—such as how bias contributed to social inequality—their attitudes toward other groups later reverted to their starting point.[3]

In another study, sociologists Alexandra Kalev and Frank Dobbin reviewed hundreds of companies and the diversity initiatives they'd introduced over a thirty-year period. These included objective hiring tests, mentorships, and specialized recruitment. They also included diversity trainings. When the researchers analyzed

---

* Since the uprisings of 2020, these approaches increasingly include anti-racism trainings.

actual rates of promotion in the companies following these initia-tives, they found that when diversity trainings for managers were mandatory, the odds of Black women becoming managers five years later decreased by 9 percent. The odds of Asian American men and women dropped 4 to 5 percent, and the odds of White women and Black men becoming managers did not change at all.[4]

While this study was correlational and does not prove that the train-ings caused these outcomes, the findings are troubling. The researchers propose that managers may resist and resent anything that makes them feel that their autonomy is being undermined. It's also possible that managers felt threatened. Indeed, one study found that when White men went through a hiring simulation, those who were told that the company was strongly pro-diversity displayed physiological signs of cardiovascular threat.[5]

The trainings might also be unintentionally sending the wrong message. For instance, unconscious bias trainings often emphasize that unconscious bias is common. Yet research suggests that when people receive messages that everyone stereotypes, they may do more of it. If people believe that stereotyping is normal, they may be less motivated to change. Moreover, trainings focused on racism that occurs in mixed-race groups—particularly if offered by an unskilled facilitator—can be exhausting and harmful for people of color, who may be asked to "teach" White people, or treated as a spectacle.[6]

This lack of clarity about effective approaches is also seen in the larger pantheon of anti-prejudice interventions. When psychologist Elizabeth Levy Paluck reviewed hundreds of interventions designed to reduce prejudice, she found that only 11 percent of the studies used experiments tested outside of a laboratory. In other words, the vast majority of studies of anti-prejudice efforts are not tested in the actual environments in which we live and work. Because the deleterious effects of anti-prejudice interventions can be profound, Paluck has asserted that proving such efforts work as intended is "an ethical imperative, on the level of rigorous testing of medical interventions."[7]

On the other hand, leaving people to their own devices is also troublesome. When confronting issues of diversity, people in majority groups often resort to asserting that they don't even see differences: "I'm color-blind," or "I'm gender-blind." The venture capitalist we met in chapter 1, for instance, defended his company's track record of not hiring women by stating, "I like to think—and genuinely believe—that we are blind to somebody's sex." Indeed, in a hierarchical society, in which "difference" is often linked to rank, not observing differences may appear useful.[8]

People may also insist on their own impartiality, stating that they themselves are objective and thus incapable of bias. Those who fear being vulnerable may prefer a shield of objectivity to admitting limitations. Members of dominant groups may, in the absence of self-examination, even believe they are being objective. But research shows that both of these strategies—belief in one's own objectivity and "blindness" to gender or color—not only are ineffective but actually make bias worse.

In one study, participants were asked to imagine themselves hiring a manager. Half the participants were first required to complete a questionnaire about their objectivity, answering whether they agreed with statements like "My judgments are based on a logical analysis of the facts" and "My decision-making is rational." They were then presented with a description of a candidate. At random, some were told the candidate was named "Lisa," others "Gary." The description was the same for both "Lisa" and "Gary": a technically proficient candidate who lacked interpersonal skills. Participants were then asked to rate the strength of the candidate's credentials and whether they'd recommend that the candidate be hired. The other half of the participants were not asked about their objectivity until after they reviewed the candidate.

Among both groups, nearly 90 percent of the participants ranked their personal objectivity as above average. But the study found that participants who had first been directed to dwell on their objectivity favored "Gary" over "Lisa." They also rated "Gary" higher in interpersonal skills, even though the description of the candidates and their skills was exactly the same. These results weren't seen in the other group. In other words, when participants felt they were objective, it led them to discriminate *more* in their decision-making. Indeed, another recent study found that people who believed that gender discrimination was no longer a problem in their field rated a male employee as more competent than an identical female employee, and also recommended an 8 percent higher salary.[9]

Another study explored the impact of "color blindness," investigating how White employees' attitudes about racial and ethnic differences affected the experiences of employees of color. Researchers surveyed nearly five thousand workers in eighteen different departments of a health-care organization, assessing the extent to which those departments practiced "color blindness" or multiculturalism.* The study found that in departments where differences were downplayed, employees of color perceived more bias and felt less engaged. By contrast, when White employees noted and appreciated differences, employees of color felt more engaged and detected less bias.[10]

There appears to be a neural basis for this: brain imaging studies suggest that when people are motivated to check biased behavior, they pay more attention to racial cues, and then work to curb their own stereotyping.[11] Curtailing biased behavior seems to require paying attention to differences—the very opposite of trying to be "blind" to them.

IN LIGHT OF THIS CONUNDRUM—MURKINESS about the effectiveness of diversity training but certainty about the harms of the status quo— approaches like the Madison workshop offer a way forward. Unlike many bias trainings, it has been rigorously evaluated. Unlike many rigorously evaluated interventions, it has been tested in the real world. It is frank but nonjudgmental. And it centers on the importance of

---

* To assess "color blindness," the survey asked questions such as whether "employees should downplay racial and ethnic differences" and "the organization should encourage minorities to adapt to mainstream ways." Questions designed to assess multiculturalism included whether "organizational policies should support diversity" and "employees should recognize and celebrate racial and ethnic differences."

making active, conscious efforts to move beyond one's habits of thinking. It does not ask people to change their implicit associations. It asks
them to notice their habits and change their behavior.

An hour into the workshop in Wisconsin, Devine rolled up one
sleeve of her blue paisley blouse and walked over to a Black student sitting in the front row. "People think, 'If I don't want to treat people based
on race, then I'm going to be color-blind, or gender-blind or age-blind,'"
she said. "It's not a very effective strategy. First, it's impossible." She held
her pale forearm next to the student's. "There's a difference," she said. Students exchanged glances. "Who's the man?" she continued, now looking
at Cox. Then she raised her eyebrows and gestured to herself. "Who's the
old person?"

If pointing out a person's skin color in a workshop on overcoming
bias seems strange, that is part of the point. Trying to deny these differences, Devine asserts, makes discrimination worse. Perceiving distinctions is something humans naturally do. Humans, after all, see age and
gender and skin color: that's vision. Humans have associations about
these categories: that's culture. Using these associations to then make
judgments about an individual—that, Devine believes, is habit. It's not
seeing difference that matters, it's reacting to difference in harmful
ways. Indeed, the student Devine had approached with her bare arm
later told me, "I was a little surprised, but I kind of appreciated it."

The workshop is conceptualized as a way to curtail biased behaviors, through those three cognitive behavior pillars of awareness, motivation, and replacement strategies. Over the course of the two-hour
presentation, Cox and Devine hit all these notes. They walked through
the science of how people can act biased without subscribing to prejudice. They gave detailed explanations of how bias works. They made
the case for just how consequential bias can be. And they were careful
to avoid fostering shame.

One study Cox showcased revealed the effect of socioeconomic status on people's interpretation of a student's abilities. Participants saw a
video of a nine-year-old White girl named Hannah. One group was led
to believe that Hannah was from a wealthy family—her parents were
professionals, she lived in a spacious suburban home, and she attended

a sleek, modern school. The other group was led to believe Hannah was poor—the video showed her in a grungy yard, amid run-down homes, attending a squat school with an asphalt playground. Next, viewers all saw the same footage of Hannah taking a test. She answered some questions correctly and got others wrong. Then the participants were asked to rate Hannah's intellectual and academic abilities and describe how they arrived at their conclusions.

The researchers found that people's impression of Hannah's social class influenced how they perceived her as a student. People who thought Hannah was privileged not only rated her performance as above grade level but invented explanations for her behavior, for instance citing her capacity "to apply what she knows to unfamiliar problems." They also remembered more of the questions she got right. Those who thought she was poor rated her as below grade level; when asked how they arrived at that conclusion, their reasons included the notion that Hannah had "difficulty accepting new information."

Her socioeconomic status even affected how people perceived the test itself: those who thought Hannah was advantaged saw the test itself as more difficult; the others saw the test as easier.[12] Everyone had viewed the same child performing the same way, but here's confirmation bias again—we tend to search for information that confirms stereotypes and ignore information that contradicts it. We nudge our interpretations of reality to fit the map of the world in our minds.

Partway through the workshop, Cox and Devine invited students to discuss how these ideas related to their own lives. Everyone had a story, which reinforced how widespread bias is. One woman recounted being steered to a sales internship despite seven years of chemistry training because she had "such a nice personality." Another woman, a teacher, talked about how differently a Black child and a White child were treated at her school. Both had autism, but when the White child sang loudly and inappropriately, teachers described him as "so cute." When the Black child behaved similarly, he was disciplined. A long-haired White man explained that people often assume he can supply them with pot. He shrugged and offered that, in the scheme of things, this wasn't so onerous.

Cox had a personal story as well. He was adopted as an infant into a multiracial family. His mother is Hawaiian Chinese, Cox and his father are White, and his four brothers and sisters are Puerto Rican, so he saw the impact of bias on his parents and siblings. But his parents are also strict Mormons, and when Cox came out to his family as gay, the prejudice he'd seen his family members experience was directed at him.

"We were on a six-lane freeway in Tallahassee, Florida," Cox recounted. "My dad took my backpack, opened the door, and threw it out of the car into a ditch. He said, 'Get out of here.' And then he drove away." The next day, Cox went to the bank to take out the money he had saved from his job delivering newspapers and found that all four thousand dollars were gone, withdrawn by his parents. After that, he slept on people's couches and lived with an older man. He finally made his way to college in Florida, then to graduate school at UW–Madison. "I lost my family to prejudice," he told me, "so I'm devoting my life to fighting prejudice."[13]

IN THE FINAL PART OF the workshop, Devine and Cox offered strategies for how to override one's own biases. Notice when stereotypes arise, they proposed, and then actively replace them with alternative images. Look for situational reasons for a person's behavior rather than assuming it comes from some inherent characteristic. Seek out and get to know people who are different from oneself.

They also suggested trying to envision the perspective of the other person. Research suggests that this approach is even more powerful when one can truly embody another's worldview. In one study, some participants were asked to imagine being color-blind; others were able to experience red-green color blindness by means of a virtual reality headset. Later, when asked to help a color-blind student with a task, those who had actually experienced color blindness spent twice as much time helping the student as those who had only used their imaginations.[14]

These behavior-change strategies were put to the test comprehensively in research by psychologist Jason Okonofua. In an effort to reduce school suspensions, Okonofua and his colleagues worked with math teachers at five diverse middle schools to bring empathy to student discipline.

In sessions spaced two months apart, teachers learned alternative reasons students misbehave in class, and how good relationships help students grow and succeed. They were discouraged from labeling students as troublemakers and encouraged to take in their perspectives, reading students' stories about feeling understood by teachers. In other words, they were asked to consider situational reasons for students' behaviors, interrupt stereotyping, and visualize students' perspectives. The training was also designed to respect teachers' autonomy and competence, treating them as experts who could help others. They were asked to reflect on how they might use these approaches in their own teaching and were told their writings would be shared to help other teachers.[15]

The intervention wasn't specifically an antibias training; it was meant to foster a context of trust, respect, and mutual understanding. And, the researchers were careful to point out, it did not encourage the teachers to avoid disciplining students or to necessarily agree with students' perspectives. But it halved the suspensions over the academic year, compared to the control group. African American and Latino students—two groups that are suspended at high rates—experienced a drop from 12.3 to 6.3 percent. Furthermore, students who had a history of suspension felt significantly more respected by those teachers who had undergone the empathy training. It had altered the way teachers behaved. A brief, similar intervention aimed at parole and probation officers led, ten months after training, to a 13 percent reduction in recidivism among adults on probation and parole. This, too, was not specifically an antibias intervention, but one aimed at fostering better relationships. And it changed the way officers thought about those they supervised.

As she wrapped up the workshop, Devine paced among the desks, making eye contact with each student. "I submit to you," she said, "that prejudice is a habit that can be broken."

Results are promising. After Devine and another colleague presented a gender-focused version of the workshop to STEM faculty at the University of Wisconsin, Madison, departmental hiring patterns changed. In the two years following the intervention, the proportion of women faculty hired in the departments that had received the training rose from 32 to 47 percent—an increase of almost half—while the proportion remained flat in other departments. In an independent survey about departmental climates conducted months after the workshop, both men and women in participating departments reported feeling that their research was more valued. They also felt more comfortable bringing up family-related issues. The intervention had, it seems, begun to improve the work atmosphere overall.[16]

In another version of the same workshop, the researchers gave hundreds of undergraduate students an intervention focused on racial bias. Weeks afterward, the students were surveyed. Those who had participated noticed bias in their daily encounters more than students who had not taken the workshop. They were also more likely to label the bias they perceived as wrong. Notably, this evolution seemed to last. Two years later, students who participated in an online forum on race were more likely to speak out against bias if they had participated in the training. The team is still trying to determine whether the race-focused interventions have an impact on the experiences of people of color, not just the perceptions and behaviors of White people. It's a big question. "If we're just making White people feel better," Devine says, "who cares?"[17]

In treating bias as a habit one must work to overcome, an approach like the Madison workshop aims to bring awareness to unexamined patterns so people can see and actively revise their behavior. Indeed, a meta-analysis of forty years of diversity trainings found that they have the greatest impact when they go beyond raising awareness to include building new skills and behaviors. It also found that in settings where people are motivated to learn, like in schools, their effects are greater.[18]

At the same time, the effects of this—or any—onetime training are limited. While the proportion of women hired by the STEM departments that had participated in the workshop rose, the overall proportion of women did not—some left. The workshop did not miraculously transform the departmental climates, either: the climate survey also asked questions about being involved in decision-making, having to work harder than colleagues to be perceived as a legitimate scholar, and many others about the experiences of women and faculty of color. The workshop didn't improve any of these other measures. The meta-analysis of forty years of trainings found they are most effective if they occur over a long period of time and are integrated into a more comprehensive, organization-wide approach to diversity. Such a training might include bias-laden scenarios specific to an organization.

Devine readily admits that a workshop is not a panacea for eliminating bias. "But if people can address it within themselves, I think it's a start. If those individuals belong to institutions, they might carry the message forward."[19]

And the workshop does raise new awareness, sometimes rapidly. In one workshop, after Cox described why it's helpful to look for situational reasons for a person's behavior, a teacher in the audience shouted out a profanity. As the other students turned to her in surprise, she explained that she had just had a realization: when her White students turn assignments in late, she asks them if there's something going on at home. When her Black students turn assignments in late, she does not ask them anything.

The workshop did have an effect on me. In the days after attending the workshop, and after spending several days in long discussions with the researchers involved, I noticed my own reactions to other people to an almost overwhelming degree. The third day I was in Madison, in the lobby of my hotel, I saw two people standing near the front desk. They were White and wearing worn, rumpled clothes, with ragged holes at the knees. A story about them flickered in my mind: they weren't hotel guests, I thought. They must be friends of the receptionist, visiting him on his break.

It was a fleeting assumption—people dressed shabbily aren't staying at a three-star hotel. But that's how bias works: it's a flicker—unseen, unchecked—that taps at behaviors, reactions, and thoughts. And this story flitted through my mind for seconds before I caught it. Maybe I was right, but maybe I was wrong. *My God*, I thought, *is this how I've been living?* This story had no real consequence, but in other contexts what starts as an undetected assumption in one person has life-changing consequences for another. Afterward, I kept watching for that flutter, like a person with a net in hand waiting for a dragonfly. And I caught it, many times. This was one step among many as I worked to end my own prejudice. Watching for it. Catching it and holding it up to the light. Releasing it. Watching for it again.

BUT WHY WOULD A WORKSHOP like this have any effect at all? After all, bias is deeply ingrained, resistant to change, and bolstered by cultural messages in 24/7 surround sound. Most of the responses in the faculty climate survey did not change at all. But some did—why?

One insight comes from an unlikely source: substance abuse prevention. Over the last few decades, substance abuse researchers have found that one way to help people stop using drugs is to pay them. In studies that have demonstrated the efficacy of this approach, people are given cash or the chance to win prizes every time they successfully pass a drug test. Often the amount of money or the value of the prizes increases with every test passed. The method has been used successfully for users of cocaine, opioids, benzodiazepines, and alcohol.[20]

For years, researchers explained its success as a result of simple economics: it makes using drugs too expensive. When staying clean comes with cash, the "cost" of doing drugs jumps to include both the money required to buy the substance and the lost value of the money or the prizes. But this interpretation does not sufficiently explain the success of this program; a simple cost-benefit analysis can't explain, for instance, why these programs work even when the reward offered is very small, with some successful programs averaging only a few dollars a week.

For this reason, neuroscientists David Redish and Paul Regier have proposed another explanation for why the method works. They suggest that what's really happening is that the offer of money flips people into a different kind of thinking. The habit of drug abuse can be so strong that people can succumb to it almost unconsciously. But when presented with the choice between using drugs or receiving a reward, people are forced to weigh two different options, which snaps them out of habitual thinking and into a different, more intentional kind of deliberation.

Habitual thinking uses parts of the brain including the basal ganglia and cerebellum. Deliberative thinking, which is slow and requires more effort, uses the prefrontal cortex, the part responsible for planning and more complex decision-making. Thinking consciously about the choice between drugs and cash, Redish and Regier argue, may cause people to give more consideration to the consequences of their actions. They may stop to evaluate potential options in light of their goals, one of which is to stay clean. Snapping out of habitual thinking creates an entry point for users to choose another path.[21]

This same shift, Redish suggests, can happen with other habitual reactions.[22] An intervention like the Madison workshop may awaken people to the fact that every time they respond in a biased way, they are making a choice. And seeing bias as a choice may help a person move from an automatic mode of thinking to conscious deliberation. As I experienced, one of the workshop's consistent effects is increasing awareness of one's own capacity to be biased. This new awareness allowed me to see my judgments as a moment of choice: Do I accept my automatic evaluation of people or do I try something new? Do I believe my first reaction or do I stop to look for further evidence? At my hotel, after spending days with the researchers, I noticed myself assuming the couple at the front desk weren't hotel guests. Then I stopped. I subjected my initial assumptions to more scrutiny. I envisioned alternative explanations. Once I shifted from simply reacting to observing my reaction, new options appeared.

But another key question about the Madison workshop is why it seems to motivate people to put in this effort. Devine's team believes it has to do with how individuals see themselves. In the late 1960s, a social psychologist named Milton Rokeach posited that the self is made of many layers, some of which are more central than others. Our values, for instance, are core to our sense of self; our beliefs or knowledge about the world a little less so. The associations and stereotypes we hold would be even more distant from our sense of identity.[23]

This hierarchy within our sense of self matters because the more central a layer is, the more resistant it is to change. It's hard, for instance, to alter whether a person values tradition, or security, or fairness. But if you do manage to alter something within one of the deepest layers, the effect can be far-reaching. "If you think of therapy, the goal often is to change processes central to how people view themselves," said Patrick Forscher, a psychologist and coauthor on many of the Madison studies. "When it works, it can create very large changes."[24] And while it may be difficult to modify a person's values—their judgment about what's important—it may be easier to alter their beliefs and knowledge about themselves and the world.

The Madison workshop zeroes in on this layer—people's belief that they don't discriminate, or that bias is not important. And this layer is actually susceptible to change. If people come to see that they themselves inadvertently discriminate, but they also value fairness and equality, that realization can be a spur to action. People want to be internally consistent.

Rokeach himself proved this in a set of experiments in the 1960s. He asked White students to rank a list of eighteen values. Students were shown that most ranked "freedom" above "equality." To induce a sense of internal inconsistency, they were then told this reflected that they cared about their own freedom much more than they cared about others' freedom. They were also presented with the suggestion that support for civil rights is support for one's own freedom and others' freedom as well. Students who were confronted with this contradiction were, months later, more likely to enroll in an ethnic studies course and respond to an invitation to join the NAACP. More than a year later, they expressed more support for civil rights for Black Americans.

Jason Okonofua and colleagues used the same approach in their intervention with probation and parole officers. As part of the experiment that led to a reduction in recidivism, officers were shown the inconsistency of viewing themselves as individuals within their group of officers, while seeing parolees as indistinct from one another. Later, these officers assigned less collective blame to adults on probation or parole.[25]

"We can't change people's values, but we can give people knowledge about how they might not be living up to their values," Cox said. "Once you have this information, you can't help but make an effort."[26]

We might call beliefs the "Goldilocks layer"—the one that's just right in terms of trying to effect change. This layer of the self is removed enough from entrenched core values that it might, with the right kind of pressure, yield to transformation. When these beliefs move, they may bring with them a torrent of other changes.

ANTIBIAS TRAININGS, EVEN CAREFULLY DESIGNED ones, are still freighted with challenges. People arrive with different levels of readiness, and at varied stages of development. One diversity consultant told me she never knows if she's helping a seed grow, watching a flower bloom, or simply watering a patch of dirt.[27] People in positions of advantage may feel threatened by the corollary idea that they have benefited from bias in their favor—assumptions of competence or innocence, or the benefit of the doubt that the venture capitalist enjoyed when he floated from journalism into venture capitalism with no relevant experience. Watching Devine, I was struck by how approachable she is when presenting. It's intentional, she said. If people feel attacked, they shut down. The message is carefully balanced. Bias is normal, but it's not acceptable. You must evolve, but you're not a bad person. Watching Cox and Devine is like watching people play the classic children's game of Operation, manipulating specific beliefs with tiny conversational tweezers, without nicking the wrong reaction and setting off an alarm.

The audience makeup may be another obstacle. As sociologists Kalev and Dobbin learned in their study of diversity trainings, voluntary trainings are much more beneficial than mandatory ones, yet on the day I attended, almost all the people in the audience were White women or people of color, some seeking a way to combat bias directed at them. The absence of White men in the group was conspicuous. Cox told me the crowd is usually more mixed, but the audience makeup points to a current reality: the most receptive audience is likely the audience most affected by bias.

That said, one analysis that Forscher conducted suggests that the ideas in the workshop may reach even those who don't attend. Forscher carried out what's known as a "network analysis" of the workshop's effect—a look at how its effect spread throughout a community. After the gender-bias intervention in the STEM departments, people in university departments who reported doing the most for gender equity weren't those who'd attended the training but those who worked closely *alongside* them. It's an unusual finding, and it's not clear exactly what this means. But it's possible that as the people who took the workshop modify their behavior, this behavior creates new norms in their environments, which then influence the people around them. This network effect was also demonstrated in a study that gave high school students a weekly class in confronting prejudice, particularly anti-fat and anti-gay prejudice. Months later, the friends and acquaintances of those in the class were more likely to sign a gay rights petition.[28] The suggestion that the workshop's benefit spread beyond highly motivated attendees is a tantalizing one. If true, it offers a potential solution to the problem of preaching to the choir.

The truth is that those who are most affected often lead the charge, a fact illustrated in another experiment, a massive one involving over three thousand employees at a global company. Participants were given an hour-long online gender bias training focused on the workplace in particular. The training explained how stereotyping operates and taught strategies to overcome it. The researchers found that the program did increase people's acknowledgment of their own bias, their support for women, and their intention to act inclusively toward women. Among groups who were the most supportive of women before the intervention, it also changed behavior: weeks later, these employees were more likely to nominate a woman to mentor over coffee.

Interestingly, it was largely women who changed their behavior—to either mentor younger women or seek out senior women for mentorship themselves. The researchers concluded that one of the training's effects might have been to highlight to women just how much harder they needed to work to gain parity. A training that included other types of bias had similar results: more minority employees were mentored over coffee or nominated for recognition, but it was minority employees doing it.[29]

THE BIGGEST LIMITATION OF AN approach like the Madison workshop may lie in its premise: the idea that most people hold authentic egalitarian beliefs but these can be hijacked by habits. What if, as the alternative view holds, people do not have separate beliefs and associations but unitary attitudes buried under layers of denial or inattention? It's possible that people *do* hold prejudiced beliefs that they have simply not closely examined. If so, a more powerful target for a training might be the false beliefs themselves. Rokeach's model would predict that changing these beliefs would have an even larger impact.

This is what happened to me. In preparing for and writing this book, I went through many phases—from feeling frustration and anger toward those who behaved in biased ways toward me to believing that bias can be a predictable consequence of living in a particular historical moment and context. As I studied and wrote and engaged with others, my own biased words and deeds were pointed out to me, and my beliefs changed again: I realized that I, too, was thinking and acting in harmful ways. I noticed my habits, but I also began a deeper process of self-examination, taking an honest inventory of my beliefs. Some, closely inspected, were odious and conflicted with my values and my conscience. As I remembered all the times I felt self-satisfied as one of few women in advanced math and science classes, I realized that I did harbor unexamined, harmful beliefs about women's worthiness, and my own. This recognition in turn prompted me to study the historical basis of these false notions, which spurred more lasting change.

Research shows, in fact, that the more we understand history, the greater our grasp of present-day prejudice. In one study, White Americans were taught the history of the government's role in creating Black ghettos through discriminatory housing policy. The participants later acknowledged more present-day racism than those who did not gain that historical knowledge. As Bob Marley sang, "If you know your history / Then you would know where you coming from." Indeed, researchers call this link between knowing the truth about the past and recognizing present discrimination the "Marley Hypothesis."[30] Learning the origin of my inherited beliefs allowed me to begin the hard work of unlearning them.

This did not happen after a two-hour workshop: it was a process that took the five years I spent writing this book, and it continues.

But the Madison workshop did offer an important step on that path—a trigger to notice the thoughts in the first place. And a short training like this may be best seen as a starting point. It makes a modest request: consider, for a moment, that you may be acting in ways at odds with your values.

Ultimately, cognitive behavior approaches alone cannot dissolve bias because we are not only individual thinking beings. We are historical beings, and interrelating beings. We are structural beings, too, operating within institutions that powerfully influence our behavior. Without refashioning these constraining structures—policies, laws, algorithms—relying on individuals is like running up a down escalator. We are also emotional beings who process complex swirls of feelings that shape our interactions. And we are physical beings who store in our bodies memories of validation and trauma, care and neglect. All these factors affect how we respond to others.

Bias is indeed often most consequential at moments where all of these currents come together: the cognitive, historical, interpersonal, structural, emotional, and physiological. In these settings, people may not slow down and engage in careful deliberation—to consider another person's perspective or the situational reasons for their behavior. In some contexts, decisions that have life-or-death results can happen very quickly, particularly when we are trained to fear one another.

The Mind, the Heart, the Moment

It was one of the warmer years on record in Minnesota, and that July evening in 2016 was cloudy and hot. In the St. Paul suburb of Falcon Heights, a thirty-two-year-old school nutrition supervisor named Philando Castile drove down Larpenteur Avenue with his girlfriend, Diamond Reynolds. They had just gone grocery shopping, and the car was loaded up with frozen shrimp, rice, chicken, and marinade—ingredients for that night's dinner. Reynolds's four-year-old daughter was in the back, strapped into a car seat. Castile was working for the St. Paul public schools that summer. He didn't have to—pay was prorated over the year—but he loved the kids. One parent called him "Mr. Rogers with dreadlocks." The kids called him Mr. Phil.[1]

Jeronimo Yanez, a twenty-eight-year-old patrol officer, had begun his shift a few hours earlier. He was on the lookout for robbery suspects. A nearby convenience store had been held up recently, and the surveillance video showed a man with dreadlocks.

Falcon Heights is about 70 percent White. Castile was Black. When Yanez saw Castile drive by in his Oldsmobile, he radioed his partner in another squad car that he was going to investigate. The driver resembled the man in the robbery video, Yanez said, "because of the wide-set nose."[2]

Yanez pulled Castile over, walked to the driver's side window, informed Castile that he had a missing brake light, and asked for his license and insurance card. Castile handed over the insurance card and then said, slowly and calmly, "Sir, I do have to tell you that I have a firearm on me." Castile had a permit to carry a weapon in Minnesota: he had undergone vetting and authorized training and was legally required to disclose its presence upon request of the police. He carried a gun for protection because he and Reynolds lived in a high-crime area.[3]

Yanez put his hand on his own gun. Castile, having just been asked to produce his license, proceeded to reach toward his right side. Yanez, having heard the word "firearm," told Castile, "Don't reach for it." Castile and Reynolds tried to explain that he was not reaching for his gun.

Don't reach for it, then.
    I was reaching for—
Don't pull it out.
    I'm not pulling it out—
    He's not—
Don't pull it out.

Then Yanez shot at Castile seven times. Castile's last words were, "I wasn't reaching—."[4]

In the days afterward, it was as though a gash had been torn through Minneapolis and St. Paul. Grieving protesters walked onto the interstate highway that links the cities. It was a foreshadowing of the summer of 2020, when worldwide protests against racism and police violence shook every corner of the world. The Twin Cities is not a very large community, and everyone seemed to know Castile or knew someone who had. A child posted a note outside the Montessori school where Castile worked. "I rilly miss you," it said. "You have rainbows in your heart."[5]

* * *

Footage of the immediate aftermath of the shooting shows Yanez in a state of incoherence. He does not put down his gun. He neither attempts CPR on Castile nor stops screaming. Reynolds explains that Castile was getting his license and registration. Even after other officers arrive, Yanez's panic does not abate. "Fuck. Fucking A.... I'm fucked up right now...." he says. "I don't know how many rounds I let out.... He was just staring straight ahead and I was getting fucking nervous and then ... fucking I told him to get his fucking hand off his gun."

Medical evidence shows that Castile's hand could not have been on the firearm when he was shot: there was a bullet wound on his finger but no damage to the gun or his pocket. A firefighter would later testify that he saw an officer reach "deep" into Castile's pocket to find the gun. Castile appears to have been complying with *both* of Yanez's orders: both the negative order not to reach for the gun and the affirmative order, just prior, to retrieve his driver's license.[6]

When state criminal investigators interviewed Yanez, he said about Castile, "His tone was very subtle and uneasy to me ... his body language appeared defensive." Yanez smelled marijuana, he said. When he shouted his instructions, Castile "turned his shoulder, kept his left hand on the steering wheel...." Yanez went on:

> And at that point I was scared and I was in fear for my life and my partner's life.... He still kept moving his hand.... I looked and saw something in his hand.... I was trying to fumble my way through under stress.... But I wasn't given enough time.... I know he had an object.... And I thought I was gonna die ... I had no other option than to take out my firearm, and, and I shot.
>
> I don't remember how many rounds I let off.

The Medical Examiner ruled the case a homicide, and Yanez was charged with second-degree manslaughter and dangerous discharge of a firearm, the death due to gross negligence. Yanez's defense was that he feared for his life. Perceived threat is the reason officers generally give for using deadly force. It can be and is used to cover up overt racist violence and has no relation to acts like Derek Chauvin's sadistic, slow-motion murder of George Floyd.[7]

Yet Yanez's hysterical behavior does suggest panic. Yanez described not perceiving Castile's words because he was getting "tunnel vision." Loss of peripheral vision is a symptom of an extreme stress and fear response, as are the inability to remember preceding details and the emotional dysfunction he displayed. Why did he panic? He thought he saw an object in his hand, Yanez said. During the trial, he claimed he saw Castile pull out a gun, which we know did not happen. "I had no other choice," he said. "I didn't want to shoot Mr. Castile."[8]

The jury, made up of ten White and two Black Minnesotans, determined that Yanez seemed honest and was not lying about seeing Castile's hand on a gun. No presumption of truth-telling was extended to Castile, who clearly explained that he wasn't reaching for the gun. The jury voted not guilty on all counts. For the second time, the city exploded.[9]

Yanez's defense attorney maintained that this incident was not about race but about the presence of a gun, as if Castile's race played no role in Yanez's sequence of responses. In fact, research shows that race undoubtedly played a critical role, not only in Yanez's decision to stop Castile, but in his perception of the encounter and in the production of his fear.

IN 2003 AND 2004, CIVIL rights attorney Connie Rice spent eighteen months interviewing Los Angeles police officers. This was after the most wide-scale police corruption case in L.A. history had prompted broad department-wide reforms. Asked to help assess whether the reforms had made a difference in the LAPD, Rice interviewed more than eight hundred officers—male and female, of all races and ethnicities. They were candid about the shortcomings of reform, describing ongoing lack of accountability for illegal behavior, perverse incentives, and adversarial mindsets toward the community, among other persistent problems.[10]

But officers also told Rice something that surprised her. They confided that they were afraid of Black men. "Look lady, I'm going to be honest with you. Black people scare me. I didn't grow up around Black people," she remembers them saying. "I grew up in Antelope Valley. We didn't have any Blacks. And I don't really know how to talk to them," or "Lady, Black men scare me—and I need help." Rice tried to keep her reaction to herself. She didn't want to lose their trust by showing her shock. Inwardly, she said, "I was stunned." Fear is not something police officers typically admit to; vulnerability is seen as a liability. But officers of all races were confessing to racialized fear.[11]

Fear, of course, does not account for cases of intentional brutality. But the anxiety these cops professed is consistent with research showing that the mere notion of Blackness conjures up the idea of crime in many people's minds. Psychologist Jennifer Eberhardt, who has spent decades studying how race influences perceptions and behavior, conducted a study in which she subliminally exposed people to Black and White men's faces. She then showed them grainy, hard-to-make-out images of knives and guns. Those exposed to Black men's faces picked out the weapons more readily than those exposed to White men's faces. They didn't just think of crime—they *saw* crime. Eberhardt also showed police officers a series of Black and White faces they were told might be faces of criminals and asked them whether each face looked criminal. Having a Black face is not a crime, as Eberhardt points out, yet her study found that the more stereotypically "Black" the face was rated in terms of facial features and skin tone, the more likely officers were to identify it as criminal.[12]

Other research shows that many Americans judge Black men to be generally more threatening than White men. One study found that Black boys are seen as older and less innocent than White boys of the same age. People who are not Black consistently overestimate how physically large Black men are, perceiving them as taller, more muscular, and more capable of causing harm than White men of equivalent size. Black individuals, too, overestimate Black men's size.[13]

A name that sounds stereotypically Black can by itself elicit a distorted perception. In one study, non-Black participants were shown pictures of sixteen White male bodies from the neck down, altered to make skin tone ambiguous. Some participants were told the figures in the pictures had names that sounded stereotypically Black like Tyrone and DeShawn; others that the figures had names like Connor or Cody, which sound stereotypically White. Participants saw "Tyrone" and "DeShawn" as taller and heavier than the same bodies labeled "Connor" and "Cody."[14]

For White Americans, another study found, faces with darker skin tones even trigger more of a response in the amygdala, a part of the brain involved in detecting threats. Feelings like threat are significant because they closely predict how people will act. When researchers analyzed fifty-seven different studies of racial discrimination, they found that people's emotions about different racial groups had *twice* as much effect on their behavior as their intellectual beliefs. Emotions can influence our behavior toward members of a group in ways as subtle as amount of eye contact or as consequential as use of police force.[15]

As noted, perceived threat is a primary reason officers give for using force; the belief that one's life is in imminent danger is, as of this writing, its precise legal justification.* And Black men are disproportionately victims of force in the absence of actual danger. An analysis of nearly one thousand fatal shootings by on-duty police officers found that compared to White victims, Black victims were nearly twice as likely to be unarmed at the time they were shot. Another analysis found that Black suspects who were reported to be totally compliant with police orders were 21.3 percent more likely than White suspects to be on the receiving end of force—even when no arrests were made.† These statistics may be underreported: the records from which the

---

* This legal threshold is, of course, subjective, and courts' reliance on officer testimony—as in the case of Yanez— makes it rare for an officer to be convicted.[16]
† This study did not account for Black individuals being singled out and detained as suspects at higher rates, making its conclusion a likely underestimation. Castile was pulled over forty-six times in thirteen years.[17]

studies are made generally come from police departments, who may be undercounting internal problematic behavior.* This lived experience places an extraordinary and impossible burden upon Black Americans: to manage officers' fears in order to protect their own lives. In his book *Chokehold*, law professor Paul Butler describes the burden as dispiriting, embarrassing, and relentless: "Every time you leave your home you are the star of a bizarre security theater."[†18]

This fear of and desire to suppress people racialized as "Black" is not simply a phenomenon of police psychology or contemporary culture: its long shadow extends back to American chattel slavery. From enslaved Africans' earliest arrival on Western shores, White slave owners lived in a phantasmagoria of fear. South Carolina's legal statutes, for instance, described the enslaved as "barbarous, wild, savage." After the 1739 Stono Rebellion, terrified Carolinians quickly signed into law the Negro Act of 1740, which sought to crush potential threats by prohibiting enslaved people from learning to write, growing food, earning money, dressing nicely, moving abroad, or meeting in groups. Three years later, South Carolina passed another law requiring all White men under the age of sixty to bring a gun to church to guard against "the wicked attempts of Negroes."

Contemporary research finds a continuing pattern of control and punishment: when Jennifer Eberhardt and colleagues analyzed contemporary sentencing, they found that in cases where a murder victim was White, defendants with "stereotypically Black" features, including skin color, hair color, and facial features, were more than twice as likely to be sentenced to death than defendants who appeared less

---

* It's not only Black Americans who are particularly at risk. Latino individuals die as a result of interactions with police at a higher rate than White individuals. For Native Americans, the rate is higher than for any other group. These numbers are also likely vastly underreported because of a failure to identify mixed-race victims as Native American, and a large number of people experiencing homelessness who may be more difficult to track.[19]

† Data on people killed by police is limited by the fact that there has been no systematic effort to collect it; participation in the FBI's new use-of-force data collection effort is voluntary. We do, however, keep national statistics on the number of people stung by jellyfish, coral, and bees.

"stereotypically Black." This was after controlling for contributing factors, including the severity of the crime.[20]

Contemporary police training often compounds fear by emphasizing what can go wrong in an encounter. Officers are taught to notice "pre-attack indicators" such as anxiety and reduced mental processing in suspects, but many of these behaviors are the same responses a person exhibits when feeling threatened. Black individuals in particular are at risk of the phenomenon known as stereotype threat: concern about being stereotyped which can affect one's actions and behavior. They may be especially prone to appear anxious and therefore, in the eyes of police, suspicious.[21]

Furthermore, "warrior trainings," increasingly popular after 9/11, teach officers that they are at war in a hostile world. In The Bulletproof Warrior, a seminar developed by former army ranger Dave Grossman, attendees watch videos of shoot-outs and are encouraged to see the world as divided into sheep and wolves. The sheep are the public, oblivious to evil; the wolves are the criminals, waiting to attack. (Police are sheepdogs, the sheep's defense.) "Every single one of you is in the front line of a live ammo combat patrol every day of your life," Grossman tells trainees. Hesitation is deadly, he says. Cops, like soldiers, must be unafraid to kill.* Grossman himself has never served in combat. One of the officers who took The Bulletproof Warrior was Jeronimo Yanez.[22]

WHEN A FORMER MINNEAPOLIS POLICE sergeant watched the footage taken of Yanez immediately after the shooting, he saw a person rendered entirely nonfunctional. In policing, the level of awareness people bring to an incident is described by the Cooper Color Code. Civilians generally operate in the "green zone"—somewhat aware of one's surroundings, but not on high alert. Police officers operate in the "yellow zone," an elevated level of awareness in which they pay close attention

---

* In 2015, a total of 53.5 million police encounters with the public resulted in a rate of shooting deaths of officers of about 0.0000008.[23]

to their environment. When officers learn that a firearm is present, they shift to the "red zone," a state of high alert. In that situation an officer should stand back, give clear instructions, and begin to ask questions.

Yanez, it seems, skipped the red zone entirely. Upon hearing that a firearm was present, he went directly to the "black zone," a state of emotional dysregulation in which, as the sergeant explained, the brain "stops working."[24]

THE MURDER OF GEORGE FLOYD in the summer of 2020 catalyzed a global movement to reimagine a public safety apparatus, one that neither fears the public nor terrorizes the people it is hired to protect. Some suggested reforms are incremental, such as banning specific maneuvers, requiring police to live in the cities they serve, and revamping their training.* Other proposals are more wide-ranging, such as breaking the grip of police unions, which can insulate officers from accountability, rebuilding departments from scratch, and creating separate traffic enforcement agencies, so armed police no longer conduct traffic stops. Had Yanez been unarmed—or removed from service entirely—Castile would likely still be alive.

Another set of voices have called for abolition of policing as we know it. As psychologist Phillip Atiba Goff points out, the debate between reform and abolition rests on the diagnosis of the problem— whether it is fundamentally a problem of policy and bureaucracy or of basic mission. Reformists aim their scope at policies; abolitionists point out that decades of reforms did not prevent George Floyd's death. A new vision for public safety will require grappling in a meaningful way with the history of American policing as an institution—an institution that was historically tasked with suppressing and oppressing Black people specifically. The precise structure of a just public safety organization remains to be seen. But the essential changes it must incorporate may

---

* In the aftermath of George Floyd's murder, the Minnesota legislature banned warrior-style training.

still be insufficient to tackle the slippery problem of unconscious bias. The principles here can support these changes.[25]

According to the cognitive perspective on how perceptions of race can influence minds and behaviors, it's plausible that what happened on the night of July 6 was this: Yanez discerned a superficial similarity and, based on his race, misidentified Castile as a burglary suspect. As Yanez approached the car, he had the armed robbery in mind. His training heightened the expectation of danger, and seeing Castile's face triggered stereotypes of criminality.* By the time Yanez arrived at the driver side door, he was, perhaps, already feeling threatened. In that state of mind, he was even more likely to apply racial stereotypes.[26]

The expectation of danger influenced how Yanez perceived Castile, even as Castile was likely trying to manage his own fear of Yanez. Castile's awareness that he could be seen through a stereotyped lens may have emerged in his expressions. Thus, Yanez projected danger onto Castile's body language ("appeared defensive"), eye contact ("staring straight ahead"), and tone ("subtle and uneasy"). When Castile mentioned he had a firearm, Yanez's sense of threat converted into panic. Now in the black zone, he ceased seeing options for what to do next. He was unable to give clear directions or take in verbal information. His perception narrowed exclusively to what he perceived as the source of the threat: Castile's moving hands. We know Yanez did not see Castile's hand on a weapon. Yet Yanez, predicting violence, shot. Mind, body, history, institution: fused together to lethal effect.

Castile was calmly reaching for his wallet. Yet Yanez was incoherent with terror. The threat was imaginary, but the fear was real. What do we do about *that*?

HILLSBORO, OREGON, IS A TOWN of about a hundred thousand people, a short train ride from Portland. Intel is a big employer. Violent crime and property crime rates are lower than the national average. Hillsboro

---

* Yanez was Mexican American; studies show that one need not be part of a dominant group in order to harbor harmful implicit associations.

has been, in the words of its police officers, a pretty pleasant place to be a cop.[27]

But around 2003, a Hillsboro police sergeant named Richard Goerling observed among his fellow officers something he calls "the asshole factor"—routine aggression directed at the public. Goerling even noticed this abrasiveness in himself, wondering at his own lack of kindness when responding to calls. Acting aggressive toward the people they'd sworn to protect was, in Goerling's words, "a performance failure," so he began reading widely about the science of performance. Specifically, he looked for approaches used by people who had to perform flawlessly, like athletes. Elite performers, he learned, used mind-body practices like yoga and meditation.[28]

Goerling had been an athlete in high school; he knew that a person's state of mind directly affected how they used their body. But the connection between mind and body wasn't something police ever discussed, and Goerling was intrigued. He began talking with a yoga teacher who had tried, at one point, to offer a free class for Hillsboro first responders—when the day came, no one showed up. Goerling invited the teacher to a class in "tactical communication," and he agreed, introducing meditation, leading cops through an exercise in mindfully eating a single raisin. Many rolled their eyes. But a handful were interested enough to join Goerling in a multiweek mindfulness course.[29]

Mindfulness, as it has been adopted in the West from Buddhist principles, has been defined and redefined over the past several decades. A working definition is this: mindfulness is a state of non-judgmental awareness of what's happening inside and outside the body at any given moment. In mindfulness training, participants practice noting their feelings, thoughts, and sensations, including when their minds begin to wander. The focus is on acknowledging what is happening now without trying to avoid, deny, or suppress it, and without getting pulled into thinking about the past or future. Through exercises like body scanning and breath awareness, people can learn to pay attention to their own habitual reactions and to bring an affectionate and even "friendly" curiosity to whatever is happening in the given moment. Legal and mindfulness scholar Rhonda Magee adds that mindfulness lets us see "the whole" and be with each other in life-affirming ways.[30]

It's hard to imagine more incongruous images than a mindful meditator and an American cop. The meditator sits monklike, with closed eyes and a relaxed, beatific smile, bathed in soft light and wearing loose, flowing robes. The militarized officer scans the environment, tense and frowning, strapped into a bulletproof vest. And sitting in his mindfulness course, Goerling found it deeply uncomfortable to do what was asked of him: to feel his emotions, his physical sensations, and his breath. Cops are trained to be alert to external events; they call it "situational awareness." They're rarely asked to look inward. The intimacy of the sessions was, to Goerling, kind of embarrassing. But maybe that was okay, he thought. Maybe there was a way to make this more palatable to cops. Because he had begun to see a link between the aggression he witnessed and what was happening in cops' minds and bodies.[31]

THE JOB OF A POLICE officer involves being sent, with technological efficiency, to some of the most difficult situations happening in a community at any given time: drug overdoses, child abuse, assaults, shootings, burglaries. Police routinely deliver the worst news a person can receive, knocking on doors at two a.m. to tell parents their child has died in a car crash. They are asked to respond to situations that the people involved feel they can no longer manage themselves. Over time, the work begins to erode officers mentally and physically. Fatigue sets in, bodily pain, sleep disorders, irritability. Officers have higher rates of heart disease than the general population and one of the higher rates of illness and injury among professions. A 2020 study of a large urban police department found that more than a quarter of officers screened positive for depression, PTSD, suicidal ideation, or other severe mental distress. In 2019 in the U.S., more died by suicide than on the job. Quite simply, Goerling said, "we're broken." These impairments are further exacerbated by a professional culture that Goerling calls "a toxic drip" of hostility, intimidation, denial of suffering, and intolerance of emotions.[32]

One year turns into five. Five turns into ten. One study found that for each additional year on the job, officers were 16 percent more likely to use force. Another found that officers who start their police career later are less likely to shoot. Clocking extra duty hours increases officers' chances of exhibiting serious performance problems, like using excessive force, unnecessarily pursuing individuals, or discharging their firearms without reason. And a feedback loop sets in: after an officer uses force in a critical number of traffic stops, the risk of future performance problems jumps.[33]

This damage directly affects how cops interact with citizens: chronic stress makes police more likely to make errors and poor decisions. Stress is linked with inappropriate aggression in officers; it can also lead to sleep disorders, which make officers more likely to explode in anger at citizens.* Chronically stressed officers may also be more likely to be involved in deadly shootings.[34]

In particular, chronic stress affects how the brain processes threats. A person's fear response involves multiple parts of the brain, including the amygdala, which helps detect salient threats in the environment and generate feelings of fear and anxiety, and the prefrontal cortex and other areas, which modulate a person's reaction to bring it in line with reality. When a person is in a state of emotional regulation, these responses are balanced. But prolonged stress increases the activity of the amygdala, facilitating the growth of neurons in this region while diminishing the strength of the prefrontal cortex. This throws off one's ability to regulate emotions. While fear has a role in policing (it's reasonable to feel terrified if one is first on the scene of a shooting by a career criminal), a chronically stressed police officer may be faster to feel and respond to fear. Because heightened amygdala activity and

---

* Anger is sometimes described as a "funnel emotion" for men: feelings of grief or sadness or shame can be converted to and expressed as anger, a more culturally acceptable way for men to express their emotions.

a weaker response from the prefrontal cortex are also associated with aggression, they could also be more prone to violence.[35]

These stresses may also affect cognitive control, the ability to override one's impulses and act with intention. People with stronger cognitive control are better able to interrupt habitual responses. But when the brain is taxed, through cognitive, physical, or emotional impairment or even time pressure, mental resources are depleted. And cognitive control is decimated by chronic stress. An overloaded mind, drained of resources, with diminished cognitive control, will more likely rely on mental shortcuts such as stereotypes. Officers in this state will likely have a harder time preventing racial stereotypes from affecting their actions. Chronic stress, in other words, creates a perfect storm for bias.[36]

Indeed, studies suggest that impaired officers do more racial profiling. One analysis of over ten thousand stops by Oakland, California, police found that when officers were stressed and fatigued, they searched and handcuffed African Americans at higher rates. A study of police recruits, too, found that those who were more fatigued showed more bias when performing in a simulation: they mistakenly identified an unarmed Black suspect as armed and decided to shoot.[37]

JERONIMO YANEZ SEEMS TO HAVE been impaired in just these ways. Yanez had been working nights for four and a half years. By the evening of July 6, 2016, he had been involved in multiple critical incidents, including witnessing a shooting and being first to respond to a murder-suicide in which a man had killed his wife and son while their other children were inside the house. The fissures were showing. Months before the night he shot Philando Castile, Yanez had pulled over a car with a broken taillight. While he was standing at the driver's side, another car drove past, coming within inches of his body. Yanez abandoned the first stop, jumped back into his cruiser, and began following the second car. When he pulled the driver over he erupted in a rage and began bellowing orders. "OUT OF THE CAR. KEEP WALKING. KEEP WALKING. KEEP WALKING," he screamed. "GET THE FUCK BACK THERE." Afterward, it took Yanez thirteen minutes to regain control of his breathing. His colleagues expressed concern. Yanez apologized, saying he had "flipped out."[38]

Attorney Connie Rice's diagnosis, after speaking with eight hundred officers from the LAPD, was a department of "crushed spirits carrying badges and guns." Goerling echoed her sentiment. "The dirty secret," he said, "is that we don't have healthy organizations and we don't have healthy people." The performance problems Goerling observed in police could often be, he determined, directly traced to their own lack of well-being, which could, in turn, be traced to the job and institution itself.* And the community bears the brunt. Few police, Goerling maintained, start their day determined to violate someone's civil rights. "We wake up and say, Fuck, it's another day. I hope I make it through."[39]

Would it be possible, Goerling wondered, to fix this brokenness? Might there be a way to systematically change officers' behavior by changing their minds and bodies? Another police officer had asked this question in the early 1990s. A Wisconsin police sergeant named Cheri Maples had faced homophobia and harassment from colleagues, and after years of job-related stresses, she found herself becoming callous and cynical. Seven years into her career, she attended a meditation retreat given by renowned Buddhist teacher Thich Nhat Hanh. In the weeks and months afterward, she thought people around her were becoming kinder. Then she realized it was she who was changing.[40]

On a domestic violence call, a scared woman told Maples that her husband would not let her fetch their child. They'd just broken up, and he was blocking her from their agreed-upon pickup. At the time, Maples's department had a mandatory arrest policy for anyone acting threatening; normally, she would have handcuffed him for disorderly conduct. But Maples asked the woman to go wait in the car and then knocked on the door. An angry six-foot, three-inch man opened it. Maples was five foot three. She explained she was there to listen and help; she told him she could see how much he loved his daughter. She then suggested the girl go to her mother while the two of them talked.

---

* The fact that no one in a supervisory position addressed Yanez's state points to another institutional failure.

He agreed. Maples sat down with him on the couch. The man started crying. Defying her police training, Maples held him in her arms.

Three days later Maples ran into the man in the community. He gave her a bear hug and told her she had saved his life. Maples later reflected that her practice changed the way she viewed the people she encountered. What she started to see "was a suffering human being who needed my help."[41]

FOR THE PAST SEVERAL DECADES, scientists have been trying to learn exactly what mindfulness and meditation do to the human psyche and body. By 2017, thousands of studies on the subject had been published. Researchers Richard Davidson and Dan Goleman sifted through all this research and concluded that only a small percentage of it was rigorous enough to be trusted. But those dozens of studies demonstrated several undeniable effects: a decrease in reactivity to stress, an increase in attention and in care for others, and lessened bodily inflammation, depression, and anxiety.[42]

Why might cultivating nonjudgmental awareness and attention have these effects? It helps to think about how mindfulness works. We can think of our typical habitual reactions as tightly linked to our perceptions. We take in sensory information, we interpret or categorize it, and we react with some combination of feelings, behavior, and thoughts. We might react with joy or relief, or fear, anxiety, or aggression. These steps happen so quickly—and we react so automatically—that they essentially function as one, like a bundle of sticks wrapped together tightly with a strap. For example, you might receive a request from a difficult coworker, immediately feel tense or angry, and fire off a curt response. You didn't weigh multiple responses and carefully decide that being curt was the best option: you felt and reacted in one fell swoop.

If the different parts of a habit are like individual sticks bound together, mindfulness seems to loosen the strap. As those sticks—the parts of the habit—are gently separated, one can examine them in turn. Neuroscientist Yoona Kang has theorized that each component of mindfulness contributes an important element to this "deautomatizing" of mental habits. Practicing awareness allows us to notice our thoughts, feelings, and behaviors as they arise. Practicing non-judgment helps us sit with those thoughts and feelings rather than turn away, even if they're unpleasant. And practicing attention strengthens our cognitive control, so we can have more influence over our reactions.

Once a habit is interrupted, alternatives become possible. Mindfulness may contribute to overall well-being because it allows a person to interrupt habitual negativity and regulate difficult internal states like anxiety. In the case of a difficult coworker, you may still experience tension in response to the person, but with mindfulness, you can pause and notice your feelings—heightened anxiety, perhaps, or tightness in the jaw. Once you notice them, you can choose what to do next. You might choose to take deep breaths to calm yourself, or try to see the situation from their perspective, or wait until the feeling passes before you respond.

Mindfulness has generally been studied as it benefits the individual; it has not been thoroughly researched as an interpersonal or social practice, so studies of mindfulness as a tool for overcoming bias are still in early stages. But early research is promising, finding that subjects who participate in mindfulness meditation show less implicit race and age bias on the IAT, and that it seems to help deconstruct automatic reactions.[43]

Some of the most powerful research on this inner work examines its capacity to open new avenues of human connection. These studies explore a related form of meditation called compassion, loving-kindness, or "metta," meditation. While mindfulness meditation focuses on seeing the present moment clearly and without judgment, loving-kindness meditation focuses on compassion for the self and others, on caring for and wishing to help those who suffer. In this practice, you are asked to close your eyes and imagine someone who is very dear to you. Then, while picturing the person, you repeat to yourself, "May you be well. May you be happy. May you be free from harm." You notice and stay with the warm feeling that arises and repeat the process—for yourself, for an acquaintance, for someone with whom you have a difficult relationship. Each time, you wish that the person be well and free from suffering. Finally, you send these wishes to everyone in the world.[44]

This deep concentration on extending care to others can create a profound feeling of interconnectedness. It can also promote a feeling of equality among people. Neuroscientist Helen Weng found that people who had trained in compassion meditation were more altruistic toward a victim of an unfair social interaction compared to a control group. When she performed fMRI scans of participants (images that help reveal which parts of the brain are handling specific functions), she found that the altruism was linked to neural systems that involved cognitive control, positive emotions, and understanding others' emotional states.[45]

Neuroscientist Yoona Kang's research suggests an intriguing possibility for how this inner work affects bias in particular. In her study, she recruited a Tibetan Buddhist priest to lead weekly compassion meditation workshops. She gave participants IATs to assess anti-Black and anti-homeless implicit associations. Over six hour-long sessions, the participants, who were not Black and not homeless, practiced extending care to widening circles of people. They were also given instructions and meditation videos for daily home practice. Another group simply learned about meditation but didn't actually do it, and a third was put on a wait list.

After the intervention, IAT scores of the meditators effectively dropped to zero: they scored as if they held no biases at all. While one might deduce that participants simply became better at overriding their reflexive stereotypical reactions, there's another possibility. In another study, Kang conducted fMRI scans of people of a variety of racial groups while they were engaged in loving-kindness meditation. Compared to people engaged in a control activity, their brains were more active in the RTPJ, a region that handles concepts related to others' mental states—the area that helps us consider what might be happening in another's mind. After a month of home practice, they were tested for implicit bias against a highly stigmatized group—people who live with drug addiction. Those who showed the greatest RTPJ activity showed the greatest decrease in bias, which suggests that this meditation may reduce bias in part by improving our ability to consider and care for another's inner experience.[46]

Other research presents another tantalizing possibility. Mindfulness scholars have long described the individual, separate self as a fiction. Enlightenment, said the thirteenth-century Zen master Dogen, comes from removing "the barriers between oneself and others." In meditation, something of this sort may be happening on a neural level. One study recruited experienced loving-kindness meditators and examined their brain responses after being shown images of themselves and images of others. Compared to a control group, the meditators' brains, over a particular region, responded more similarly to the two kinds of pictures. In other words, they differentiated less between self and other. Loving-kindness meditation may, it seems, diminish the firm distinction between you and me.

Buddhist teacher Thich Nhat Hanh calls this "interbeing": we exist with and through one another. These neural changes provide a startling insight into one type of meditation's capacity to erode bias. Bias requires a firm division between the self and the other. But if the distinction between the concept of "I" and the concept of "you" is not so clear, bias loses its meaning. If the separation between us starts to dissolve, how can we rank—or hurt—one another?[47]

RICHARD GOERLING, THE HILLSBORO OFFICER, knew mindfulness would not solve all the problems of policing. But, he thought, perhaps it could improve behavior so that, like Cheri Maples, officers could encounter the public with compassion and awareness. Maybe mindfulness could help counteract what he believes officers bring to every encounter: fear, anger, and ego.

Goerling tried to bring it up to his Hillsboro supervisors. The response was curt. "Are you fucking kidding me?" they asked. "False religion," said some. "Devil worship," said others. But Goerling persisted. He tried broaching the idea with the SWAT unit, then the K-9. He carted around studies that showed how mindfulness could improve sleep, pain, and injuries. He got certified as a trainer through UCLA and started exploring the idea of running clinical studies on the impact of mindfulness on cops.

Then Hillsboro PD faced a tragedy of its own.

On January 20, 2013, Hillsboro 911 received a call about a man who, that night, had turned violent. He'd shot the family cat and was now holding his wife and children hostage in the bathroom. The man was also a Hillsboro police officer, one who had been having problems for years. He'd struggled with alcohol; he'd had conflicts in the department. When officers arrived, he pointed his gun and began shooting, at one point switching to armor-piercing bullets. More than a hundred rounds were fired. By the time he surrendered, he'd attacked eleven of his own colleagues. The officer was sentenced to ten years for attempted murder. The police chief resigned. The new chief contacted Goerling. "Go ahead," Goerling remembers him saying. "Try out this mindfulness thing."

Goerling and a small team, including the yoga teacher and researchers from a nearby university, began running eight-week courses—mindfulness training retrofitted for police officers. Goerling named the course Mindfulness-Based Resilience Training. The curriculum included standard sitting and walking meditations and practice with attention and focus. To make it more culturally acceptable, there was a little less interpersonal sharing and a little more scientific evidence. Officers practiced bringing mindfulness to the simple task of putting on their uniform and paying attention to the sensations of stress in their bodies while listening to 911 calls. Over the next year, dozens of officers from around the state of Oregon enrolled in trials the team ran to find out whether these practices would make a difference in their bodies—and their minds.

As this experiment was taking place, another sergeant a few hours away was on a parallel path. Brian Beekman is a police lieutenant in the city of Bend, Oregon, a 90 percent White town on the eastern edge of the Cascade Mountains. Over his fifteen years in the department, Beekman had seen police face every kind of personal struggle—cardiac failure, drugs, alcohol, divorce, suicide. But the police culture treated these as a given. "You're a cop. That's what happens. Emotional stress? Shut it down, turn it off."

Beekman had started out his career following that advice. Early on, he responded to a call in which a child had been killed by a day-care provider. He'd been one of the first on the scene. The feelings that arose were overwhelming, and he had no way to manage them. So he just began closing himself off. "I remember putting up those walls and just feeling cold to the whole thing," he said. It seemed like the only option. But over time, he noticed himself hardening, deadening. "That's how those calls change you," he said. "They start to chip away at your humanity."

In another call, a man had come home to find that his wife had shot herself and their two children. One of the kids was still alive when Beekman got there. Beekman rushed the kid out of the house and brought the father to the police station. He sat there with this man whose wife had just murdered his child. But calls from dispatch were stacking up, and Beekman was still on duty. Soon he found himself adjudicating a dispute between two neighbors arguing over the exact location of their property line.

"I remember thinking to myself, *This is crazy. I just went to a death scene. There was a child that was just killed an hour before, and now I have to deal with you?*" In that moment, the stable, dispassionate state of mind he needed to make wise decisions morphed into pure anger. "If you're an officer with a badge who's very angry," Beekman said, "you see how that's a problem. What position is that officer in to make the decision to lawfully use force?"[48]

Beekman and a sergeant named Scott Vincent had already ushered various wellness programs in to the Bend PD, including a physical fitness program, yoga, and mindful movement classes. When, in 2013, a family member showed him a magazine article about Goerling and his work, Beekman hightailed it to Hillsboro to meet with him. They quickly began collaborating. They put together three-day mindfulness retreats for police—a condensed version of the eight-week course—that they offered in a rustic lodge just north of Bend.

Beekman began sending his Bend cops to the retreat, and he added a regular, daily mindfulness practice session to police shifts. Like other officers, Bend police had originally rolled their eyes at the idea. But they'd started feeling better physically from yoga, and they trusted Beekman, so they were willing to try it. For fifteen minutes on the day shift, cops began gathering together in a plain conference room, turning the lights off, and following a guided meditation practice.

In 2015, Goerling and the team published the results of their research. After eight weeks of mindfulness training, officers improved in nearly every aspect of mind and body health studied. They reported less anger, less fatigue, and lower levels of burnout. They were less impulsive when stressed. They reported finding it easier to manage their emotions. A replication confirmed the findings, also noting that officers became less aggressive and more psychologically flexible.[49]

These were small studies, and it still wasn't clear how all these changes would actually affect the way cops treated the community. But in Bend, the ongoing approach to officer wellness—the yoga, movement, and meditation classes—was over the years starting to change officers and their community relationships. Injuries and medical costs went down, and performance began to improve. In the six years after 2012, citizen complaints as a proportion of all calls dropped by 12 percent. The Bend Police Department also reduced its use of force. Compared to 2012, the number of times force was used in 2019 as a proportion of all calls decreased 40 percent.[50]

IN SPRING IN BEND, A cutting wind snaps over the cold, dun-colored ground. Clear, hard buttons of resin stud the tall conifers, scattering sunlight like crystals. The Cascade Mountains rise in the distance— North, Middle, and South Sister. On the west side of town, an unbroken line of bicyclists in neon Lycra round the curves of winding, manicured streets. Friendly joggers chug past houses in the Oregon vernacular style—arts and crafts in rich, earthy colors like ocher and forest green.

Last April, I traveled to central Oregon to join a couple dozen first responders for one of Goerling's three-day retreats. Officers were women and men, and of all races and ethnic backgrounds, all from western U.S. states. Some were there voluntarily, including a few Bend officers who had not yet attended a retreat. A large portion were there under duress; they were the last stragglers in a department out of state that now required it of everyone. The oldest attendee was a portly, graying fifty-something; the youngest was a skinny kid from California just out of training, wearing a nubby, fleece-lined sweatshirt. Blinking uneasily, he looked like a lamb.

At dinner, I sat down with a few cops, asked about their jobs and stressors. Drugs, one said. Kids get hooked on their parents' pills, then switch to heroin because it's cheaper. The rest nodded. "I see things no one should see," said Patty, a community officer. A sergeant from California explained that no matter what you see, you can't let anyone know how it affects you. You show up to a motorcycle accident, and the driver is ripped in two, barely alive—that driver is looking to you for hope. You can't show your horror. "Feelings come up, you shut 'em down," he said. A detective named Tony shook his head. "Nothing frustrates me," he said. "I'm good at compartmentalizing." He was one of the crew who was required to take the training. He didn't need this, he averred. He had plenty of other ways of managing stress.[51]

After dinner, we adjourned to a pine conference room with soaring ceilings. Goerling stood in front. He is tall and eager, White, with a shock of white-gray hair and an open, unlined face. As he spoke, he sometimes jerked his shoulder, as if nagged by an old injury. "We're not here to be weirdos and burn incense," he said. "We're here because what we're doing isn't working well."

Next to Goerling stood his coleader, a lean mindfulness teacher from Los Angeles. The cops wore sweatpants and sweatshirts, their chairs pulled into a scraggly circle. A mustached lieutenant wore a T-shirt that said THROAT PUNCH. There was Big M, with a brush of black hair, and another in a red hooded sweatshirt, lolling back in his seat, eyes focused on the ceiling. A patrol officer wore a T-shirt showing a grim reaper handing a little boy a skateboard.

Goerling began by making the case for mindfulness. He ran through the stats on police injuries, illnesses, trauma. Hearing about the ways police were damaged, Tony and the lieutenant traded annoyed glances. "Your skepticism is welcome," Goerling said. "Your cynicism is welcome. If you think this is a waste of time, that's okay. I'm quite convinced it's not." Some officers who had been forced to attend were quick to make their irritation clear. One sergeant said, plainly, "I don't want to be here." An officer who'd been on patrol for twenty-five years said he'd rather be home with his cats. (When I expressed surprise at the length of his tenure, he smiled. "I exfoliate," he said.) A few denied that this training would have any utility for them. They already knew how to be resilient: it was to be "bulletproof"—that word again. Immune to damage. When asked to share their experiences of stress, the mustached lieutenant said flatly, "I don't get stressed. I give stress."

Others were intrigued. Patty had just had a baby, she said, and had no idea how to be a parent. Maybe, she thought, this could help her be a good role model for her son. The fifty-something just wanted to feel better in his body. "You may think you're fine now," he said to the younger attendees, "but believe me, it adds up." A 911 dispatcher with dark-lined eyes said her stress manifested in headaches so painful she needed prescription medication. She had been skeptical, but now she wondered whether this could help her get off the meds.

The next morning, we ran through breathing exercises and practiced paying attention. Goerling interwove the practices with studies, explanations, detailing the physiological benefits of mindfulness—the decreased inflammation, the lower stress levels. When we practiced body awareness, I noticed an ache in my neck and a chill across my skin. We worked on focus and attention, mentally counting up from 1 and down from 30 at the same time, alternating: 1, 30, 2, 29, 3, 28. Patty closed her eyes. Big M fell asleep.

The crew began asking questions. "What if you see something horrific? Will this help?" a young man wanted to know. "We're not meant to see that." His voice was urgent; he was thinking about something specific, but he didn't share what it was. No one asked. Goerling nodded and offered a quote from a trauma expert: "Your body has been reset to interpret the world as a terrifying place." This practice, he said, allows people to be in trauma and recover from it, and also grow in compassion and empathy. Later, someone asked if mindfulness could hurt. Could someone become so self-aware they freeze? Goerling tried to reassure that these skills would help them envision a greater set of potential outcomes when entering a situation, so they would act with more agility, not less.

We practiced being aware of body sensations by holding an ice cube in one palm. Goerling asked us to close our eyes and describe the sensation to ourselves. If we label and describe what is happening to us, he said, we can step outside it and see it more objectively. We can see reality as it unfolds instead of being trapped inside it. The cubes melted in our hands. Afterward, we shared our observations in small groups.

Some texted and smirked like fourteen-year-olds in detention. Others described the progression of pain from tingling to burning to numbness.

During an afternoon break, Scott Vincent, the sergeant who had helped set up the wellness program in Bend, arrived with some dim, mood-setting floor lamps. He, too, had at first been skeptical of mindfulness and inner-oriented practices. "I wasn't into namaste, I wasn't into yoga. And I'm a pretty hardheaded guy." But, he said, the practice changed him. And it changed how he saw the public. After decades of this work, you can start to see people as objects, he said. "*You're* a crackhead, *you're* an asshole, *you're* a doper," he added, pointing to phantom suspects between us. "But it grinds you. It grinds both ways."[52]

Later that night, some of the cops mutinied, lashing out at Goerling. "You're talking down to us," a sergeant said. "You don't know me, you don't know what I've been through. You're saying cops are unhappy, we're unhealthy, you're saying there's something wrong with us." Goerling listened. Here was everything we'd been speaking about: the anger, the anxiety, the defensiveness and threat. I could see Goerling trying to slow this encounter down and breathe, just as he'd been trying to coach the officers to do.

"I didn't mean to suggest there's anything wrong with you," Goerling said slowly. "But that's what you're *saying*," the sergeant insisted. He was getting angry now, and his voice turned hard. He stared at Goerling. "I'M AS HAPPY AS I CAN BE!" he yelled. The words echoed against the high ceiling. Then the lieutenant said loudly, "I think if you need this, you're weak. Go be a teacher."

The air felt combustible. "Okay," Goerling said. "Let's take a break." If they didn't feel they were getting any benefit from the sessions, they were free to leave. I asked the young man in the fuzzy jacket what he was going to do. He didn't know, he said. He looked miserable. He eyed the lieutenant who had said mindfulness was for the weak. Then he followed him out the door.

The majority of the attendees didn't return. Some played pool at the lodge later, debriefing after the explosion. It was insulting, they said. They felt they'd been talked down to and not given credit for the work they've already done on themselves. The sergeant who had angrily shouted about

how happy he was told me his department had lost a colleague to suicide. A bunch of officers had sought therapy. Later, I learned from some
officers who stayed to finish the training that they found the techniques
helpful and were planning to use them. But those who left could not even
countenance the notion that something was wrong in their minds or bodies. What was meant as a life raft was perceived as an accusation of failure.

A FEW DAYS LATER, I met up with Goerling in Hillsboro. We drove the
quiet streets in his black SUV. He's used to resistance, he told me, but he
did not expect the level encountered at the retreat. He traced it to the fact
that many officers had been forced to be there, but also to a culture that's
so hardened, so traumatized that it can't even allow itself to be helped.
That was toxic police culture on display, he said, flatly. Ego, fear, and anger
are what officers bring to every encounter, he said, and there it was, all of it.
There was also a fear of looking weak, of seeming insufficiently masculine.
And it doesn't have to be that way.

He drove us past a teenager in camo shorts, carrying a plastic bag
and scowling.

"Take this kid across the street. I can say, 'He probably stole those
shoes because how could he afford those bright shiny shoes?' I could
just be a completely biased person. Or I can be like, 'Yeah, he's got some
stuff from Target he bought. He's this young man trying to find his place
in the world and how he walks and how he dresses is about power, and
he's reaching for some.' I can understand that and not be offended by it.
How he walks up to me isn't an affront to my authority."

IN AN IDEAL WORLD, THIS work would be done preventively, woven
into training early on in careers, before the cynicism and toxic buildup
that we saw on display had time to accumulate. At the retreat, Scott
Vincent mentioned that incorporating mindfulness practice into the
ongoing training and job of policing could turn some cops off from the
profession entirely. But, said this thirty-year veteran, that might be a
good thing: change the entire culture.

In Bend, this shift may have already started. One afternoon I met with Eric Russell, a thirtysomething with sparse eyebrows and arctic blue eyes. Before moving to Bend, he served overseas in the Marines. After returning to civilian life, he joined SWAT in Hillsboro. "I was that young, excited cop," he tells me. "I wanted to crush crime. If people were yelling and screaming, I had to match their tone."

Russell was guarded as he told his story. He, too, had thought mindfulness was absurd. He ate well, went to the gym, took care of himself. He didn't need this mumbo jumbo. But he liked Goerling, so he signed on to a course. And then, he found himself using some of the practices. One day, he was sitting in a van, right before a SWAT response, "packed like tuna in a can." They were waiting for a suspect, and the tension rose as they readied for the moment they were going make the arrest. Then, Russell told me, something happened to him physically.

"So I'm sitting there, and I just kind of leaned forward, took a deep breath, and felt, you know, the rise and fall of my chest against my big armored vest." He centered himself, and he consciously refocused on his purpose: "I'm trying to protect the citizens around us, I'm trying to take this person into custody with the minimum kind of force response." He breathed, and he gained control of his emotions. "That was my big aha moment."[53]

Now Russell practices mindfulness regularly. Sometimes he'll put a sticker on his rearview mirror as a reminder to stop and take ten seconds to focus. On a call, he says, he'll monitor his own reactions and pay attention to his breath before he even gets out of his car. He'll take time to notice his heart rate. Because he's aware of what he's experiencing, he can then use breathing to regulate his response, even in a dangerous situation.

And because his emotions are in check, he has the mental resources to stay calm and avoid jumping to conclusions. Emotional regulation also helps him communicate precisely. As he puts it, a disorganized mind leads to confused speech. "Maybe I'm stuck in my own delusion. I'm telling him the same command over and over. And I'm yelling blue but it's coming out green."

Yelling blue but it's coming out green. That's what Yanez did: he yelled a garbled instruction. "Don't reach for it," he said, meaning, "Don't move at all."

Russell explains that he's had to unlearn much of his police training. He doesn't need to match anger with anger or stress with stress. He doesn't take any interaction personally. "When I show up, they're not mad that Eric's here, they're mad that someone has been introduced to their problem." If you show up angry, he says, you may not even notice that a person is complying. You're going to miss important indicators.

Again, I think of Yanez. *He was just getting his license and registration.*

The fear, anger, and ego that Goerling sees officers bring to every interaction have all been shaped into something more responsive, attuned, and flexible. Mindfulness, Russell says, helps him resist the impulse to predetermine what's going to happen and be open to a variety of possible outcomes. *I had no other option. I had no other choice.*

Crucially, he can also remain compassionate. In Bend, police are often called to respond to someone in a mental health crisis. Before, Russell would label the person "crazy" and be determined to "fix" the situation. Now, he thinks, "How can I help this person? I don't want to hurt this person I signed up to protect." In his nine years as an officer, Russell has gotten into a physical altercation only twice.

For Beekman, too, mindfulness has remade his style of policing. Like Maples, like Goerling, Vincent, and Russell, he's a changed person, he says. We met one afternoon in a busy, light-filled café on the west side of Bend. Dressed in a neat cotton shirt and glasses, he could be mistaken for a high school guidance counselor. Five years into the job, Beekman explained, he'd be called to a house for domestic violence, maybe the twentieth time to the same house. "I was full of judgment before I even stepped out of the car. I'd think, 'What's wrong with them?' That puts you in a certain space, when you're walking up to deal with something."

But two weeks before we met in Bend, he'd been called to the imme-diate aftermath of a suicide. When he arrived at the scene, he noticed his feelings. Instead of shutting them down and walling them off, he let himself feel angry at the man who'd left behind his son. He recognized that he didn't know the full story. As he surveyed the environment, he kept repeating to himself, *Non-judgment. Present moment.* He observed his own thoughts, regulated them, and fulfilled his role without either burying his emotions or succumbing to anger.

In some cases of force, Beekman wonders whether an officer's mind was healthy. "A lot of these overreactive actions by officers—are they calm, confident, present, seeing all the cues, reading the person?" Beekman asks. Again, I thought of Yanez. *His tone was very subtle and uneasy. His body language appeared defensive.* "Are those observational skills serving them well?" *I thought he was reaching for the gun.*

Mindfulness might also help manage other harmful emotions officers bring to an encounter. Research finds, for instance, that police officers who feel threatened in their masculinity—who, for instance, find it stressful to have a woman boss, or to express feelings of hurt or fear—may be more likely to attempt to restore that masculinity by using greater force on Black men, who are stereotyped as masculine. In officers, even fear of appearing racist—otherwise known as stereotype threat—is linked to greater use of force against Black civilians.[54] As mindfulness brings greater self-awareness and regulation, it may help retrain these unacknowledged yet dangerous states of mind.

But Beekman, too, has had a hard time persuading police to join him. When he taught a professional development class, he started passing around a book on emotional intelligence. The officers looked at the cover and refused to even open the book. "Feelings?" they said. "No."[55]

For those who begin this inner work, mindfulness seems to do two things. It provides a set of specific skills to use on the spot, in real time, in each interaction and each decision. It allows them to feel fear and anger arise, and then notice and regulate these so perceptions are not skewed by unregulated emotion. When fear comes alive, says Goerling, it's possible to embrace it and be with it, so it doesn't get between him and the encounter. *I was scared to death*, Yanez said. *I was in fear for my life.*

Over time, mindfulness also seems to reshape an officer's inner being, the source from which behavior springs. The mind and body become less like a brittle shield and more like a deep well, a reservoir that can absorb and work with anything that enters. Compassion and humanity are preserved. In the best cases, the mindful officer can see the person on the other end of the encounter not as an adversary but as another suffering human being. When Cheri Maples, the Wisconsin sergeant, attended her first meditation retreat with Thich Nhat Hanh, she was introduced to a set of five ethical principles meant to guide one toward wisdom and compassion. One was reverence for life. Maples asked how she could possibly fulfill this goal of nonviolence while also being armed. Who other than a person acting mindfully, the teacher asked, would we want carrying a gun?[56]

What would have happened had Yanez shown up differently on the night of July 6, 2016? What if he had arrived with a mind that was calm and nonreactive? Perhaps he would have been able to recognize Castile's body language as threatened, not threatening. He might have had the presence of mind to step away and slow down. He might have spoken clearly. He might have prevented panic from creating its deadly distortions.

The truth is we don't know yet whether mindfulness training will reduce racial bias in police behavior. The studies are not yet complete. But we do know this: chronic stress and the other impairments cops experience lead them to act more aggressively and use more force. These impairments are also known to increase bias. Mindfulness, on the other hand, diminishes such impairments. It also allows individuals to observe their habits of mind and begin to change their reflexive reactions.

THESE INTERNAL PRACTICES OF MINDFULNESS and compassion can be tools not just for police but for all of us working to transcend our own biases. One challenge of this work is recognizing all the ways our culture signals and enforces the human hierarchies that underlie bias and the extent to which we are engaged in the project of maintaining them. For those of us who have privileged aspects of our identities, this recognition is difficult in two senses of the word. First, it is practically hard. Privileges—with regard to gender identity, sexual orientation, race, ethnicity, or beyond—can create a sort of blindness to others' suffering: when the world unfurls its red carpets for you, it is challenging to see the shards of glass others must traverse. This lack of insight with regard to race, philosopher Charles Mills points out, yields the ironic outcome that "Whites will in general be unable to understand the world they themselves have made." Moreover, these cognitive shortcomings are psychologically and socially useful. Avoidance of reality, misunderstanding, and self-deception were required for colonization and enslavement, Mills points out, just as they are required today to maintain the status quo.[57]

But cultivating awareness, attention, and non-judgment—fostering mindfulness—enables us to perceive realities to which we may not have had access before. Practicing mindfulness, Kang told me, makes it more possible to perceive the "luminous, clear equality that everyone has rather than the labels that create prejudice and stereotyping." A psychologist working in an inpatient psychiatric facility described to me how his mindfulness practice helped dissolve the instinctive fear he had previously felt around some patients.

"Because I know my own heart," he said, "I can relate to others' hearts."[58]

Law professor Rhonda Magee, who has taught mindfulness to thousands of students, describes how the practice allows people to engage deeply with difficult questions of race and racism, such as whose suffering one has and has not been conditioned to see. Some students have shared that the skills brought by mindfulness allow them to move from understanding that race is a construction to actually deconstructing it in real time. One White student, for instance, gained a particular type of self-knowledge: he had always insisted that he didn't "see" race, but with mindfulness, he uncovered that in fact he had been actively turning away from the subject. This kind of shift in awareness might seem subtle, Magee says, but it can change a person's entire trajectory.[59]

Seeing these realities can not only be practically difficult but can also stir up painful emotions. Contending with systemic bias and racism honestly means confronting history. For many Americans, that means grappling with pain this country has never addressed—the brutality the world saw in the video of George Floyd's murder, multiplied and magnified backward over centuries, history like a haunted house one has barely entered. We may newly perceive the harms we have caused or not prevented; those of us in a dominant group may see our own barbarous indifference to others' past and present pain. For a person uncovering one's own biases, these truths can be deeply distressing.

Prominent among the painful feelings is shame, which has been described as the internalization of scorn—the feeling that not only has one done bad, one *is* bad. Shame can lead to a host of dysfunctions: defensiveness, anxiety, a desperate need for external validation. In the context of race, these reactions in White people are often labeled "White fragility" and defined as an inability to tolerate racial stress. But what is perceived as "fragility" may be an unskilled response to horror and shame. And a harmful cycle can emerge in which some resort to bullying and denigrating others: shaming others, out of shame, for feeling shame. Shame has also been described as the stress response turned inward, such that "fight, flight, freeze" becomes self-criticism, isolation, and self-absorbed rumination.[60] All of these can halt the process of deeply examining and transforming one's own biases.

Here, too, mindfulness can point the way forward. One acronym often invoked with mindfulness, Magee points out, is RAIN: recognize, allow, investigate, nurture. In building awareness, mindfulness helps us recognize what is present in our thoughts and our emotions. We learn to allow what is rather than deny or avoid it. We can then investigate what emerges: Why am I feeling this shame, or anxiety, or fear? What is beneath this reaction? If we do this in a spirit of nurturing or compassion, we can transform these difficult feelings. Thich Nhat Hanh prescribes meeting our own suffering as a mother embraces a child. Our enemy is not ourselves or any person, he says: our enemy is the violence, ignorance, and injustice in us and in others. Compassion is shame's antidote.

Now we can look deeply at our own dysfunctional patterns, deeply enough to uproot them: the lies we may have unwittingly absorbed, the harms we may perpetuate. When we inadvertently cause others harm, as Magee writes in her book *The Inner Work of Racial Justice*, mindfulness practices also help us "develop confidence to repair our relations with others."[61]

Mindfulness does not let anyone off the hook; in contrast, it makes reality impossible to ignore. While it alone cannot magically dissolve bias, it can facilitate uncovering what lurks in our own hearts and minds and tolerating what we see and feel. In fact, Kang conducted a study in which she found that engaging in compassion meditation made individuals react less defensively to potentially threatening messages—and change their behavior more. Her study looked at people with obesity who were presented with messages about healthy behavior, but defensiveness is one of the biggest barriers to people grappling with their own biases, too; the use of compassion meditation may be a crucial tool in creating an open, receptive response to information that people might otherwise reject.[62]

Indeed, if those of us who experience defensiveness can witness our own response, we may be able to see that, like any emotional reaction, it is a sign pointing in an important direction: a gift of information from the body to itself. In regard to this country's racist history, defensiveness is perhaps cellular-level recognition of moral responsibility. As philosopher Armen Marsoobian points out, we need not have personally caused a harm to be responsible for it. Collectives, whether a racialized group like White people or institutions like policing, have identities over time. Belonging to them, he argues, involves a moral obligation: if these entities have abused in the past, then current members have a duty to repair them. This nullifies the "bad apples" argument about poor police behavior: every member of an institution is responsible for its integrity.[63]

This repair can begin with improving present-day conduct. But in any institution or collective, reducing bias requires much more than relying on individuals to change. This is nowhere more clear than in policing, in which structures and incentives shape behavior, and change means reimagining an institution.

The Watts Jigsaw

"I need your help" was not what civil rights attorney Connie Rice had expected to hear from the chief of the Los Angeles Police Department when she picked up the phone one day in 2003. For nearly fifteen years leading up to that call, she had battled the department and the county sheriffs. She had, as part of legal coalitions, sued them for racial discrimination against Angelenos, for beatings and brutal interrogations of suspects, and for civil rights abuses like using dogs to illegally attack Black and Latino teenagers. She had sued the department for bias against its own officers—for discrimination in promotions and the mistreatment of women, gay, Black, and Latino police.[1] She knew the LAPD's dodges and weaves by heart, the way a boxer can anticipate the moves of a familiar opponent. The most likely request from this longtime foe? That she leave them alone.

But when Bill Bratton, the chief, asked for assistance, she said yes. To understand why, it's helpful to know something about the history of the LAPD and about Connie Rice. Rice was the daughter of an air force colonel and a biology teacher, and she had grown up on military bases around the world. When she was a child, her mother tried to prepare her for racism, explaining that some people had a brain sickness that made them think that the Rice family was inferior because they were Black. It was an effort at inoculation, a way to immunize Rice against a lifetime of attempts to diminish her. When White children taunted her or when White adults marveled at how "articulate" and "well-mannered" she was, Rice remembered they were ill, and she pitied them.[2]

But as the family moved between England and Japan, Arizona and Washington, DC, Rice also saw stark differences between her own life within rarefied military circles and the material privations many of her classmates faced, like the first grader she befriended at her DC elementary school, whose clothing reeked of urine—a boy she defended from bullies and later thought of as her first client—and the children of migrant farmworkers at her Phoenix junior high, who lived in tin-roofed shacks at the edge of town. By the time she graduated from Harvard and NYU Law School, Rice had developed a keen interest in advocating for marginalized people. Doubtful her vision would always align with her employers', on the first day of every job, she drafted a standing letter of resignation.[3]

In 1990, Rice landed at the Los Angeles office of the NAACP Legal Defense and Educational Fund, a practice founded by U.S. Supreme Court Justice Thurgood Marshall and focused on civil rights. (Its most famous case: *Brown v. Board of Education*.) When she arrived, she crisscrossed the city, asking residents about their biggest concerns. "Everyone," she told me, "from upper-class African Americans to underclass gang members said LAPD is the scourge of our existence."[4]

At the time the LAPD was in the midst of what Rice called a decades-long campaign of abuse, humiliation, and degradation of Black and brown Angelenos. Over the years, the department had evolved into what was essentially a paramilitary organization, and its brutality was aimed at communities of color. This was the department that, after the rebellions of the 1960s, pioneered the use of SWAT teams— Special Weapons and Tactics—to deploy explosive devices against the city's own residents. At one point, officers looking for drugs used an amphibious armored truck and a fourteen-foot steel battering ram to tear open the wall of a house. In another raid, police wrecked two apartment buildings, beating residents, destroying furniture, pouring bleach on clothing, punching holes in walls, and shattering toilets with hammers. Officers threw a dining room table out a window; one mauled a living room wall with a homemade battering ram after finding a suspicious packet of white powder. The powder turned out to be flour. The goal had been to send a message that drugs and gangs would not be tolerated. In the end, the police found fewer than six ounces of marijuana and less than an ounce of cocaine. Twenty-two people were left homeless.[5]

To see what was happening for herself, Rice would drive around L.A. at night, disguised: her hair wrapped, her pearls and tailored St. John suits swapped for street clothes. She'd park her Honda Civic and watch officers pulling Black men out of their houses and lining them up under klieg lights to take pictures of their tattoos. "No warrants, not even reasonable suspicion, just everybody Black, they'd pull them out of their house," she said.[6]

Yet when Black and Latino Angelenos were themselves victims of crime, police responded with less urgency. The result was a police apparatus that was simultaneously too much and not enough. It suppressed communities of color with aggressive arrests and "search and destroy" tactics, keeping them from the wedding-cake chateaux of Beverly Hills, and at the same time failed to protect those communities from harm.[7]

This mix of brutality and indifference was put on display for the world shortly after Rice arrived in Los Angeles. In 1991, four White officers were taped beating and tasering Rodney King, an unarmed Black man, while more than a dozen others watched. When a jury acquitted the officers a year later, the uprisings that broke out burned and decimated Black, Hispanic, and Korean communities. But officers withdrew from the violence; while the fires mounted, then chief Daryl Gates attended a fundraiser in the tony neighborhood of

Brentwood. More than fifty people died and more than two thousand were injured. A post-unrest investigation found damning evidence of widespread departmental racism: officers had used codes like "NHI" when responding to calls from Black and other marginalized communities. "NHI" stood for "no humans involved."* Rice and her colleagues kept filing lawsuits against the LAPD. And then one day the phone rang.[8]

Bratton, the chief asking Rice for help, had been brought in to clean up a mess. The department had been upended by a disaster known as the Rampart scandal. Dozens of officers were implicated in a pattern of blazing and criminal corruption. "Anti-gang" officers had in effect become their own gang, planting guns, stealing and selling cocaine, beating and shooting suspects, participating in a bank robbery, and framing civilians. They even colluded with immigration services to have witnesses deported from the country. Supervisors let it happen. The scandal nearly upended the city's entire criminal justice system. The corruption had tainted thousands of criminal convictions and prompted more than 140 civil lawsuits against the city; the U.S. Department of Justice threatened its own lawsuit unless the city be placed under a "consent decree," which meant being supervised by a federal judge. Under the consent decree, the department kicked off reforms: tighter officer discipline, use of force changes, better investigation of complaints.[9]

Bratton wanted to understand whether the reforms had actually addressed the underlying causes—and why the disaster had happened. Rice agreed to help because she had come to a realization. Litigation was powerful—even exhilarating—but it was limited. After hammering the department with lawsuits, she saw that courts alone could not change police behavior. They could only, as she put it, curb the "screaming neon of abuse." They could enforce constitutional compliance; they could not enforce kindness or decency. They

---

* The attitude can be traced to the top: Gates attempted to explain the deaths of Black residents during police choke holds that constrict blood flow to the brain by saying, and this is a verbatim quote, "In some blacks when it is applied, the veins or the arteries do not open as fast as they do on normal people."[10]

couldn't even limit all abuse, because not all of it was unconstitutional. They certainly couldn't influence those decisions left to an officer's discretion.[11]

If Rice wanted to change how police behaved, she realized, she'd have to understand how they thought. To do that, she'd have to go inside.

It was in the eighteen months after that phone call, in 2003 and 2004, that Rice interviewed more than eight hundred officers—Black, Latino, White, and Asian American cops, women and men. The police culture that had made the corruption possible, she found, was defined by insularity and a warrior mentality. The department was also marked by callousness, to the public and to one another. "One epiphany for me," she said, "was that they treat each other so cruelly."[12]

Crucially, Rice saw that the officers lacked a relationship with the communities they were supposed to protect and serve. There were well-meaning cops—she'd witnessed them herself—but they went into neighborhoods only to get information and make arrests; they lacked meaningful connections. This lack of familiarity led them to stereotype entire communities. Watts, for instance, in South L.A., is a historically Black (and now largely Latino) area that suffered from overcrowding, lack of jobs, and struggling schools. It also suffered from high levels of violence, primarily stemming from gangs. Often, police stereotyped Watts residents as potential gang members.

This dynamic led to a deep, pervasive, racialized fear on the part of the police. It was not only the instant mind-stopping panic of the kind we saw in Jeronimo Yanez, whipped up when emotional dysregulation meets a racial stereotype. There was an ongoing baseline of mistrust. Officers treated everyone as a possible menace.

"If you're afraid of someone," Rice said, "you objectify them. They're not a human being, they're a threat. So, everything about them is scarier than the actual facts." Fear can distort and pervert the threat assessment. The way police operated fostered a lack of familiarity, which led to fear and dehumanization.[13]

This was the limit of litigation, Rice realized: the law can prevent illegal abuse, but it can't alter the way a person feels. "A court can't order a cop to love a poor Black kid," she said. "A court can't order police to change their fear of Black people."[14]

This deeper work would require meaningful connections. How to begin? "You hang around with folks. You get to know them," Rice said. If she could orchestrate new interactions between police and the communities they served, perhaps she could neutralize this fear.

What would replace it, Rice didn't know. Familiarity and respect were essential. Trust she could hope for. Relationships that could begin to mend what decades of discriminatory policing had broken—maybe. Love? Love was an unreasonable aspiration.

IN 1954, A PSYCHOLOGIST NAMED Gordon Allport published a book called *The Nature of Prejudice*. It was the first social-scientific effort to analyze the phenomenon systematically. In it, he described the landscape of bias in the 1950s in terms that sound eerily familiar. Refugees "roam in inhospitable lands," he wrote. Many people of color around the globe "suffer indignities at the hands of Whites who invent a fanciful racist doctrine to justify their condescension." Diaspora Jews find themselves "surrounded by anti-Semitism." The book is mostly diagnostic, but Allport also ventures some ideas for reducing prejudice expressed by people in dominant groups. One approach is replacing ignorance of others with deep knowledge through a specific type of interpersonal contact. Casual, meaningless contact doesn't work, he writes, because it can fail to offer new information, adding only more mistrust and suspicion. Interpersonal contact will only decrease prejudice, Allport asserts, if people from different groups can join together with equal status to work cooperatively. They must have common goals. And the efforts should be backed by an institutional authority.[15]

This idea became known as the "contact hypothesis," and it has an intuitive appeal: if one group of people can get to know another group as equals, perhaps false beliefs can be replaced with complex, realistic perceptions. And it's been born out in many contexts. Toward the end of World War II, the U.S. Army integrated several combat divisions. White soldiers who fought beside Black soldiers underwent a "significant change in their racial attitudes," a later survey found. Among White officers and enlisted men, prior prejudice against Black soldiers was replaced with admiration and respect. "They were the best platoon in the regiment," wrote one officer. "I wish I could get a presidential citation for them."* In a study from the 1980s, students spent time in a substantive conversation with a gay man; later, those students expressed significantly less homophobia in survey responses† than those who had not been in conversation with him. A program that integrated people with disabilities into a workplace also improved the prejudicial attitudes of their coworkers.[16]

Of course, there are limitations to the power of contact. Gordon Allport noted that contact between groups is less potent without the approval of a powerful authority and, indeed, contact alone cannot override other social and political forces. The twentieth century is rife with examples of intimate relationships that failed to counteract prejudice—Germans denouncing Jewish coworkers, Rwandan Hutu turning against Tutsi family members.

As is the case with much prejudice research, this approach tends to focus on the attitudes and experiences of people in positions of advantage—how *they* change, how *they* benefit. The army did not ask the integrated Black soldiers how this experience affected them; no one asked the gay man how these conversations affected him. And the toll can be great, especially if the less advantaged group is further objectified, becoming an instrument for others' self-improvement. Meaningful contact as Allport prescribes it is often an effort to counteract the unequal distribution of power. Here is an additional demand placed on the less powerful.[17]

Further, each group may experience interpersonal contact quite differently, as one study of Black and White conversation partners found. When Black individuals expected their White partner would be prejudiced, they worked harder to connect and also felt worse

---

* The resulting report, based on the surveys of hundreds of White officers and enlisted men, was never made public out of fear of alienating segregationist congressmen and encouraging the NAACP to push for further integration of the military.

† Survey questions included whether students would be comfortable with gay coworkers or teachers and whether they would want to have friends who were gay.

about the interaction afterward. But in this same scenario, the White individuals liked their Black partners more. One encounter, two different impressions—a tiny example of the massive complexities inherent in connections between groups. These are real risks. While there is a growing number of rigorous real-world experiments that test how and why and when contact works, the specifics of what is required for contact to change people are yet to be fully understood.[18]

BUT FOR CONNIE RICE, IT was clear that the lack of meaningful contact was creating unacceptable levels of danger. By the time Bratton asked her to help analyze the LAPD's response to the Rampart scandal, she was running an offshoot law and policy group called the Advancement Project that focused on the problems facing the most marginalized communities. Rice's focus was trained on safety. In the public housing developments in Watts, for instance, Rice witnessed family strength and love, creativity and resilience, and she also saw people terrorized and traumatized by violent, multigenerational gangs. She was winning her discrimination cases, but her clients were losing their lives. If her clients were cut off from the civil right of basic safety, no other rights—the right to get work, be promoted, or go to school—mattered. And through anti-gang work, she saw how completely entwined gangs were with status quo policing.[19]

LAPD's aggressive approach, she saw, wasn't only failing to promote safety but was actively undermining it: the destruction of trust between police and the community created a vacuum that empowered the gangs, who were responsible for the majority of the violence. Lack of trust and police's failure to protect witnesses meant people were reluctant to report crimes or help police solve them. This allowed gangs to operate with impunity.

In 2007, Rice and a colleague, a lawyer named Susan Lee, along with dozens of subject matter experts released a thousand-page report detailing the ways in which Los Angeles's approach had led to disaster. The city, they concluded, had failed to see gangs as a public health epidemic. It used disconnected, tactical approaches against what was a strategic threat, arresting individual gang members instead of executing a large-scale coordinated effort to change behaviors, norms, incentives, and opportunities. The approach was akin to addressing a viral pandemic by only hospitalizing individual cases instead of widespread testing, masks, and contact tracing.[20]

And the city's failures were evident in the data: a nearly thirty-year "war on gangs," had not dented the problem. From the 1980s to the early 2000s, one hundred thousand people were shot and fifty police officers were killed. In many years, homicides in L.A. County had surged beyond a thousand a year.[21]

What was needed was a comprehensive strategy to address the conditions that allow violence to emerge, one that would make communities healthy and strong enough so that young people would not be drawn into gangs. This included strengthening schools, supporting families, eliminating the inequalities embedded in the prison system, bolstering opportunities, and creating role models. In short, opening viable alternative paths. As a priest who worked for decades with gang members said, he had never met a child with hope who joined a gang.[22]

To Rice, the first order of business was day-to-day safety. To secure it, LAPD needed to have trusting relationships with community members. And that meant fundamentally changing officers' behavior.

In a small pocket of L.A., this was already happening. A volunteer-led group called the Watts Gang Task Force had formed in 2006. Founded by people who had lost loved ones to gangs and gang violence, and former gang members who had become professional gang interventionists, the group began meeting to build community-led efforts. Phil Tingirides, then a police captain in the division that included Watts, attended early meetings. Tingirides is in his fifties, White, and grew up in part in South L.A. When he sat in on these meetings, the community shared their anger over past police abuses. He listened to their rage and hurt, and he didn't offer justifications or shirk blame. Instead, he owned the history. Tingirides used the word "we" when describing police violence: when *we* shot that kid; when *we* locked that man up. In response to the community's frustrations, Tingirides began investigating crimes in the neighborhood with renewed urgency. He reassigned officers who had acted disrespectfully and coached others to behave with more compassion. Soon, Tingirides began working with a sergeant in the Southeast Division, Emada Castillo (later Emada Tingirides—they married), to connect to schools, bringing community-focused officers to read to children during the school day.[23]

But this was an isolated effort. Sustainable change needed support from the top. Institutions are not typically excited about sweeping changes, and political leaders, Rice realized, often don't focus on the marginalized unless they're forced to. No one loses their job when a poor child is unable to walk to school for fear of safety. And the structural reform Rice and Lee were proposing was deep.

Then, an opening appeared. In 2010, Rice's team was finishing another report, this one on improving safety in Watts housing developments. On the day the report was released, a family was robbed and assaulted by a gang as they were moving into their new apartment in one of the same housing developments described in the report. A maintenance worker heard screams and intervened just before an attempted rape.[24]

Rice and Susan Lee asked for an immediate meeting with the new chief of police, Charlie Beck. They could submit report after report, they realized, but none of those doorstops were going to force necessary changes. The attack was proof that police in Watts were failing to provide basic safety, they asserted. Then they proposed the creation of a new police unit.[25]

The new unit they outlined would consist of around fifty officers dedicated to three public housing developments in Watts and another in East L.A., all areas wounded by multigenerational gang violence and trauma. The officers would patrol on foot, not in cars. They would be required to make a five-year commitment to the unit. They'd receive more pay. And, crucially, they would be beholden to different incentives. They'd be instructed to shift their focus from making arrests to building relationships, and they would earn promotions based not on the number of arrests but on their ability to demonstrate that they'd established trust in the community.[26]

Rice and Lee were, in effect, asking to overturn decades of standard police practice.

"Okay," Beck said. "Let's try it."

The program was dubbed the Community Safety Partnership.

Emada Tingirides was put in charge of its administration. As she read applications from LAPD officers, she looked for people whose life experiences would allow them to understand the community. Tingirides herself had lived in Watts as a child, right across the street from one of the housing developments. Her mother had been fifteen when Tingirides was born, and she had spent time in foster care, a personal history that allowed her to powerfully relate to life experiences of the people in the community. People going through hardships would say to her, "You don't understand," and she'd say, "I do understand." She also saw what was special about the community: pride, she said, and generations of families who survived and wanted to make it better.[27]

Tingirides sought out Black and Latino officers, particularly those who were bilingual and had ties to the neighborhood. She looked for cops with skills in conflict resolution who were open to new ways of policing, or who demonstrated promise in those areas. While officers would still be empowered to arrest people who were behaving violently or to bring in outside resources if a situation became dangerous, arrests would be a last resort. For forty-five slots, CSP received nearly four hundred applications.

Rice and Lee then began a door-knocking campaign, interviewing residents about what the community needed. As Rice presented the vision for CSP, she recalls, she didn't sugarcoat the magnitude of what it would demand of them: to cooperate with the police department that had inflicted violence on their own families, that had killed or abused or wrongly jailed their sons and grandsons. "I'm here to ask you to do something I don't think I could do myself," she told people. "I'm asking you to work with these cops because we have to change them. And they can't change without your help."[28]

Responses ranged from incredulity to anger. "You're going to pay for cops to knock on my door?" furious residents asked. "Are they going to harass us for getting together on our porches?" But others quietly revealed that they thought this kind of investment in the community was long overdue.[29]

The program had other ambassadors, too: gang interventionists who had worked in the community. Initially, they, too, were dubious. Andre Christian, a gang intervention worker, said the initiative had all the

makings of another kind of suppression, a prettified version of surveillance and control. He knew it well: he had grown up in the neighborhood
and had once been choked by an officer for calling him "man" instead
of "sir." But he knew Rice personally through their anti-gang work, and
he trusted her. The requirement of a five-year commitment meant it was
serious, not another Mickey Mouse effort that would vanish after a year.
When Rice and Lee asked Christian and other gang workers to help spread
the word about CSP, they agreed. They had already earned credibility in
the community by defusing tension and preventing retaliatory shootings;
when they showed up at someone's door to talk about CSP, Christian said,
"It wasn't like someone from the moon coming in." Said Susan Lee, "It's
one thing for a short Asian woman to show up. It's another thing for a
gang intervention worker who they respect as a peace builder."[30]

Once recruited, the officers also underwent a new kind of training. Beyond courses in weapons handling and tactical breathing,
they learned dialogue and context. Rice and Lee brought in experts
to teach the history of Watts and answer one of the officers' biggest
questions: "Why does this community hate us?" Younger police, Rice
said, often didn't understand that their uniform represented a legacy
of abuse.[31]

So they were introduced to the history of slave patrols. While the
history of American policing typically taught in training academies
traces its origins to northern police agencies, these were not formalized until the mid-nineteenth century. Slave patrols, the first publicly
funded police organizations in the South, date back to the early 1700s
and, as one Southern governor wrote, were "intimately connected
with the good order and police of the state." These police were tasked
with searching the structures where enslaved people lived, breaking
up their gatherings, and controlling their movement. They inspected
the "passes" slaves were required to carry in public that showed their
destination and reason for traveling. And they violently beat slaves
for infractions such as traveling without a pass. They roamed "beats"
under leaders called "captains." Rice was not shy about pointing out
the resemblance between modern police badges and those worn by
slave patrollers.[32]

Melvyn Hayward, another gang specialist, taught officers about the different cultures in the three developments—while the population was now majority Latino, Black residents still held most of the leadership roles, he noted, and communicated the most with law enforcement. The Latino residents were more reticent. Multigenerational Chicano communities preferred to keep their concerns private; immigrants who had recently come from countries where police were entirely corrupt largely avoided interacting with law enforcement. Others feared deportation.[33]

This cultural understanding was essential, particularly because the proposed vision centered on human connection. "When you're out there," Rice told the recruits, "I want you to see these brown and Black kids as your own children. I want you to protect them as if they were your own children." Rice remembers the officers sitting in silence as they absorbed this mandate. "You're no longer in the arrest business," Rice said. "You're in the trust-building business."[34]

To that end, Rice and Lee invited community leaders to collaborate on a project with the officers, right in the training room. These were pastors and block presidents, elders everyone went to for advice. Lee had blown up giant maps of the neighborhood and handed out colored stickers, asking the community leaders and the officers who knew the neighborhood to identify "hot spots" where crime and safety were concerns.[35]

Some community members balked. It seemed like snitching. But eventually someone stood up and added a sticker to the map, like an alley that needed attention. Someone else added a sticker to a corner where kids would get shaken down for pocket change. It was an early hint that cooperation was possible. Police listened and offered ideas. The community watched the police, in real time, focusing on community safety, not just making arrests. And the police who had preconceptions about Watts heard leaders speaking authoritatively about the challenges they faced and the resources they had already marshaled.[36]

There was a shared goal. There was cooperation. There was a sense of partnership. Each side began to see something new in the other. At the very best of these meetings, the two groups exchanged phone numbers on their way out the door.

THE COMMUNITY SAFETY PARTNERSHIP LAUNCHED in November 2011, beginning what one CSP officer called "community policing on steroids." Phil Tingirides, the captain who had connected with the Watts Gang Task Force, oversaw the Southeast Division that included most of the CSP. He exhorted officers to use their imagination to come up with programs that would attract kids. "If it's legal and moral," he said, "try it."[37]

At the formal kickoffs, the officers introduced themselves and brought food trucks and barbecue. They joined baseball games. Over the following weeks, working without a blueprint, the police set out to make connections. They focused on listening. They apologized for past wrongs and asked for forgiveness. They encouraged people to be truthful about their feelings toward police and thanked them for the opportunity to listen.[38]

The interactions did not always go well. When CSP officers dropped in unannounced on a fathers' group in the Jordan Downs housing development, they stood around awkwardly, holding their walkie-talkies. The fathers, who were in the midst of sharing personal stories about their children, were bewildered by the police presence. One officer began talking about troublemakers in the community and referred to them as "knuckleheads" and "motherfuckers" until a facilitator explained that profanity was not allowed in their meetings. Later, the policeman scribbled an apology and a smiley face on his business card, and handed it over in an effort to make amends.

One of the fathers pointed to a different officer. "I remember you. You messed me up. I wasn't doing anything, and you came in and handcuffed me up in front of my kids and locked me up," he said. "The next thing I knew I was serving thirteen years in the pen because of you." Then he ran out of the room. The other fathers looked at one another in silence. At the next meeting, the officer returned, explaining it couldn't possibly have been him, he hadn't been on duty the day the man had been arrested. The father was mistaken. But he had not returned.[39]

CSP officers found themselves having to unlearn habits that were deeply ingrained over years—sometimes decades. Some had been known for their past callousness. Andre Christian, the gang worker, said that one of the CSP officers used to jump out of his patrol car and act so aggressively, he imagined he had never been hugged as a child. After CSP started, that cop chased a man down, and, using force, arrested him. Tingirides berated him: arrests were to be used as a last resort. If the officer kept policing in the old way, he'd put the credibility of the entire program at risk. Christian said he could always tell when officers had been chastised—the next day they'd look sheepish.[40]

But slowly, the CSP began finding how to be helpful in ways the community requested. When elders in the Jordan Downs development asked that an alley filled with old mattresses and used for drugs and prostitution be cleaned up, the officers brought hoses and trucks to haul away the trash. They set up camping trips, helped create the first Watts Girl Scout troop, and played basketball with residents. They ran school backpack giveaways and health fairs. They paid special attention to the elders of the community, bringing the bifocals they needed for everyday mobility.[41]

The point, Tingirides said, wasn't to come in "on my shining horse to save the community." Often the CSP officers were amplifying existing efforts and coordinating resources. In the Ramona Gardens development in East L.A., an underground farmers' market had developed as a source of fruits and vegetables in a food desert. Officers helped residents get permits, tarps, and stands to operate the market as a legal, viable business. In Nickerson Gardens, under the aegis of CSP, a resident named Lupita Valdovinos and an officer named Jeff Joyce started a youth mentoring program, with soccer games, a homework club, and Saturday classes on stress and healthy eating. Soccer coaches came from the community, with Joyce coaching alongside.[42]

Changes in the officers became noticeable. Mothers noted the difference in the way they treated their children—instead of harassing or intimidating them, they would talk with them kindly. "This is a different look, a different feel," said Melody Culpepper, one mother. A resident named Lucelia Hooper, who moved to Nickerson Gardens in the 1970s, said, "They show me the love and respect that I deserve." When one man started having a seizure, a CSP officer began giving him mouth-to-mouth resuscitation. Because the man had vomited, she suctioned vomit from his mouth with her own. A neighbor approached her afterward, surprised. "Even I wouldn't have done that," he confessed.[43]

Officers' fear diminished, too. At the first baseball games, the officers had worn their bulletproof vests and kept their guns holstered. A year into the program, they played in shorts and T-shirts and left the guns in their cars. When, one day, a young child ran toward the cops in Nickerson Gardens holding what looked like a nine-millimeter gun, the police didn't reach for their firearms. They had spent so much time getting to know the children, they didn't assume the boy meant harm. The child remained safe. The gun turned out to be a toy. Had twelve-year-old Tamir Rice been judged the same way, he would likely be alive today.[44]

And the community's attitude shifted. Every positive interaction, Christian said, sent ripples out into the community. Word traveled that officers were treating people with respect and care. People began nodding or waving to cops. They'd shake hands, talk about sports. Phil and Emada Tingirides would, off-duty, visit locals who were in the hospital because they had by now become friends. Officers started inviting children over to their homes for dinner, and residents invited them back to quinceañeras, graduations, funerals, or to be *padrinos* at baptisms. The community is "getting overwhelmed with so much goodness," said Christian. "It's still coming. It won't stop."[45]

CSP officers still made arrests for violent behavior. Gang units were called in at times. But so far the unit has not resorted to the traditional wide-scale suppression methods of the past. While it's difficult to show causality, an independent analysis suggests CSP is decreasing arrest rates. In Jordan Downs, Nickerson Gardens, and Imperial Courts, arrests rates at the end of 2019 were about 50 percent of their levels in the year leading up to CSP. At one point, when there was a shooting suspect in one of the developments, the police didn't kick down doors to find him. Instead, they went from apartment to apartment for two hours, talking with people about what was happening. An hour later, multiple people called with information on the shooter's location.

Officers made the arrest, but without inflicting additional trauma on the next generation of children.[46]

Safety for residents increased, too. The independent analysis compared CSP sites to similar developments and found that CSP cut violent crime by nearly a quarter compared to expected levels. Homicides in particular have dropped precipitously. In the ten years prior to CSP, there had been seventy homicides in the three most dangerous CSP developments, including twenty-five homicides in Jordan Downs alone. In the nine years since the start of CSP, these developments combined have registered twenty-one homicides—a decrease from an average of seven per year to about two. According to Tingirides, most were also solved within two weeks,* in one case because a resident called a CSP cop on his cell phone and simply told him who had committed the crime. Susan Lee recalls LAPD leadership having difficulty accepting that arrests and crime could decrease in tandem, but these results support what many scholars maintain: crime reduction requires neither incarceration nor aggressive policing.[47]

The connection between police behavior and crime is worth emphasizing. Research by legal scholar Tracey Meares, sociologist Andrew Papachristos, and others suggests that when people see police and the legal system as legitimate, this alone can reduce crime. The researchers have found that past offenders who have positive perceptions of the police in turn hold stronger beliefs in the legitimacy of the law; those who believe more strongly in the law's legitimacy are more likely to comply with the law. Because people's perceptions of police and their legitimacy improve when police treat them fairly, respectfully, and honestly, officers' fair treatment may alone serve as a form of crime prevention. The community members and former gang members add this: if you have hope, and resources, and the certainty that someone cares about you, you don't turn to crime.[48]

And it's not the case that crime is simply being displaced to other

---

* In law enforcement, a "clearance rate" refers to the percentage of crimes in which an arrest is made or a suspect is identified. In LAPD, the typical clearance rate for homicides is 60 or 70 percent, about the national average. In the division containing the three original Watts CSP developments, the rate at one point hit 87 percent.

neighborhoods. An independent analysis by UCLA released in March 2020 found that CSP reduced crime without simply shifting it elsewhere: while Los Angeles experienced an overall reduction during the years since CSP was introduced, the CSP housing projects outpaced the city's overall trend. The analysts estimate the program alone was responsible for preventing approximately 221 violent crimes over a six-year period, saving the city over $90 million. In over one hundred interviews and twenty-eight focus groups, people shared that they felt safer going outside, mingling with neighbors, using green spaces—that their lives had been changed in concrete ways. Officers said that community members' feeling safer was the most gratifying part of the job.[49]

The CSP unit possessed some of the characteristics suggested by the contact hypothesis. There was a spirit of partnership, and cooperation. There was a common goal of community-wide safety. It all took place with the support of authorities, both that of the local leaders and of the LAPD. And there was another characteristic, not mentioned by Allport, but essential to the process. It had to do with the homework club.

ABOUT FIFTEEN YEARS AFTER THE *Brown v. Board of Education* decision, a social psychologist named Elliot Aronson was asked to help the city of Austin, Texas. The schools had finally, belatedly, been desegregated, and Black and Mexican American children were being bused to schools attended by White children. Within a few weeks, fights broke out in elementary classrooms, and the playgrounds had become tense battlefields organized along racial lines.[50]

When Aronson came to the district, he saw that the competitive nature of the classrooms only exacerbated the problem. Children were always working against one another; nothing in the daily routine encouraged mutual understanding or respect. He wondered whether the classroom could instead become a setting in which students could heal distrust and even grow to like one another. So he tried an experiment.[51]

In some rooms, he arranged the children into teams of about six students, with each team comprised of children of different racial and ethnic groups. He adjusted the children's school projects so that they

had interlocking parts, like a jigsaw puzzle. Every student was responsible for a key contribution. If the subject was gardening, one student
focused on flowers, another on soil, another on vegetables, and so on.
The flower students were all given material related to their topic, and
they met together in "expert groups" to discuss what they learned; so did
the other specialties. The experts then returned to their teams and taught
one another what they had learned. Students had to listen and ask meaningful questions to help the experts share their ideas. The team's overall
success was dependent on each member's contribution, and each expert
contributed something essential for the others.

Aronson and his team found that after six weeks, the children working
on the collaborative projects had higher self-esteem. They liked school
better. They were better able to envision others' emotional perspectives.
What's more, the fighting stopped. Students started to like the members
of their team more, including those in other racial or ethnic groups.
They also liked the rest of their classmates more than they had before
the experiment and began seeking out classmates who they had previously distrusted. After the experiment, Aronson climbed onto the roof
and took aerial photographs of the playground. The photos showed that
children from different racial and ethnic groups had become friends.[52]

In a jigsaw classroom, every student was recognized as contributing something important. Shy, less confident students were bolstered
by the help of their expert groups and grew to be seen as essential team
members. Students grew to perceive one another as trusted resources,
and they succeeded by helping one another. The jigsaw method met
Allport's conditions for the contact hypothesis: equal status, cooperative
work toward a shared goal, the support of an authority or institution.

The jigsaw approach has been found to reduce stereotyping and
prejudice in a variety of schools from Germany to Australia. In fact,
half a century before Aronson formalized the approach, a young Black
woman named Juliette Derricotte tested it as a way to help integrate
midwestern universities. In the 1920s, Derricotte and a White colleague,
both working for the YWCA, met with White and Black students.
Prefiguring Milton Rokeach and his demonstration that confronting
people with their own inconsistency prompts change, they pointed

out the hypocrisy of the students calling themselves Christian while refusing to speak with their classmates. Derricotte then assigned them joint research projects on the history of slavery, racism, and Black cultural contributions. Over the following year, they worked in interracial teams to answer the questions she had posed. The cooperative study was so successful that the teams began sharing their findings across the university and creating public programs of Black poetry and music.[53]

In Austin, Aronson and the researchers also stumbled on something else, Aronson told me: the way giving changes the giver. Jigsaw students served one another with the gift of their hard work and expertise. In Aronson's view, when you serve a person, having negative stereotypes about them creates an internal conflict. To reduce the conflict, you change your idea of the other person. "If I'm stereotyping you, and then I go out of my way to help you with something, I'm working hard for you," he said. "I downplay negative thoughts about you and emphasize everything I see about you that's positive."[54]

Gifts can also change the receiver. In his classic book *The Gift*, Lewis Hyde writes that gifts "carry an identity with them, and to accept the gift amounts to incorporating the new identity. It is as if such a gift passes through the body and leaves us altered." Gifts also infuse social connections with new emotions, creating, as Hyde says, a new "feeling-bond between two people."[55]

The presence of meaningful interdependent contact, supercharged with the power of giving, may help explain why prejudice dissolves so rapidly in soldiers fighting side by side in war. In combat, soldiers are willing to give their own lives for one another. It may also explain why sports can be such an effective tool for reducing prejudice. Here, too, players come to depend on one another and also offer the gift of their skill or, at times, the ball. In one recent study, a cricket league was set up to test whether playing on the same team for eight months could reduce prejudice among men of different castes in India. In another, a soccer league in Iraq made up of Christian players was integrated with Muslims. In both cases, some forms of intergroup prejudice dropped as a result of the experiment. The Christian soccer players on the integrated team were more likely to later sign up for a mixed-religion team and to vote for a Muslim to receive a sportsmanship award. The Indian players who had teammates from other castes later reported more cross-caste friendships outside the team, compared to a control group. They also chose more men from other castes as future teammates.[56]

The CSP program is, in some ways, a jigsaw program. The community and the officers collaborate, in a spirit of cooperation, on the shared goal of community safety. Each party contributes something of value. On the Nickerson Gardens soccer team, officers and residents coach the children together. In another, Safe Passage, officers and parents form walking teams to ensure children can safely get to school. By performing acts of service, the CSP cops invest in the community. "When I serve you, I invest in you, I imbue you with new value and new esteem," Rice said, echoing Aronson. The homework club, the camping trips—these acts of service changed the police.[57]

WHEN FACING A COMPLEX DECISION, wrote psychologists Amos Tversky and Daniel Kahneman in a 1974 paper, people often use mental shortcuts. These shortcuts, or heuristics, are quick algorithms we use to make predictions when an actual answer is uncertain to us. For instance, if you're speculating about whether you're likely to be in a car accident, you might use the mental shortcut of "availability," making an estimate based on people who have been in accidents who quickly come to mind. Heuristics are useful and efficient: we rarely have time to understand and process every aspect of a situation. But, crucially, Tversky and Kahneman maintained, these shortcuts can also lead to errors. Indeed, a stereotype is a type of heuristic: it's a quick prediction about another person, and it can often be totally wrong.[58]

Sometimes, however, people are able to make predictions quickly *and* correctly. In his book-length study of decision-making, *Sources of Power*, psychologist Gary Klein uncovered myriad cases of people using mental shortcuts to make essential, even lifesaving, split-second decisions. Nurses in neonatal intensive care units use such shortcuts to identify which two-pound baby in the ward is about to develop an infection, even before it shows up on a test. Experienced firefighters sense when a roof is about to collapse just by gauging how spongy it is underfoot. These decisions, sometimes called "intuition," can often be right.[59]

Why are some quick predictions right and others wrong? In a jointly-authored paper (subtitled "A Failure to Disagree"), Klein and Kahneman propose that features of the thing or situation we are assessing are a major factor—it's easier to make good predictions, for instance, about predictable environments. But we also have to have the opportunity to learn deeply about those environments. We have to develop the ability to truly recognize what we see.[60]

Take, for example, learning to recognize a cat. When we are children, learning what a cat is, we first see a variety of different members of that category—alley cats and Persians, Garfield and barn kittens. Each time we see an individual cat, sensory information about the animal enters our brain. Our brain looks for commonalities, storing the essential traits that cats have in common—furry, whiskered, smaller than a breadbox—as a pattern of neural activity. Our memory for this category—as for any category—is in fact a specific pattern of neural activity, ready to fire whenever something catlike comes into view. You might think of it as a neural "signature" that the brain recognizes as a cat.[61]

As we go out in the world, our brain laps up sensory input—the glint of light in trees, the chiaroscuro shadows between buildings, the warm face of a friend. All this input begins activating patterns of neural activity. But, because the brain is geared toward predicting what's about to happen, a pattern need only be *partially* filled before the brain decides what it is perceiving. If, as we walk down the street, we see a flash of fur in our peripheral vision, the brain will recognize this as "cat," even with only some of its "cat" pattern filled in.

What's happening here is called pattern completion, and it happens in a couple hundred milliseconds. Rather than wait to take in every bit of information available, the brain absorbs just enough to compare the object of perception with the existing categories stored in memory. As soon as its neurons start firing in a pattern that's more similar to one category than another—at least, more like "cat" than the next closest category—the brain recognizes that as its best guess and the pattern auto-completes. The brain literally rushes to judgment. When compounded with stereotyping, this process has far-reaching, even lethal, consequences: eye-tracking research finds that non-Black people viewing an armed Black person in a simulation make the decision to shoot in self-defense before their eyes fully process the gun.[62]

In everyday contexts, the ability to sort what is perceived into correct categories depends on our ability to differentiate neural patterns—both those that we've cultivated in our minds deliberately and those that we've received passively through cultural osmosis. This allows us to perceive what a given thing shares and doesn't share with another thing. We can see the challenges arise in the case of language. Sounds, too, create neural patterns in the mind, and one's ability to decode other languages depends on one's ability to recognize the sounds. Native English speakers often have difficulty distinguishing among the soft and hard "l" of Turkish, the aspirated and unaspirated "t" of Hindi, or the three different sounds in Korean that all sound like "k" to the untrained, English-speaking ear. Many English speakers haven't had enough exposure to these sounds to have developed the ability to differentiate them from one another.

The brain's facility with different neural patterns—or lack thereof—also helps explain why people can have trouble distinguishing among faces of those of other racial or ethnic groups. An inability to recognize individual faces is at least partially rooted in an inability to distinguish among different neural patterns. This "cross-race effect" is particularly present in individuals who have had little exposure to other groups. Korean adults raised in Korea, for instance, have more difficulty distinguishing White faces from one another than Asian faces, while Korean adults adopted as children by White European parents experience the reverse.[63]

Individuals who are mistaken for someone else of their same racial or ethnic background can feel unseen and erased; they may also be wrongfully convicted. Recent analyses suggest about 30 percent of wrongful convictions in the United States involve eyewitnesses' misidentification of individuals of another race, with about one third of unjustified convictions rescinded through DNA evidence involving the misidentification of Black people by Whites. Indeed, a person convicted of murder in the United States is 38 percent more likely to be exonerated if they are Black.[64]

Humans are, in a sense, experts at recognizing and categorizing people from the groups with which they have the most familiarity. In segregated societies, they tend to be most familiar with their own race. Research shows that as people increase their exposure to other races—as they develop expertise—the problem of the cross-race effect decreases. The more experience one has distinguishing among different members of the same group, the stronger one's ability to see individuals. With expertise comes more sophisticated ways of handling sensory input.

We may think of an expert as someone with a well-developed ability to differentiate among and categorize the neural patterns triggered by the things of the world. A novice has a more limited and crude ability. I, a soccer newbie, might watch a penalty kick and simply see a ball hurtling toward a goal. A World Cup goalie, on the other hand, may perceive the angle of approach, the location of the player's feet, the direction of the player's hips and head, and be able, in milliseconds, to recognize—categorize—the kind of kick and likely trajectory. The World Cup goalie, the NICU nurse, and the firefighter all use heuristics to make quick decisions. But, because they're experts, they can quickly and accurately perceive reality.[65]

But the importance of being able to differentiate extends beyond the basic perceptual processes we've just discussed. It also applies to the way we think about people. Psychologist Ellen Langer demonstrated this with sixth graders in the Boston area in the 1980s. One class of students was, over several days, shown images of people with disabilities. Another class was shown slides of people without disabilities, and every child was given a booklet of questions about the slides. Some booklets contained questions prompting careful attention to the particulars of each person—for instance, students were asked to list how a person in a wheelchair might drive a car, or multiple reasons a blind musician might or might not be good at their profession. Other booklets provoked less attention.

When, days later, the researchers assessed the children's responses to images of people with disabilities, the results were striking. First, they found that the students who had been required to think deeply saw the individuals with disabilities more accurately. When asked who they'd choose as a partner in a game like a wheelchair race or blindfolded Pin-the-Tail-on-the-Donkey, they were more likely to pick people whose specific disabilities would give them a special advantage—considering the blindfolded game, for instance, they chose the blind person to help rather than a child without a disability. But the children who had been asked to think carefully about people with disabilities also became more inclusive: they were more likely to want someone with a disability to join them on a picnic and were more than twice as likely to want a child who was blind or in a wheelchair paired with them in any game at all.[66]

Developing the ability to correctly differentiate—becoming an expert—makes a lot possible. Because one can more easily distinguish among the various members of a group, one can more accurately perceive what makes them special. Thoughtfully attending to distinctions, Langer's study suggests, also illuminates the absurdity of applying a broad stereotype: there are simply too many differences among members of a group for the stereotypes to be correct. This may explain why the children not only saw people with disabilities more accurately but were more welcoming toward them, too. Negative stereotypes no longer held.

This process of expanding and refining these skills has helped police in some cities change their interactions with one group at great risk of experiencing police violence: those with mental health challenges. They belong to one of the most stigmatized groups in society, but police may lack familiarity with their experiences and resulting behaviors, often receiving fewer than eight hours of relevant training. Without familiarity, police may misperceive actions as dangerous, when in fact they signal a personal crisis. Someone shouting and brandishing a knife might appear threatening; in reality, that person could be acting out of terror of their own delusions. Some analyses show that this group accounts for a quarter of police shooting deaths, including many killed at home and many who aren't in possession of a firearm. In Minnesota, 45 percent of people killed by police between 2000 and 2015 were in severe mental and emotional distress, or had such a history.[67]

A number of departments have begun to address this problem through an approach called Crisis Intervention Team (CIT) training, which is designed to improve the police response to such cases. One essential component is training officers to recognize and distinguish among different emotional and behavioral states. Officers learn the signs and symptoms of post-traumatic stress disorder, mania, and psychosis, as well as traumatic brain injuries and Alzheimer's. They may learn, for instance, that an individual with autism may avoid eye contact, while a person in psychosis may be unresponsive because they're unable to interpret commands.[68]

In the process, the officers' categories for behaviors that would normally be deemed threatening are multiplied and refined. A novice might see someone muttering and waving a knife and decide that he is dangerous. A CIT-trained officer might be able to identify a psychotic break and see that the knife is being wielded in self-defense against imaginary dangers.

As police expertise increases, their reactions may shift to better fit the circumstances. One study of more than a thousand calls relating to individuals suspected of having behavioral issues found that the officers trained in CIT were more likely than the others to rely on verbal negotiation. According to another study of more than six hundred police calls, CIT-trained officers resorted infrequently to force, even when facing a high threat of violence. Officers in Bend, Oregon, attribute some of their own decrease in use of force to their crisis intervention training. As Eric Russell, the patrol officer in Bend who keeps a mindfulness sticker on his mirror, recounted, he used to label a person "crazy." Now he focuses on how he can help. More effective still may be crisis-trained workers who are not police at all. In Eugene, Oregon, 911 calls that have a behavioral health component are routed to crisis responders who do not carry weapons. They do possess advanced training in mental health crisis and de-escalation. In 2019, the crisis team responded to 24,000 calls; only 150 needed police backup. In thirty years, the team has never been responsible for a serious injury or death. They are armed not with weapons but with expertise.[69]

THIS TRANSFORMATION—FROM NOVICE TO EXPERT—IS also what happened in Watts. "The problem," Rice said, "is when police aren't fluent in a culture. They can't tell what's a real threat." When officers saw the child with the toy gun in Nickerson Gardens, they knew they didn't have to draw their weapons. They were able to perceive the boy accurately and respond appropriately. They had moved beyond seeing only perpetrators and victims. They didn't see a criminal; instead, they saw a child. They could distinguish between a real threat and a projection, a true danger from an illusion.[70]

Jorja Leap, the lead researcher of the UCLA analysis, found that before CSP, officers had divided residents into one of two categories: perpetrator or victim. But in the subsequent years, that false binary disintegrated. Watts did suffer from violence, but even in the most dangerous places, only a small fraction of people actually pose a danger. Fluency with the place allowed CSP officers to see the community as multifaceted and nuanced. Along the way, a new category developed in their minds: community partners.[71]

The same transformation happened for the residents. Before CSP, everyone in a badge looked like an abuser and a threat. Now, they began to recognize officers as people, as various as they were. Not only were there more categories, there were also individuals—a category of one.

"You know," Rice said, "once you know somebody by their first name—even if it's Pookie—and even if Pookie is a gang member, you know what, once you sit down and talk to Pookie, once you have breakfast with him every now and then, once you start hanging out with him and play basketball a couple of times a week, guess what? You're going to become really relaxed around him and you're not going to fear him anymore. Because really, while Pookie is a gang member, he's not one of the bad guys, and you can begin to make those distinctions. And surely and slowly that's how we did it."[72]

One of the strategies for breaking down bias is to imagine another person's point of view. The power of meaningful contact with people from another group is that you don't have to imagine their perspective: you actually know them. You don't have to mentally replace stereotypes because you see that people don't fit them. And you don't have to guess situational reasons for a person's behavior because you witness the situation.

Ultimately, to grow and deepen one's perception of another group is to see them as human. "A lot of what we did," Phil Tingirides said, "was humanize the community to the police, and the police to the community." As one resident put it when describing the CSP officers, "Once they started interacting with the community, they changed."[73]

Lupita Valdovinos, the cofounder of the youth mentoring program in Nickerson Gardens who now runs the homework club, hopes this nuanced perception of the community can grow beyond the neighborhood. "You can show everyone that doesn't believe in Watts," she said, "that Watts brings out good flowers."[74]

The program is not without challenges. The UCLA report uncovered that some residents and officers remain confused about the program's ultimate goals. While some CSP officers see themselves as liaisons between the community and other LAPD units called into the area, others remain uncertain about managing their dual role of building relationships and enforcing the law. Some residents are happy with the community-oriented programs CSP offers (one parent said CSP had delivered more than it had promised), but others feel left out, or lament a dearth of opportunities for older adolescents at risk of joining gangs. The other evaluation found "uneven implementation across sites," and CSP officers shared worries about ensuring that all recruits to the program are the right fit. Some residents were enthusiastic, while others expressed frustration and a desire for more law enforcement activity, noting officers' lack of visibility and inconsistent responses to crime. Some residents still observed police stereotyping Black residents or harassing them during outdoor get-togethers.[75]

After the upheavals of the summer of 2020, Los Angeles created a bureau to make the program more cohesive and clear. Emada

Tingirides, now deputy chief, was tapped to lead it. But the problem of police legitimacy has found its way into CSP; Valdovinos told me that after George Floyd was murdered, some residents withdrew from the CSP officers they had grown to trust. One strong criticism is that the roles CSP has carved out for officers are better suited to social workers, or the kind of crisis workers who are so effective at handling behavioral health calls in Eugene, Oregon. Tingirides has acknowledged that it might be beneficial to transition some of the CSP officers' roles to others.[76]

But criticism of the program by some in the community is telling. The UCLA analysis found that residents wanted *more* of a presence of CSP in their communities, not less—more collaboration, more programming. Some prefer CSP to other social programs; one mother said she felt safe and secure when her child was with officers. Others asked for more CSP programs for at-risk children. "They're the ones that need the CSP officers," one resident said. Nine years in, the majority of residents—at least in the two communities polled—want CSP to stay. In Ramona Gardens, the figure was nearly 80 percent.[77]

CSP is perhaps best seen as a transitional approach that can build trust, change police conduct, and reduce crime in high-crime areas. Regardless of how the institutions responsible for public safety evolve, what the program demonstrates is that meaningful, sustained connections can change people: fear and mistrust can give way to human relationships. In a larger sense, the program demonstrates the role structural change plays in individual transformation. By changing the incentives, goals, rules, and mandates of their jobs, CSP changed the way officers behaved, felt, and showed up to work. Said Jeff Joyce, who cofounded the Nickerson Gardens mentoring program, "Maybe I saw people as human before, but now I can treat people as human."[78]

Rice's original goal was for cops to develop fluency with the community and understanding. She hoped that with greater familiarity, mutual fear and mistrust could transform into respect and collaboration. The seeds were there. She recalls opening the paper back in 2011 to a story about Phil Tingirides and another officer and some kids from Watts learning to surf. Tingirides realized these L.A. kids had never seen the ocean, so he organized a bus and took them to the beach. "I knew what I was looking at," Rice said. "I was looking at love."[79]

Jeff Joyce, for his part, lives southwest of Nickerson Gardens, in a house with two living units and a pool. A couple of years ago, some Nickerson residents gathered there for a Fourth of July barbecue. "Wow, he must really trust me," a teenager recalled thinking at the time. The relationships deepened. In spring of 2020, when the pandemic meant graduations around the country were canceled, students and their families held a socially distanced ceremony in the backyard of Joyce's house. Graduates stepped up to a wooden podium built by students from the soccer program. There were tears, speeches, and proud parents sitting on the lawn. By that time, the main unit of the house had become available—Joyce lives in a small bachelor pad above the garage—and one of the Nickerson families had moved in. They have dinner together Sunday nights now, taking turns making quesadillas or fettucine Alfredo, cocreating a new kind of family none of them expected.[80]

PART III

Making It Last

Designing for Flawed Humans

I met Chris in my first month at a small, hard-partying Catholic high school in northeastern Wisconsin, where kids jammed cigarettes between the fingers of the school's life-size Jesus statue and skipped mass to eat fries at the fast-food joint across the street. Chris was a sharp-witted high school junior with a low, amused laugh; I was a bright-eyed frosh, gothy but looking for friends. Chris and her circle perched somewhere adjacent to the school's social hierarchy, and she surveyed the adolescent drama and absurdity with cool, heavy-lidded under-standing. I admired her from afar and shuffled around the edges of her orbit, gleeful whenever she motioned for me to join her gang for lunch. She was clever but unfailingly kind, a combination of earnestness and ironic amusement endemic to the Midwest. When she laughed, I relaxed. Everyone did.

After high school, we lost touch. I went east; Chris stayed in the Midwest. To pay for school at the University of Minnesota, she hawked costume jewelry at Dayton's department store. Once, she lost a bet with a coworker and had to pay up by donning all of the most lurid gold chains, slinking around the jewelry department like a slender, gimlet-eyed Mr. T. She got married to a tall classmate named Adam and merged with the mainstream—became a lawyer, had a couple of daughters. She'd go running at the YWCA and cook oatmeal for breakfast. When I ran into her years later, the rebelliousness was subdued, but the old kindness was there. She asked me about my life and watched me carefully when I told her. And she held on to her taste for the absurd. When she and her girlfriends got their nails done, she'd choose the gaudiest, most sparkly shades and grin.

Then in 2010, at age thirty-five, she went to the ER with stomach pains. She struggled to describe the pain—it wasn't like anything she'd felt before. The doctor told her it was indigestion and sent her home. But the symptoms kept coming back. She was strangely tired and constipated. She returned to the doctor. She didn't feel right, she said. *Of course you're tired*, he told her, *you're raising kids. You're stressed. You* should *be tired.*

Frustrated, she saw other doctors. *You're a working mom*, they said. *You need to relax. Add fiber.* The problems ratcheted up in frequency. She was anemic, and always so tired. She'd feel sleepy when having coffee with a friend. *Get some rest*, she was told. *Try sleeping pills.* One doctor mentioned a scan in passing but said it probably wouldn't detect anything.

By 2012, the fatigue was so overwhelming, Chris couldn't walk around the block. She'd fall asleep at three in the afternoon. Her skin was turning pale. She felt pain when she ate. Adam suggested she see his childhood physician, who practiced forty minutes away. That doctor tested her blood. Her iron was so low, he thought she was bleeding internally. He scheduled a CT scan and a colonoscopy. When they revealed a golf ball–size tumor, Chris felt, for a moment, relieved. She *was* sick. She'd been telling them all along. Now there was a specific problem to solve. But the relief was short-lived. Surgery six days later showed that the tumor had spread into her abdomen. At age thirty-seven, Chris had stage IV colon cancer.[1]

HISTORICALLY, RESEARCH ABOUT THE ROOTS of health disparities—differences in health and disease among different social groups—has sought answers in the patients: their behavior, their status, their circumstances. Perhaps, the thinking went, some patients wait longer to seek help in the first place, or they don't comply with doctors' orders. Maybe patients receive fewer interventions because that's what they prefer. In fact, one analysis found that in thirty years of research devoted to reducing health disparities, 80 percent of efforts were focused on changing patients themselves or their communities. For Black Americans, health disparities have long been seen as originating in the bodies of the patients, a notion promoted by the racism of the nineteenth-century medical field. Medical journals published countless articles detailing invented physiological flaws of Black Americans; statistics pointing to increased mortality rates in the late nineteenth century were seen as evidence not of social and economic oppression and exclusion but of physical inferiority. How could doctors reconcile this supposed debility with the fact that the United States was largely built by slave labor? Easy: it was freedom that caused disease. Wrote physician John Van Evrie, the Black American's risk of dying "accelerated or diminished in exact proportion as 'impartial freedom' was thrust upon him."*[2]

In this century, research has increasingly focused on the social and environmental determinants of health, including the way differences in access to insurance and care also change health outcomes. The devastating disparate impact of COVID-19 on communities of color vividly illuminates these factors: the disproportionate burden can be traced to a web of social inequities, including more dangerous working conditions, lack of access to essential resources, and chronic health conditions stemming from ongoing exposure to inequality, racism, exclusion, and pollution. For trans people, particularly trans women of color, the burden of disease is enormous. Trans individuals, whose marginalization results in high rates of poverty, workplace discrimination, unemployment, and serious psychological distress, face much higher rates of

---

* The zeal with which nineteenth-century scientists tried to extract their preferred conclusions from contrary evidence knew no bounds: the low rate of suicides among Black people was also taken as evidence of psychological weakness.[3]

chronic conditions such as asthma, chronic pulmonary obstructive disorder, depression, and HIV than the cisgender population. A survey of nearly 28,000 trans individuals in the United States found that one third had not sought necessary health care because they could not afford it. As a White, fully employed cisgender woman with health insurance and other socioeconomic advantages, Chris, by contrast, had access to many social benefits as she made her way through the medical maze, including the resources to seek and secure care from another doctor when hers proved inadequate.[4]

More recently, researchers have also begun looking at differences that originate in the providers themselves—differences in how doctors and other health-care professionals treat patients. And study after study shows that they treat some groups differently from others.

Black patients, for instance, are less likely than White patients to receive pain medication for the same symptoms, a pattern of disparate treatment that holds even for children. Researchers attribute this finding to false stereotypes that Black people don't feel pain to the same degree as Whites—stereotypes that date back to chattel slavery and were used to justify inhumane treatment. Spurious ideas about biological differences between Black and White people promoted in medical journals were an underlying assumption of the inhumane twentieth-century Tuskegee syphilis experiment—in which treatment was withheld from Black patients—that ended in 1972. A 2016 study found that half of White medical trainees hold at least one false belief about racial differences, such as the idea that Black people have thicker skin than Whites.

The problem pervades medical education, where "race" is presented as a risk factor for myriad diseases, rather than the accumulation of stressors linked to racism. Black immigrants from the Caribbean, for instance, have lower rates of hypertension and cardiovascular disease than American-born Black people, but after a couple of decades, their health dips downward toward that of the U.S.-born Black population, a result generally attributed to the particular racism they encounter in the United States.[5]

Black patients are also given fewer therapeutic procedures, even when studies control for insurance, illness severity, and type of hospital. For heart attacks, they are less likely to receive guideline-based care; in intensive care units for heart failure, they are less likely to see a cardiologist, which is linked to survival. These biases also affect the quality of interactions in clinics: many studies find that doctors use more negative language with Black patients and dominate discussions more. Studies routinely find that even when differences in access to care, hospital qualities, and disease severity are accounted for, a Black person with vascular disease is more likely than a White one to undergo an amputation as opposed to surgery that would spare the patient's leg.[6]

In a striking example of disparate care, at least two studies of patients within the Veterans Administration system found that Black patients were less likely than White ones to receive a safer, minimally invasive form of surgery. In emergency departments, Black patients are more likely to be seen as capable of becoming violent, and more likely to be put in restraints. Researchers even believe racist stereotypes are one reason the opioid crisis affected the White population first: doctors avoided prescribing opioids to Black patients because they stereotyped them as "drug-seeking."[7]

The list goes on. Doctors spend less time and build less emotional rapport with obese patients. Transgender people face overt prejudice and discrimination. The survey of nearly thirty thousand trans individuals found that in the year leading up to the survey, a third of respondents had had a negative encounter with a health-care provider, including being refused treatment. Almost a fourth were so concerned about mistreatment they avoided necessary health care. Transgender individuals can therefore face a dangerous choice: disclose their status as trans and risk discrimination, or conceal it and risk inappropriate treatment. Meanwhile, because sexual orientation is typically not recorded in medical records, it is unknown how much bias people of different sexual orientations face.[8]

Women as a group receive fewer and less timely interventions, receive less pain treatment, and are less frequently referred to specialists. One study of nearly eighty thousand patients in over four hundred

hospitals found that women having heart attacks experience dangerous treatment delays and that once in the hospital they more often
die. After a heart attack, women are less likely to be referred to cardiac
rehabilitation or to be prescribed the right medication. Critically ill
women older than fifty are less likely to receive lifesaving interventions
than men of the same age; women who have knee pain are twenty-two
times less likely to be referred for a knee replacement than a man. A
Canadian study of nearly five hundred thousand patients showed that
after adjusting for the severity of illness, women spent a shorter time in
the ICU and were less likely to receive life support; after age fifty, they
were also significantly more likely to die after a critical illness.[9]

Women of color are at particular risk for poor treatment. A recent
analysis of their childbirth experiences found that they frequently
encountered condescending, ineffective communication and disrespect
from providers; some women felt bullied into having C-sections. Serena
Williams's childbirth story is by now well known: the tennis star has a
history of blood clots, but when she recognized the symptoms and asked
for immediate scans and treatment, the nurse and the doctor doubted
her. Williams finally got what she needed, but ignoring women's symptoms and distress contributes to higher maternal mortality rates among
Black, Alaska Native, and American Indian women. Indeed, Black
women alone in the United States are three to four times more likely to
die of complications from childbirth than White women.

And this health burden cannot be solely attributed to differences in
socioeconomic privilege: the gap between heart disease rates for Black
and White women is greatest at the highest levels of education. Racist
stereotypes about Black women's sexuality, long promoted in the field
of medicine, have also historically influenced diagnoses. Early in the
last century, a medical journal propounded the "African's birthright to
sexual madness." In the second half of the twentieth century, a Black
gynecologist found that 40 percent of his Black patients with endometriosis had been misdiagnosed with pelvic inflammatory disorder. The
disease is sexually transmitted.[10]

For all women, one reason for this inferior care is that their expression of pain and suffering is more likely to be viewed as an untrustworthy overreaction. Women have long been stereotyped as overemotional, agonized creatures with outsize responses, "hysterical" beings whose physical symptoms were manifestations of psychological problems. Research shows that adults watching a child express pain perceive the pain as less intense when the child is described as a girl.[11]

There's also a structural reason for inferior care: women have historically been excluded from much of medical research. The reasons are varied, ranging from a desire to protect childbearing women from drugs that could impair fetal development to notions that women's hormones could complicate research, to an implicit judgment that men's lives were simply more worth saving. Many landmark studies on aging and heart disease never included women; the all-men study of cardiovascular disease named "MRFIT" emerged from a mindset that male breadwinners having heart attacks was a national emergency, even though cardiovascular disease is also the leading cause of death for women. In one particularly egregious example, a 1980s study examining the effect of obesity on breast cancer and uterine cancer *excluded women* because men's hormones were "simpler" and "cheaper" to study.

Basic to these practices was an operating assumption that men were the default humans, of which women were a subcategory that could safely be left out of studies. Of course, there's a logical problem here: the assertion is that women are both so complicated and different that they can't be included in research, and also so similar that any findings should seamlessly extend to them. In the 1990s Congress insisted that National Institutes of Health–funded medical studies include women; earlier, many drug studies also left out women, an exclusion that may help explain why women are 50 to 75 percent more likely to experience adverse side effects from drugs.[12]

It's important to note that the sexual categories of "male" and "female" have boundaries that are not tidy but fluid, as the numbers of intersex, trans, and nonbinary people confirm, and also that there's a danger in overemphasizing any category. As sociologist Steven Epstein points out, medicine often starts with categories that are socially and politically relevant, but these are not always medically relevant. Categories that are not politically relevant may have a medical meaning. (Parkinson's disease, for instance, is associated with having red hair.) Relying on categories like race risks erasing the social causes of health disparities and may entrench the false and damaging ideas that are inscribed in medical practice. Patients racialized as Black, for instance, are presumed to have less baseline lung capacity, which may lead to their lung disease being unrecognized. At the same time, ignoring differences such as sex is perilous: as a result of their exclusion, women's symptoms have not been medically well understood. Doctors were told, for example, that women present with "atypical symptoms" of heart attacks. In fact, these "atypical" symptoms are typical—for women. They were only "atypical" because they hadn't been studied.[13]

Women and men also vary in their susceptibility to different diseases, and in the course and symptoms of those diseases. They respond to some drugs differently. As one concrete example, women secrete less gastric acid, so drugs that require an acidic environment may be less effective. Women's kidneys filter waste more slowly, so some medications take longer to clear from the body. A blockbuster antihistamine called Seldane was taken off the market after it was discovered that it made women in particular susceptible to a fatal arrhythmia, because they have on average a longer "QT interval" than men—the time it takes for the heart to reset between beats.[14]

The reason for these phenomena still isn't well understood. Different fat percentages and hormones play a role. But the fact is that every cell in the body, whether they're part of the reproductive system or not, has an XX or XY set of chromosomes, or in some cases XXY, XXX, or XO (only one X chromosome). This affects how cells behave, yet research has typically treated them as functionally equivalent. Scientific understanding about what this distinction means outside the sex organs is in infant stages. One study, for instance, found that "male" and "female" cultured cells responded differently to stress, even absent exposure to sex hormones. These cellular differences may contribute to different disease susceptibilities, like the fact that women are more likely to develop multiple sclerosis, lupus, and rheumatoid arthritis.[15]

This dearth of knowledge about women's bodies has led doctors to see differences where none exist and fail to see differences where they do. As journalist Maya Dusenbery argues in her book *Doing Harm*, this ignorance also interacts perniciously with historical stereotypes. When women's understudied symptoms don't match the textbooks, doctors label them "medically unexplained." These symptoms may then be classified as psychological rather than physical in origin. The fact that so many of women's symptoms are "medically unexplained" reinforces the stereotype that women's symptoms are overreactions without a medical basis, and casts doubt over all women's narratives of their own experiences.* One study found that while men who have irritable bowel syndrome are more likely to receive scans, women tend to be offered tranquilizers and lifestyle advice. In response to her pain and fatigue, my friend Chris was told she should get some sleep.[16]

The doctor who finally ordered the right tests for Chris told her that he'd seen many young women in his practice whose diagnoses had been delayed because their symptoms were attributed to stress. Indeed, studies show that women around the world experience delays in receiving a correct diagnosis for myriad diseases, including Crohn's, Ehlers-Danlos syndrome, celiac disease, and tuberculosis. A U.K. study of more than sixteen thousand patients also found delayed diagnoses for many types of cancer—bladder, gastric, head and neck, and lung cancer, and lymphoma, for instance. Women also experience delays for colon cancer. In Chris's case, she had been having symptoms for years. She had even had pencil-thin stools, which is a classic sign that a tumor is blocking the colon. No one had asked about them.[17]

---

* Ovarian cancer, as Dusenbery points out, was for years considered a "silent killer"—a disease with no known symptoms. In fact, women had been reporting symptoms for years, but these had been dismissed. A *JAMA* article that cited case studies to prove that ovarian cancer "defies early discovery" included the story of a "Mrs. M" who had had "no symptoms" at all, only "constipation for the past year and considerable bloating." Constipation and bloating are now recognized as signs of ovarian cancer.[18]

As Dusenbery argues, the problem is exacerbated by the fact that doctors rarely receive feedback about their misdiagnoses. They never learn where they went wrong.[19]

I asked Chris's husband whether her original doctors ever learned about their mistake. No, he said, she never returned to them. After she was correctly diagnosed, she was focused only on her recovery. She had surgeries, finished chemo, and had a clean scan in January 2013. Then more tumors appeared. She underwent a procedure called HIPEC (hyperthermic intraperitoneal chemotherapy), where the abdomen is bathed in heated chemo. After six months, she recovered. She christened herself a warrior. She bought cancer-kicking cowgirl boots for herself, her daughters, and her husband, and she started offering pro bono legal advice to other cancer patients. In July 2014, Chris stood in front of an advocacy crowd in a navy dress with tiny white dots and called herself a survivor.

Chris's cancer came back in December. She got sicker and weaker. The pain became uncontrollable. She began planning her funeral. She wanted people to wear black, she said. She wanted crying and the "Ave Maria." "This isn't a celebration of life," she said. "This is a tragedy." When hospice wheeled a bed into her house, she deadpanned, "I have a literal deathbed in my living room." Chris gave her best friend a pearl ring and said if she didn't wear it, Chris would become a ghost and haunt her. As she lay in bed, they flipped through photos from a recent nail salon visit. Chris looked at the pictures longingly. "I want sparkles," she said.

Chris died on June 3, 2015. She was forty years old. The day of her funeral was brilliant and blue, everything blooming. There were framed photos of Chris as a kid and a newlywed and a lawyer and a mom. There was the wailing and singing she wanted, and her sister carried the ashes, and I still see her in my mind at sixteen, every path still unfurling before her, the future a beautiful question.

DIAGNOSTIC ERRORS, IT IS ESTIMATED, cause eighty thousand deaths a year. Cognitive factors are estimated to play a role in some 75 percent of cases. What could have been done in Chris's case? Certainly, it's essential that doctors increase their awareness of their own capacity for biased decisions and their motivation to overcome it. The inner work of mindfulness and emotional regulation can also help, as biases are more likely to arise when people are mentally taxed. Meaningful, collaborative contact with those in other social groups can also help. One study found that when male emergency room physicians work with more female doctors, their female heart attack patients have greater survival rates. Another found that when non-Black medical residents had more connections with Black individuals during their residency, they expressed less racial bias—measured in both implicit and explicit ways—years later.[20]

But there's another approach to reducing bias that can support all these efforts, providing another layer of protection against the risk of interpersonal bias.

ELLIOTT HAUT IS A TRAUMA surgeon at Johns Hopkins Hospital in Baltimore. Affable and baby-faced, he looks happiest when talking about safety. The desk in his office is scattered with books about preventable deaths. A note taped over his computer reads REDUCE SYSTEM ERRORS. In other parts of the country, the trauma unit might see farm accidents or motorcycle crashes. At Hopkins, many trauma patients are victims of gunshots or stabbings. One patient arrived with the shard of a beer bottle still lodged in his neck, the entire word "Budweiser" perfectly legible along the length of jagged glass.

About fifteen years ago, Haut was tapped to oversee efforts to improve the Hopkins trauma department. The goal was to create better outcomes for the patients by improving the performance of the doctors. When Haut dove into the hospital's data, he found that patients were developing blood clots at a strikingly high rate.

Blood clots—the condition that threatened Serena Williams's life when she was in the hospital giving birth—are gelatinous globs of stuck-together blood cells that can travel through blood vessels and block blood flow to the lungs. They kill about one hundred thousand people a year—more than breast cancer, AIDS, and car crashes combined.* Many of these clots are preventable, if doctors prescribe the right clot prevention. In some cases, this means blood thinners; in others, mechanical "squeezy" boots that inflate and deflate around the legs to get the blood moving. But at Hopkins, only a third of the highest-risk patients were getting the right blood clot prevention, Haut found. "We'd get a patient into surgery—a routine surgery—and a week later they'd die of a pulmonary embolism," he told me as we sat in his office in East Baltimore, near a pile of wooden puzzles. And this problem wasn't specific to Hopkins: at hospitals around the country, patients were getting proper clot prevention only about 40 percent of the time—a problem that the American Public Health Association was calling a crisis.[21]

Haut wasn't sure why doctors were failing to prescribe the right interventions. Maybe, he thought, they overestimated the risks of blood clot prevention because patients who had developed complications from blood thinners sprang to memory more easily than those who were treated successfully. Haut wasn't thinking about disparities; his goal was to improve blood clot prevention for everyone.

To do so, Haut and his team sought out an approach that had been developed by Peter Pronovost, another Hopkins doctor, whose own father had died because of a cancer misdiagnosis. Pronovost had formulated a technique for improving medical care by adapting an approach used in aviation: the humble checklist. A checklist is just what it sounds like, a reminder of all the mandated steps a clinician should take. It plugs memory holes and hangs a safety net under human errors so they don't add up. Proper ICU care, for instance, requires nearly two hundred separate actions each day. Complications can arise from missing even one or two.

Pronovost showed that using a checklist in intensive care units reduced infections simply by ensuring that doctors adhere, each time, to a predetermined set of tasks. In one trial, a five-step checklist reminding

---

* Being immobile in a hospital bed increases the risk because blood that isn't moving is more likely to clot. Trauma also increases the risk because it changes the chemical makeup of the blood, so it's more likely to clot.

workers in over one hundred ICUs to do things like wash their hands and clean the patient's skin with antiseptic led to a 66 percent drop in catheter-related bloodstream infections. The drop held steady over the eighteen months of the study. A study of patients at eight hospitals showed that after checklists were introduced, complications from surgery dropped by 36 percent, and death rates dropped by 47 percent.[22]

Haut and his team decided to try developing a checklist for blood clot prevention. In their version, whenever a health-care provider admitted a patient to the hospital, a computerized checklist would pop up on-screen. The checklist would walk the doctor step by step through risk factors for blood clots and for bleeding from blood-thinning medication. After the checklist was complete, the system would suggest a recommended treatment—a blood thinner, for instance, or a mechanical squeezy boot to move the blood. If doctors didn't choose the suggested treatment, they had to document their reasons.

The approach worked. After introducing the checklist, the percentage of patients getting the right clot prevention surged, and preventable clots in trauma and internal medicine were close to eliminated. One study of a month of hospital admissions found that the number of internal medicine patients who returned to the hospital with blood clots within ninety days of discharge fell from twenty to two. And after the introduction of the checklist, the rate of fatal pulmonary embolism was cut in half.[23]

That could have been the end of the story. But Haut's office was, at the time, two doors down from the office of Adil Haider, a doctor who studies gender and racial disparities in health care. Their conversations prompted Haut to wonder whether there had been disparities in blood clot prevention. The team hadn't sliced the data that way, but when they went back over the numbers, an alarming pattern appeared. While 31 percent of male trauma patients had failed to get treatment, the rate was 45 percent for women. In other words, women had been nearly 50 percent more likely to miss out on blood clot prevention than men, and in greater danger of dying of this particular cause.

It's possible that factors other than gender might have been at play. Most patients who arrive with gunshot wounds, for instance, are men; perhaps doctors prescribed more prevention for more severe injuries. But as the researcher who analyzed the data put it, the disparities in treatment fit a consistent, large, and well-established pattern of women receiving suboptimal care.[24]

Looking at the numbers after the checklist was introduced, Haut and the team found that it had eliminated the gender disparities. Women and men received the right clot prevention at exactly the same rates. The gap had disappeared.[25]

IN 2008, THE UNIVERSITY OF Chicago economist Richard Thaler and legal scholar Cass Sunstein coined the term "choice architecture" to describe a powerful phenomenon: the context within which we make a choice has a profound influence on the way we choose. Just as the design of a physical environment can influence our behavior (as seen in coffee shops that skimp on electric outlets to discourage people from sitting with laptops), the design of a process can also shape our behavior.[26] It, too, can be thought of as a kind of architecture.

For instance, University of Minnesota researchers discovered that they could coax students into eating more vegetables simply by redesigning their lunchtime routine. In a typical lunch line, students encounter carrots next to more tempting options like french fries and pizza. Instead, the researchers gave kids a cup of carrots the moment they arrived in the cafeteria, when they were at their most hungry. It worked: kids ate a lot more carrots. The key was to put the carrots "in a competition they actually can win"—a contest not against french fries but against being really hungry. To change how students ate, it wasn't necessary to sell them on the virtues of vitamin A. What changed was the choice architecture.[27]

The Hopkins checklist is a kind of choice architecture, too—a way of shaping a doctor's behavior not through persuasion but through design. It doesn't ask doctors to think more carefully about their biases; it simply interrupts the process by which they make decisions. The Hopkins checklist forces doctors to disentangle the thinking that goes into a medical decision. In a way, it acts like a prism, reverse-engineering a holistic judgment into its constituent parts, the way a prism separates white light into its rainbow colors.

The checklist also supports that human judgment. It is meant to remind doctors of steps they might forget, but bias isn't really about forgetting. It's about using assumptions to judge and evaluate, without necessarily being aware of the presence of those assumptions. Some doctors resist the intrusion, pointing out that these mandated checklists aren't perfect. As one hospitalist told me, they may not take into account the full range of factors a doctor might consider. While a clot checklist asks questions to assess risk at one moment in time, an experienced doctor might note that a patient with pain could have a procedure tomorrow that could change the risk profile over time. The checklist does not have the capacity to account for this nuance. As the medical scenarios become more complex, the checklist may be best considered a fail-safe for decision-making, not a substitute.*[28]

But checklists have been shown to reduce bias elsewhere. After a structured decision-making tool was introduced in the state of Illinois, the disparities between psychiatric hospitalization of young, low-risk Hispanic and Black patients and White patients shrank. When the Mayo Clinic instituted a system of automatic referrals for cardiac rehabilitation after heart attacks, the gender gap between men's and women's referral rates disappeared. The effects extend beyond health: when psychologist Jennifer Eberhardt worked with Oakland police to create a one-question checklist—"is there intelligence linking this person to a specific crime?"—traffic stops of African Americans fell 43 percent. Her checklist approach also slashed racial profiling on a social media platform.[29]

Using principles of behavioral design to reduce bias dates back to 1952, when the Boston Symphony Orchestra began changing the way it auditioned musicians. Instead of having musicians play in full view of a panel of judges, a screen was set up to divide them. Women musicians

---

* It's important to note that any system, including the kind of algorithm used here, can encode bias. Algorithms in artificial intelligence, for instance, are only as good as the data on which they are trained, data that might, for instance, be oversampled from one population and undersampled from another. Researchers have uncovered bias in facial recognition software, such that Black women are classified incorrectly more than other groups.

were asked to remove their shoes, so clacking high heels wouldn't be a tell. Instead, a man standing onstage created fake footsteps with his shoes.

Over the following decades, curtained auditions rippled through American orchestras—a heavy cloth was hung from the ceiling, or a room divider was stretched like an accordion across the stage. By the 1990s, most had adopted the practice. When economists Claudia Goldin and Cecilia Rouse studied the differences between orchestras that did and did not use this approach, they found stark evidence that masking gender changed judges' assessment of women's skills. Analyzing thousands of orchestra records, they found that concealing musicians' identities increased women's odds of advancing to the next round of auditions by 50 percent. This alone, they conclude, is a large part of why women's proportion increased. Today, they make up almost 40 percent of orchestras.[30]

Of course, the masked approach isn't possible in medicine, which usually depends on face-to-face interactions between doctor and patient. Structured decision-making like checklists is a close cousin. But in contexts where face-to-face contact isn't required, blurring out a person's social identity can shield a person's evaluation from harmful stereotyping. Blocked from using assumptions and preconceptions, those in positions of power are forced to rely on official criteria alone. It is a blunt tool for reducing bias, to be sure. But it, too, can unleash powerful changes.

CYNTHIA PARK KNEW SOMETHING WAS wrong as soon as she ran the numbers. It was 2001, and Park, a genial Texan with a faint accent, was working as a statistician for the schools in Broward County, Florida. Her team analyzed things like student attendance and teacher evaluations. But before she'd landed the job of data geek, Park had been a teacher of students she calls "severely gifted" back in her home state. Park had fallen in love with these students—their quirkiness and wild questions. So she was curious about the academically unusual kids in Florida—who they were, how they were doing. No one on the stats team was looking into them, so Park decided to run her own queries. What she found was alarming.

In Broward, a school district of a quarter million students—the sixth-largest school district in the United States—the kids in the gifted programs looked nothing like the general population. White children were a minority of the student population, but they made up almost 60 percent of those labeled gifted. Students of color made up the majority of those attending Broward schools, but only 28 percent of the kids in the gifted programs. In over a dozen elementary schools made up primarily of students of color, no one had been identified as gifted.[31]

Park had grown up in a family of eccentrics (her father was a socially awkward rocket scientist) and understood that academically gifted children faced challenges, too. Her Texas students could form coherent arguments about time travel but would routinely walk into the tetherball pole outside the school. One profoundly talented art student said it felt like a brick was on top of his head every time he sat down to write, so Park let him dictate his essays instead. Another struggled with social skills and came to school with matted tangles in her hair. Park brought a ruler to show the girl exactly how far to stand from people to avoid making them uncomfortable. At recess, she brushed the girl's hair. Park sewed little velvet pillows for students who were easily distracted, so they'd have something to fidget with to help them focus. One child preferred sandpaper to velvet and focused so hard his fingers bled.

Park understood that these kids needed just as much special attention and resources as those with traditional learning disabilities. They, too, were challenged: being this unusual could be isolating, even painful. The students didn't fit into the mainstream school culture and were often teased mercilessly. They needed a teacher who understood them, who accepted them and could help them. In Broward, the "gifted" label would ensure years of extra attention.

This made the Broward stats especially concerning. If the numbers were this lopsided—and this statistically unimaginable—that meant the county was full of children who needed resources and attention and weren't receiving them.

Two years later, Park was put in charge of the gifted program. She decided she needed to make her concerns visible and undeniable. Because the school board was filled with people who weren't particularly savvy with statistics, she needed to make her data sing.

Park decided to make the numbers visual. First, she printed out a large map of the county. Next, she placed a single red dot on the home address of every public school student who'd been identified as gifted. Broward County is a wide rectangle wedged near the southern tip of Florida, with Fort Lauderdale and Hollywood on the Atlantic coast, and suburban communities like Coral Springs and Weston farther inland. Wealthy families on the coast tend to put their kids in private school, while those inland tend to enroll them in public schools. When Park completed her map, it looked like a tide had swept red dots from the coast and deposited them inland on the tony, manicured areas where wealthy White families lived. The gifted label, it was plain, was concentrated in the Whitest communities.[32]

At that time, the process for identifying gifted students in Broward began in first or second grade. If teachers judged that a child might qualify for the gifted program, they would refer that child for a test with the school psychologist. Parents could also hire a private psychologist to perform testing. Any child whose IQ was measured to be 130 or above would then be evaluated for inclusion in the gifted program. Because of the well-documented depressing effects of early childhood poverty and limited English language proficiency on standardized test performance, low-income and ELL (English language learner) students had a lower cutoff of 115.

But the county's IQ scores looked funny. Instead of having a normal distribution, the scores dipped to zero at exactly 129, and then spiked sharply at 130, the minimum cutoff to be labeled "gifted." Many of the students were miraculously reaching the exact score required for the gifted label. And none—not a single child—had the disqualifying score of 129.

As Laura Giuliano, an economist who later analyzed the data, told me dryly, "It appears that there was a market for high IQ scores." The private psychologists had been ostensibly hired to evaluate children but, at several hundred dollars a pop, they had in practice been hired to ensure the gifted label. Giuliano told me that when her own kids neared school age, other parents whispered to her about which psychologists were the "good" ones. "Good," it seems, meant mastering the art of "uncovering" an IQ of 130.[33]

This IQ-for-hire was one clue to understanding the disparity: wealthy, mostly White parents were essentially buying slots for their children. But it still didn't explain the low absolute numbers of Black and Latino kids, or English language learners, or low-income students with the gifted label. Even if wealthy White children disproportionately received the label, that shouldn't depress the numbers for anyone else.

Park suspected there might be a problem with the first step in the process, when teachers and parents referred students for testing. So in November 2003, she showed her map to the school board and proposed a new way of identifying gifted students. Broward, Park said, shouldn't rely on anyone's individual judgment about who should be tested. The county should screen every single child. Faced with the blinking red map of inequality, the school board unanimously voted yes.

In 2005, Broward County rolled out universal screening. Staff were paid overtime to test every single one of the county's twenty thousand second graders. Because IQ tests and other standardized tests are notoriously plagued by bias, the test chosen was a nonverbal cognitive test that minimized these risks.* Instead of relying on words or images related to any particular culture, it measured general problem-solving skills.[34]

After students took the test, Park's team hand-delivered permission slips in Haitian creole, Portuguese, and Spanish to every school, so parents could approve the next steps. They fielded calls from parents who worried that this written notice about their "exceptional student" referred to a behavior problem. The team reassured them that it was just the opposite, that this was good news.[35]

When the process was complete, the results were striking. After universal screening, the number of Black and Hispanic kids who qualified as gifted tripled. Of the hundreds of additional kids who qualified for gifted programs over the next year, 80 percent were low-income or English language learners. Many of these students had scores that were significantly above the cutoff, meaning that even the most unusually

---

* There is some evidence that the test, the Naglieri Nonverbal Ability Test, may avoid biased outcomes, with roughly the same proportion of children from White, Black, and Latino groups performing at the 95th percentile.

gifted kids had previously been left behind.[36] The problem wasn't that these children weren't gifted. The problem was that no one had bothered to find out.

The changes didn't end there. Broward rules mandated that if a school had even a single gifted child in a given grade, it had to create a special "high-achiever classroom" for that student, with a specially trained teacher and more advanced coursework. That classroom was then filled out with students who had come close to the cutoff—if, for instance, a grade had four gifted students, and a high-achiever classroom had room for twenty-four students, that meant the next twenty students by score were also given spots in that classroom. These students, too, would benefit from the faster pace, greater enrichment activities, high teacher expectations, and peer support. As Giuliano and her colleague David Card found, Black and Hispanic "high achievers" who learned in these special classrooms saw huge boosts in their math and reading scores. Before being placed in these classes, these students had performed worse in math and reading than White students with the same IQ scores. Afterward, the gap disappeared. More of these students then became eligible for continued accelerated courses, sending them on a new academic path.[37]

In fact, the entire data profile of Black and Hispanic students changed. It turned out that before universal screening, not only were these students less likely to be screened for giftedness but they were also *more* likely to be screened for learning disabilities. And this over-screening was reflected in the overall distribution of IQ scores. After universal screening, the graph of Black and Hispanic students' scores lined up with the scores of White students.

Another group benefited, too: girls. Park hadn't noticed that girls had been systematically overlooked, but even parents aren't immune to gender bias. (Parents google "Is my son gifted?" at two and a half times the rate they google "Is my daughter gifted?")* After universal

---

* Parents also ask, "Is my daughter overweight?" at twice the rate as "Is my son overweight?" In reality, a greater percentage of boys are overweight.

screening, girls went from being underrepresented in the gifted program to overrepresented.[38]

Rather than relying on adults' assessments of children, the Broward system rerouted decision-making to bypass judgment entirely. Instead of confronting teachers' and parents' biases, or even convincing them that they *had* biases, Park identified the moment in which human judgment might introduce an error and then designed a work-around.

This approach to removing bias—changing processes rather than relying on changing people—is spreading. Academic journals remove names when evaluating submissions. The committee that acts as a guardian for the Hubble Space Telescope, selecting which astronomers have access to its mysteries, recently began concealing applicants' identities. An analysis of more than fifteen thousand applicants over sixteen years found that prior to masking, men's proposals were accepted at a higher rate than women's. After names were erased from applications, the disparity actually reversed.[39]

Workplaces are using behavioral design among structural efforts to reduce bias as well. The hiring process is ripe for bias because humans intuitively use "culture fit" to make hiring decisions; this means interviewers often instinctively favor candidates who resemble themselves. The phenomenon of homophily (literally, love of the same) means that we often gravitate to people who are just like us. Research by sociologist Lauren Rivera has shown at a truly granular level how homophily affects employment decisions. Her research into investment banks, law firms, and consultancies showed that some people making hiring decisions were not only more likely to choose someone who played a college sport if they themselves were college athletes, but they were more likely to hire someone who played the *same* sport.[40] Homophily is an example of bias's double whammy: not only is there bias against people who belong to stigmatized groups, there's bias in favor of those most similar to the people making the decisions.

Research also shows that when standardized criteria for choosing a candidate are lacking, people often redefine them to match the qualities a favored candidate possesses. In one study, for instance, subjects were shown résumés for male and female police chief candidates, Michael or

Michelle. The police chief candidates were also given one of two sets of credentials, either extensive experience "on the street" or formal education but little experience. The streetwise applicant was tough, worked in rough neighborhoods, and got along well with coworkers. The educated applicant had administrative experience and was politically savvy and adept with media.

In the experiment, the researchers manipulated the profiles so that sometimes the educated candidate was Michael and sometimes Michelle. They then asked the subjects which criteria were important for the position of police chief and who should be hired. Interestingly, the subjects saw the formally educated Michelle as just as smart, communicative, and politically savvy as the formally educated Michael. But when Michelle possessed these credentials, the subjects decided these were less important hiring criteria. When Michael possessed them, they were seen as critical for the job. Even stereotypically feminine traits, like being family-oriented, were seen as more important when they described Michael. In the end, the subjects chose the educated Michael for chief: they adjusted their requirements to favor him. The technical term for this is "redefining merit"—mentally rejiggering criteria to fit the already favored candidate. The researchers concluded that to reduce job discrimination, organizations need "the establishment of standards of merit prior to the review of candidates."[41]

To combat homophily, redefining merit, and other forms of bias, companies have begun redesigning their interview processes. In tech companies, the "whiteboard interview" is a classic step in the hiring process. Candidates are asked to solve a coding problem in real time in front of evaluators—typically, in the words of one engineer, "three neckbeards in a room tapping their feet and looking at their phones." There are a few negative consequences of this approach. Bias can easily creep into the evaluation. Tracing one's thought process with a dry-erase marker in front of a live, skeptical audience can also create extra stress for the candidate. Interpersonal phenomena like stereotype threat—in which people from stigmatized groups spend mental energy grappling with the prospect that they'll be seen as confirming negative stereotypes—can lead women and others from underrepresented groups with the same skills to perform more poorly. A recent study found that this type of coding interview and its attendant anxiety can reduce performance by half.[42]

One tech company used an approach modeled on masked orchestra auditions: candidates were given a problem to complete at home or in the seclusion of the company's office. All personal identifiers were wiped from each candidate's work, which was then evaluated according to set criteria. The company also standardized interview questions ahead of time so that each candidate faced fair, equal questions designed to unearth the specific characteristics it was seeking for that position. Additionally, the company changed the language in its job descriptions to ensure they appeal to a wide audience, using phrases like "care deeply" and "lasting relationships" instead of "wickedly" and "maniacal." These changes resulted in a boost in numbers of women and underrepresented employees in technical and management roles.[43] Changing the choice architecture made for less biased choices.

Would structured decision-making have helped my friend Chris? A doctor who is aware of their own capacity to be biased, draws on inner resources to combat it, and uses supportive diagnostic checklists as a fail-safe, we can hope, would not have told her that she just needed to relax. But we'll never know. Ultimately the decision is left to the physician. And the truth is, lack of medical knowledge combined with stereotyping still doesn't fully explain the inferior treatment women receive. As Sharonne Hayes, a cardiologist at Mayo Clinic explained, even when the diagnosis is unequivocal (as in a heart attack) and the guidelines are clear (as in which medications to give patients afterward), women *still* receive worse care.[44] This points to a troubling reality more fundamental still.

One unusual study looked at a thousand Canadian patients who were admitted to a hospital for heart problems. The researchers examined how quickly patients received imaging, medication, and an angioplasty. But they also had the patients fill out detailed personal questionnaires about their occupations, lifestyle, and personalities. While women generally experienced longer delays, the delays were especially pronounced for both women and men who possessed stereotypically feminine personality traits—who were nurturing or gentle—or were more responsible for housework. In other words, the more traditionally feminine the patients were, the worse they fared.*[45]

What this suggests is that despite clinicians' best intentions, they may simply be devaluing that which is associated with women. As Hayes puts it, these problems of disparate treatment won't go away until we start seeing women's lives as important.[41] Years ago I wrote a story about gender bias in medicine that ended with a figure: the number of women whose lives could be saved. An editor added that this translated into saving mothers, daughters, and grandmothers, as if their lives' worth was not inherent but needed qualification.[46]

This, then, points to the inherent fragility of choice architecture as a means of overcoming bias. The approaches are only as robust as the larger value systems within which they take place. The Broward schools saw successes, but when budget cuts later scaled back and at one point suspended the program, the rates of Black, Hispanic, low-income, and ELL students identified as gifted drifted toward their previous levels. A well-known European media company tested a pilot program of masked auditions to diversify new hires, and while this nearly tripled the proportion of ethnic minorities on the hiring short list, the pilot failed to find support from management and was ended.[47] The tech company that used structural changes likewise increased the diversity of its employees, but in some cases failed to retain and promote women of color in particular. Choice architecture helps, but structural adjustments like these cannot overpower more fundamental forces that tend toward maintaining the status quo. The question is—what can?

---

* Every additional point on the "femininity scale" led to a 31 percent decrease in the chance patients received an intervention in a timely fashion. Every point on the masculinity scale, by contrast, led to a 62 percent increase in timely care.

Dismantling Homogeneity

In spring of 2013, a twenty-eight-year-old Brazilian-born, U.K.-dwelling journalist named Caroline Criado-Perez began what she thought would be a quiet campaign to keep women on British currency. The face of prison reformer Elizabeth Fry—the only woman on a banknote—was to be replaced by Winston Churchill, so Criado-Perez started an online petition to reverse the decision and its "message that no woman has done anything important enough to appear." Soon, women were protesting outside the Bank of England dressed as historic figures, like Boudicca, warrior and Celtic rebel queen.

When the Bank of England announced that it would now feature Jane Austen on the ten-pound note, Criado-Perez began receiving threats on Twitter—a torrent. At one point, she was receiving nearly one rape or death threat per minute. When she complained to Twitter, the company suggested she report the tweets. This required completing a nine-part questionnaire for each threat. The police finally intervened, and two people were ultimately arrested and jailed. Under pressure, Twitter rolled out a "report abuse" button.[1] This provided a different route to the same cumbersome nine-part questionnaire.

But the abuse and harassment didn't stop—not for Criado-Perez, and not for untold thousands on the site. A 2020 analysis found that of the tweets women congressional candidates in the United States receive, 15 to 39 percent are abusive, compared to on average 5 to 10 percent of those received by men. Abuse escalates for women of color: Amnesty International's review of tweets received by women journalists and lawmakers in the United States and U.K. found that as a group Black, Asian, Latina, and mixed-race women are 34 percent more likely to receive abusive tweets than White women. Black women alone are 84 percent more likely to be harassed than White women on Twitter. Of the twenty-six thousand abusive tweets sent in half a year to women in Parliament in the U.K., half were directed at a single Black woman MP. In the United States, the candidate with the highest rate of abuse on Twitter—at 39 percent of all messages—was Ilhan Omar, a Somali American congresswoman.[2]

The problem has been endemic at Twitter from its very beginning in 2006. In 2018, Twitter CEO Jack Dorsey finally admitted that the company had failed to anticipate how it could become a breeding ground for abuse. But according to former employees, the founders also inadvertently built the mechanisms for abuse into the product architecture. Anyone can tweet to anyone; likes, retweets, and the hierarchy of visible responses to tweets have all optimized Twitter for harassment.[3]

"Twitter is good at two things: real time information and abuse," Leslie Miley, a former engineering manager at Twitter, told me. "They both are disseminated in the exact same way. The vectors that allow news content to go viral allow trolls and abuse to go viral."[4] Twitter's "algorithmic timeline" prioritizes the content it shows according to what it deems most engaging and relevant. People engage highly with negative content that arouses anxiety and anger, whether they're reading news or harmful comments.[5] While news is amplified by readers and broadcast to a wide audience, abuse can be amplified through an army of trolls and aimed at a very narrow audience—an audience of one.

"The platform was built in a way that made the weaponization of information easy to do and hard to stop," Miley says.[6]

But why would Twitter have been unable to foresee this major problem?

ONE ANSWER CAN BE FOUND in an observation made by computer scientist Melvin Conway in the early days of computing. Conway noticed that the structure of a piece of software always reflected the structure of the organization that created it. If a product was developed by four distinct teams, its final version would have four distinct parts. Wrote Conway, "The very act of organizing a design team means that certain design decisions have already been made, explicitly or otherwise.... There is no such thing as a design group which is both organized and unbiased."[7] Conway's insight reflects a larger truth: software always exhibits essential features of the group that built it.

One feature of Twitter's founding team was homogeneity: all four cofounders were young White men. This homogeneity, according to many former senior employees at the company, produced a critical blind spot.[8] Twitter's creators didn't anticipate abuse because online harassment—the vindictive, physically threatening, terrifying kind directed at Criado-Perez—had never been part of their online experience.

When Twitter's cofounder Evan Williams described his baptism into the world of connectivity to an interviewer, he recalled the thrill of using online bulletin boards as a kid from Nebraska in the early 1990s. He'd dial in and suddenly be connected with people around the world. This showed him, he said, that the world was full of minds, and the minds were full of ideas, and that "the Internet is this big machine that takes certain bits from these minds and puts them in other minds."[9] Linking all these minds would become, in time, the focus of his career. His dream of transmitting thoughts seamlessly from one brain to another found its purest expression in Twitter.

But for anyone from a marginalized group, the threat inherent in connecting all these minds would have been obvious from the beginning.

Instant thought transmission might be acceptable if the ideas are kind; if those thoughts are hate-filled and violent, the prospect turns nightmarish. Leslie Miley had a very different experience of the early, pre–social media Internet than did Williams. Because of his name, people often assumed Miley was a woman. The results were alarming. "I was often asked to come into private chat rooms. If I didn't want to, I was called 'bitch,'" Miley told me. A person who had had this experience would have anticipated the ways Twitter could be used for harm. "If you had had different people at the table, they would have asked, 'How can we give people tools to protect against abusers, because this has happened to me,'" Miley noted. "Someone like me would have said, 'Let's really think this through.'"[10]

Even after Twitter's utility as an abuse machine became apparent to the founders, they failed to act. An early user and friend of the founders, Ariel Waldman, began receiving harassing tweets shortly after Twitter launched, but after she reported it, the founders didn't take the problem seriously. When Twitter introduced the "follow" button, Evan Williams told an interviewer, the team mused about whether it was "creepy." Was it stalking? "We thought about that," he said. "We had lots of jokes about that." The psychic distance between the founders' experience and that of stalking victims was so vast that they saw it as a joke.* A decade and a half after its founding, abuse persists; of a sample taken in 2020, 16.5 percent of the tweets received by Congresswoman Alexandria Ocasio-Cortez were abusive. Early attempts like banning specific offensive words missed the point and demonstrated the leaders' incomprehension. Said Miley, "They didn't understand that it was about context, not content"—that the specific words used were less important than the overall atmosphere the abuse created.[11]

Consequences multiply. In 2018, a team at Facebook was working on what the company calls "fake engagement"—for instance, inauthentic

---

* Twitter was eventually used for real-world stalking. In one instance, after being repeatedly harassed by a man she had accidentally "followed" on Twitter, a British woman came home to find the man at her door. She called the police. At the trial, he claimed they were in a relationship and used the fact that she had followed him on Twitter as proof. [12]

accounts making comments. A young data scientist named Sophie Zhang proactively uncovered extraordinary manipulation by political figures in countries ranging from Azerbaijan to India to Honduras. But the company, based and founded in the United States, steered resources to the United States and western Europe. Zhang escalated her concerns, assuming leaders would address them—global politics was not her purview. Everyone agreed this was bad, but no one assumed responsibility. Zhang soldiered on solo, sometimes working eighty hours a week. Civil unrest erupted in Bolivia after she found but could not attend to manipulation in that country. Dozens of deaths followed. "I have personally made decisions that affected national presidents without oversight," Zhang wrote in a damning 2020 memo. Zhang, stretched thin and told to focus elsewhere, was fired for performance issues. The factors at play here are numerous, but had stakeholders from these countries been decision makers, how might the perceived urgency of these crises have changed?[13]

When faced with homogeneity like that of Twitter's founding team (tech is currently 74.8 percent male, with women making up 18 percent of software developers), many shake their heads and attribute this to the "pipeline," affirming that there simply aren't enough women and people of color entering these fields. But if the pipeline were the problem, then leadership in fields like medicine, law, and business would be filled with women: women have been entering the pipeline in these fields in equal numbers for decades. As of this writing, 50 percent of the top ten Fortune 500 CEOs are White men over six feet tall, a combination of traits that rarely occurs in nature.[14] In fact, only 4.65 percent of the adult U.S. population fits this description; is the CEO pipeline flooded with these towering creatures?

And top universities are turning out Black and Latino computer science majors at more than three times the rate these companies are hiring them. While these students earn nearly 18 percent of bachelor's degrees in math, computer science, and electrical engineering, they make up around 4 percent of the technical workforce at Google, Microsoft, Facebook, and Twitter. After twelve years, half of women in STEM leave the field, often citing bias and the social environment as important reasons. While equal numbers of Asian men and Asian women receive engineering PhDs, there are only a tenth as many Asian women as men in the National Academy of Engineering.[15]

Bringing new people into a field is, of course, essential, as is expanding access at a young age. But because the forces that entrench homogeneity are so strong, it's crucial to attend to the existing pipeline as well. To ensure that those already in a field are not thwarted, there need to be structural changes to the pipe.

SHORTLY AFTER SHE'D UPROOTED HER life and moved to California, physicist Peko Hosoi's phone rang. It was 2002, and Hosoi had moved to Pasadena to begin teaching at Harvey Mudd, a science-focused liberal arts college. An expert in fluid mechanics, the study of how liquids and gases flow—blood through a capillary or gas during the expansion of the universe—she had arrived so recently that the office was practically empty. One desk, no computers, and a landline, which was now ringing, disconcertingly. Who would possibly be calling? No one even knew she was there.

"I'm looking for Anette Hosoi," said the voice on the other end. This was strange. Everyone who knew her called her Peko, a nickname she'd been given in childhood by her Japanese grandmother. "This is Rohan Abeyaratne from the mechanical engineering department at MIT. I'm calling to offer you a job." *What?* Hosoi thought. *This must be a prank.* She had applied for a faculty position in that department two years earlier and had never heard back—she hadn't even received a rejection. Now, the department head was calling, with no warning, to offer a job over the phone? It wasn't possible.

Hosoi told Abeyaratne she had to think about it. Harvey Mudd was a dream—bright students, wonderful colleagues, an always-sunny location. But if she accepted this mysterious offer and moved to MIT, she'd have the chance to collaborate with graduate students, too, and pursue research at one of the top mechanical engineering departments in the world. She was just starting her career. This kind of decision could change it.[16]

What Hosoi didn't know was that that phone call was part of an experiment. And it had been undertaken in a moment of crisis.

MIT WAS, AROUND THAT TIME, a fairly strange place. I know: I was there for my first two years of college. The students in my freshman dorm inaugurated the year by dropping a block of pure sodium into the Charles River to watch it explode on impact. We doused the hallway carpet in isopropyl alcohol and then played floor hockey with a tennis ball we'd lit on fire. We spent our entertainment budget on a single Fresnel lens.

But as much as MIT was a bastion of intellectual and bodily adventure, it was also a fairly vertiginous experience for women, even though by the time I arrived we made up about 40 percent of the undergraduate population. In an odd attempt at making us feel welcome, the school offered "bathroom tours"—guided viewings of the once-rare ladies' rooms on campus. Academically, the departments were fairly segregated: women gravitated toward biology and chemistry, and engineering skewed male. In mechanical engineering, the department that would eventually make Hosoi an offer, only around 24 percent of the undergrads were women. Nationally, the field attracts fewer women than petroleum engineering, nuclear engineering, and mining.

"It was isolating," remembers Djuna Copley-Woods, a mechanical engineering major at that time. She had come to MIT in search of a path to economic stability after living on food stamps as a child. In lab classes, where she was often the only woman, she would typically be paired with the person no one else had selected as a partner. She didn't protest, she told me, because she didn't want to "make anyone feel not cool" for being paired with her. Once, she accidentally overslept on the day of an important test, waking up only minutes before it was over. She threw on her clothes and bolted to the classroom, asking the professor if she could take the test in the remaining minutes. He suggested instead that she drop the class. She did. "There was no sense that the professors saw me as a person," she said. While MIT students have always survived the notoriously brutal classes by banding together to do homework, Copley-Woods did most of it alone. "I didn't know any alternative," she said.[17]

Other women mechanical engineering majors from that time remember that the machinists teaching in the machine shop would show the male students how to use the machines. When a woman needed help, they'd just do the task for her. Women students remember wanting to downplay their differences, even feeling uncomfortable befriending female classmates. My own freshman year, before choosing to major in physics, I took an introductory course in materials engineering, a field adjacent to mechanical engineering. I was impressed by the magisterial excellence of the professor—he gave gripping lectures, piped in music that worked thematically with each class, and at one point demonstrated the difference between shear and tensile stress by ripping a phone book in half with his bare hands. I was acing the class, and the subject had ignited an interest in the field. One day midsemester, as I walked across campus, I saw the professor. I summoned my courage and told him that I really enjoyed the class. "Oh," he said, batting his eyes, "you say that to *all* the boys." I never sought out the professor again.

Later, I met with another professor, one who was working with piezoelectrics, materials that can convert mechanical signals into electricity. He offered me a position on his research project, supervised by an upperclassman. But the meeting location kept changing. Somehow (this was before cell phones) the information didn't get to me, and I kept showing up to the original location. Eventually, I stopped going. I changed tactics, approaching a professor in a different field, this time at the MIT Media Lab. When we met for an interview, he was twisting bits of tape into loops and didn't look at me during the conversation. I never followed up.* That was my experience as a White woman. A classmate of mine who is Black looked back on her experience as "horrendous."[18]

This sense of exclusion was also present for women faculty. In the few years prior to the call to Peko Hosoi, MIT had been rocked by explosive, data-driven reports on the status of women professors that had been prompted by women science faculty. In the School of Science and

---

* This professor went on to become the president of another university.

School of Engineering, the reports revealed, women achieved tenure at the same rate as men, but over time, women faculty's working lives got harder as men's got easier. Women, who made up 8 percent of the faculty, received fewer resources, lower salaries, less access to collaborations, and even—as carefully measured with tape measures—lab space. They were asked to change teaching assignments more than men, weren't invited to present at important meetings, and were excluded from committees that provided the necessary experience for academic leadership. One woman faculty member reported that colleagues dismissed the value of her work, even though it was essential to their own research.[19]

Even an effort designed to help women ended up hurting them. To help new parents care for their children, the university had given them an extra year to achieve tenure. But while women used the year as intended, men used the year to accelerate their careers, traveling around the world to give seminars, starting new companies,* and promoting their research. One study of this tenure policy in economics departments found that these "gender-neutral policies actually increase the gender gap at research-intensive universities."[20]

When women began as faculty, they thought they just had to be excellent and that their success would only be hampered by greater family responsibilities. What they found instead was a slow and steady marginalization. It was not, as Antonin Scalia would have insisted, expressed overtly or directed from above, but its net effect was systematic exclusion. The fact that women were still advancing toward tenure at the same rate as their male colleagues was remarkable. But the toll it took was so severe, some women faculty feared they were, for women students, "negative role models."[†21]

It is not surprising, perhaps, that new women faculty who were

---

* Indeed, a 2016 study found that pausing a tenure clock has opposite effects on men and women faculty: improving men's chances of tenure while decreasing women's.

† The findings were so damning that the MIT president publicly acknowledged that while he had always thought that gender discrimination was "part perception," he now saw that "reality is by far the greater part of the balance." By today's standards, this seems like a modest, condescending admission. At the time, the science community was shocked by MIT's admission of discrimination. Some worried MIT would get sued. No one sued.

offered jobs at the school turned them down at the rate of 40 percent. Thus, a cycle: the isolating and unwelcoming environment drew few women, which in turn made the isolation more intense. Biologist Nancy Hopkins, who had initiated the School of Science report, asserted that it was essential to improve the culture for the existing faculty. The report on the School of Engineering concluded that changing the culture would require a critical mass of women.[22]

The mechanical engineering department recognized itself in this analysis. It was a department of seventy faculty and had hired five women in its entire history. By the time the reports came out, only one remained. Faced with this reality, the department had a choice. It could try to improve the environment and then wait patiently for women faculty candidates to trickle in. Or it could try to proactively assemble that critical mass of women.[23] It was a choice between setting a dinner table and waiting for people to hear about a party or tracking down guests, hand-delivering invitations, and sending chauffeurs to go fetch them.

Debates ensued. Some faculty members, like Royan Abeyaratne, advocated for the active approach, arguing that without aggressive recruitment they'd be waiting forever. But others balked at the idea of preferentially adding women, arguing that these actively recruited faculty who didn't battle it out through the existing process would struggle to gain tenure. Then all the effort would be for naught. One vocal opponent made an impassioned argument: if women received any special treatment, other faculty would see them as less qualified and devalue their work. The very program meant to help women could serve to hurt them.[24]

The opponent? Mary Boyce, the sole woman professor hired by the department who was still at MIT. A titan of the field, Boyce has transformed scientific understanding of the behavior of materials. An advocate for women, she was likely trying to protect them from even more unfairness. She had, in fact, been a lead author of the damning report about the School of Engineering.

SINCE THE ADVENT OF AFFIRMATIVE action, its use as a means of correcting the inequities of the past and present has been a source of debate. Today, the practice has many definitions; for the purpose of this discussion, I take it to mean choosing a qualified woman or underrepresented minority candidate over an equally qualified male or White candidate. Defenders argue that it uncovers candidates who might not otherwise apply, corrects past and present injustice, and helps create the critical mass necessary to mitigate isolation for an underrepresented group. Skeptics argue that it is unfair, ill-advised, and divisive.

Another claim from the skeptic camp is that the practice may add to the harms inflicted on those already experiencing discrimination. Echoing Boyce's worry that preferential hiring for women would give male faculty another tool to demean them, economist Thomas Sowell has lamented that because others will doubt their achievements, Black students who are admitted to universities that have affirmative action graduate with their credentials "under a cloud of suspicion." Law professor Richard Sander has argued that students admitted through affirmative action suffer a "mismatch" between their skills and the demands of the school. Like the faculty members at MIT who believed that women preferentially hired would never achieve tenure, Sander claimed that these students are placed into a school where they will not succeed. Others argue that being a beneficiary of affirmative action creates an internalized stigma, a sense that one does not deserve to be there.[25]

But are these imagined harms real? Sander's hypothesis has been debunked. A study of twenty-seven years of affirmative action at the University of Michigan law school found that students admitted under affirmative action graduated at nearly the same rate (96 percent versus 98.5 percent) and had similar earnings. They also did significantly more community service than their White counterparts. In their landmark study of affirmative action in higher education *The Shape of the River*, sociologists Derek Bok and William Bowen found that contrary to the notion of "mismatch," the more selective the school, the lower the dropout rate among Black students. Moreover, achievement differences are not fixed but pliable. When Georgia Tech found there was a performance gap between Black and Hispanic students and White students, it created an intensive summer program; within two years these Black and Hispanic students were outperforming White students.[26]

Bok and Bowen also did not find evidence of internalized stigma in students who benefited from affirmative action. While some students feel uncomfortable about race-sensitive admissions, Black students at elite colleges actually felt more positively about their college experiences than White students. Philosopher Anita Allen—herself a beneficiary of affirmative action—points out that people who have historically been prohibited may perceive affirmative action as a necessary remedy for past exclusion, a form of reparations and a long overdue "extra chance." When lawyer Ashley Hibbett surveyed fellow Harvard Law School students about stigma and affirmative action, their responses were so varied she admitted that she had made "one of the most common mistakes that African Americans often accuse white America of committing": assuming she could find one clear pattern for the group by canvassing a handful of members.[27]

When I was a student at MIT, I did sense the "cloud of suspicion" that Sowell describes. Schools are cagey about how they make admissions decisions; at the time, MIT's official affirmative action policy included "intensified" recruitment of women and minority students.[28] I arrived confident of my own abilities, despite middling high school preparation. My freshman advisor revealed that I'd arrived with higher test scores and grades than the men in my advising group. But a sense that MIT preferentially admitted women undergraduates hung in the air like a haze.

At the times I sought out tutoring, the room was often, to my chagrin, filled with women. Upon entering, I felt a sinking sensation I now understand as shame, and I wondered what the male tutors thought. Were we confirming negative stereotypes? Later, I learned that the school's fraternities (which housed more than 50 percent of male undergraduates) had "bibles" of exams and problem sets from previous years that members could study in tight-knit study groups. Unfairness was built into the residential structure of the school.* But at the time, I

---

* It would be years before I understood that my own privileged upbringing was itself a version of those exclusive fraternities.

felt others' misperceptions as my personal burden as well as an ongoing anxiety that I might prove them right. Freshman year, after I failed a physics exam, I studied with such frenzy that Maxwell's equations spun behind my eyelids as I fell asleep.

Psychologist Madeline Heilman studied the effect of affirmative action on women's performance evaluations in the workplace by asking hundreds of managers to review a supposed employee. When they were told the person was an "affirmative action hire," the manager saw her as less competent unless there was unequivocal evidence of her success. Even now, women students at MIT encounter some version of the idea that they are less able. Today, MIT's policy is to look "holistically" at a student's background and qualifications. MIT does admit a higher proportion of its undergraduate female applicants than male. It's also true that those women graduate MIT at a higher rate and with higher GPAs, both metrics of academic excellence.[29] One explanation for the higher admit rate is that women applicants possibly hold themselves to higher standards in order to even apply. Perhaps, their performance, too, is propelled by fear of confirming others' stereotypes.

Stigma against marginalized groups, it seems, exists with or without affirmative action.* When law professor Angela Onwuachi-Willig and colleagues analyzed the experiences of students at seven elite law schools—four with affirmative action policies and three without—they found that Black and Latino students felt equally stigmatized regardless of their school's affirmative action policy. Indeed, California banned affirmative action twenty-five years ago, yet students continue to be demeaned by the insinuation that they do not deserve to be there.[30] For women at MIT, it's possible that the devaluation that stems from others' negative stereotypes would exist with or without preferential treatment.

---

* Meanwhile, a long-standing form of affirmative action for wealthy White students goes largely unchallenged. A stunning 43 percent of White applicants accepted at Harvard are legacies, athletes, and children of donors and faculty. In a recent analysis, researchers concluded that roughly 75 percent of these students would have been rejected without those traits. To the best of my knowledge, Sander has not developed any hypotheses about whether these students are "mismatched."[31]

After all, the devaluing of women faculty so thoroughly detailed by the scorching reports took place in the absence of any affirmative action designed to support them.

MARY BOYCE LOST THE DEBATE. The department began strategizing about how to increase the number of women faculty. The problem was, in a sense, an engineering one: how to achieve a desired outcome, within the constraints of the real world. First, it was essential to keep to MIT's hiring practice, such that anyone brought in as junior faculty be able to reach tenure. That means being among the very top investigators in one's field *in the world*, so the department had to believe that everyone it hired had the potential to reach that level. (Not all universities hire this way, but MIT hires faculty it expects will get tenure.) Second, there were far fewer women than men with the needed qualifications: PhDs in fields including engineering and physics.

To manage these two constraints, the department introduced a critical change in its hiring. Up until then, the mechanical engineering department hired for the subfields it wanted to develop—not just mechanics, say, but the mechanics of polymers. It then invited top candidates in the subfield to apply. But because the pool of women in mechanical engineering was already small, the likelihood that a top candidate in any specific subfield was female was microscopic. So the department changed the search criteria from a subfield to a much broader field. Mechanics of polymers became mechanics; nanoscale sensing became nanoengineering in general. Instead of trying to find women working in specific subspecialties, the department decided to locate spectacular women candidates and then make their research agendas the departmental priorities. In this way, they opened the search up to a much wider array of candidates, dramatically increasing the applicant pool without changing the standards at all.

They also changed *how* they searched. Before, the department posted job ads and waited for applications and suggestions. Now it aggressively sought out women. Professors called colleagues and department heads around the country asking for recommendations. They scoured past applications for people they had turned down or overlooked—people like Peko Hosoi. And when applications came in, the dean of engineering personally reviewed every one from a woman. If departments turned down a good candidate, they had to explain why.[32]

Hosoi said yes to MIT's offer. So did Yang Shao-Horn, a Chinese metallurgical and materials engineer who was a postdoc in France, working on lithium-ion battery innovation before it was widely accepted as a topic of research. Over the next several years, the mechanical engineering department hired a total of six women with impeccable credentials, including a mechanical engineer who was deriving basic principles to guide the design of products and a young engineer working on nanostructures. More women followed: an electrical engineer focused on nonlinear systems, and an applied mathematician studying the laws of human pathogen transmission—the physics of the sneeze.

These professors made fundamental discoveries. Hosoi arrived as an expert in fluid dynamics, but graduate students' interest in robots inspired her to merge these interests. Her work has powered the field of "soft robotics"—robots that can more safely interact with humans, compressing and expanding like octopi. She has also fused biology and design, creating a digging robot inspired by razor clams and uncovering how bats use the hair on their tongues to drink nectar. Hosoi, now tenured, also became a dean of engineering, a leader at the school-wide level. This engineer who had been turned down by the department has succeeded in her role in every measurable way.

Other women faculty hired in these years developed systems for extracting drinkable water from desert air and made breakthroughs in the development of biological cells that function like electronic circuits. Of the six women faculty initially hired, four received tenure. MIT's general rate is 47 percent; this cohort reached that milestone at a rate of 66 percent.[33]

As one long-standing mechanical engineering faculty member admitted to me, the outstanding accomplishments of these women faculty demonstrate how flawed the prior process was. "We were being biased before," he said plainly. "The only thing we changed about our selection process was to make it harder not to hire them."[34] Preferential hiring wasn't bringing in less qualified people; it was removing barriers to outstanding ones.

Other significant changes followed, too. Mary Boyce became head of the mechanical engineering department, and she assigned the teaching of core introductory classes in the department to women; as a result, students were exposed to women faculty early. The department also modernized the curriculum, offering a new mechanical engineering degree that allowed students to specialize in newer fields like robotics, nanotechnology, and sustainability. As a result, students were able to merge interests in engineering with broader social contexts, allowing students to study, for instance, not only physical properties of a new fuel but the social implications as well.

And something else happened—something the department did not anticipate. The proportion of women undergraduates majoring in mechanical engineering began to rise. The change was visible: more and more women appeared in hallways, as though precipitating out of air. Rohan Abeyaratne remembers walking past Hosoi's office in those years. "The number of female students who were going to her office hours or coming in and out or gathering there was striking," he said. "You could see that these role models were really important."[35]

The year Hosoi was hired, women made up 32.5 percent of mechanical engineering majors at MIT. Over the years, as the number of women faculty grew, the number rose steadily. As of this writing, of the MIT students majoring in mechanical engineering, 50.4 percent are women. Women are now overrepresented in the major compared to the general undergraduate population. Notably, this is not a national trend; at peer institutions like Georgia Tech or Caltech, the proportion of women majors is 21 and 30 percent, respectively. Yet at MIT, some courses in the department are completely dominated by women. In 2016, in classes such as "Mechanical Vibrations" and "Mechanics of Structures," over 70 percent of the students were women.[36]

When I asked Hosoi if preferential hiring caused her to question whether she belonged at MIT, she said no. There are so many qualified people in these fields, she said, that for anyone to be hired at a place like MIT, you have to be good and you also have to be lucky. Everyone who ends up there, of any identity, has been lucky in some way. Having extra attention paid to one's application, she feels, is one form of luck. Another woman faculty member felt differently: the self-doubt it engenders is real, she told me. Moreover, women can rise to the top of the candidate pile without a vast network, which makes things more fair, but also means the men who land on the top of the pile tend to be politically savvy and well-connected, which gives them an advantage going forward. The solution may not be cost-free.[37]

But as a blunt instrument, in this case, its effects were unexpectedly large. The mechanical engineering department had set out to make the faculty less homogenous, and it put in a massive structural change to accomplish this. But it had not set out to pull women undergraduates into the major. It didn't recruit undergraduate women to mechanical engineering or load the pipeline by requiring that all freshmen take courses in the department. It didn't shower them with encouraging messages. Instead, by diversifying the faculty, it showed students flesh-and-blood examples of people like them who already were.

What do role models do? Psychologist Nilanjana Dasgupta set out to understand their impact by pairing incoming women engineering majors with upperclassmen or upperclasswomen as mentors. (A control group received no mentors.) The mentors met individually with the women for one year, developing personal relationships and providing advice. Over that year and the following, after mentoring had ended, Dasgupta found that the women in both mentor groups grew concerned about their coursework's difficulty. Both described their mentors as equally available and dedicated.

But over the course of the study, only those with women mentors continued to feel they belonged in the field, and only they maintained a sense of their own confidence and talent as engineers; for the women with male mentors or no mentors, belonging and confidence plummeted. And it was these feelings—not grades—that determined their persistence in the field. For women with male mentors or no mentors, many switched to another major by the end of their first year in college.

Of the women students with women mentors, 100 percent remained in engineering.

Dasgupta describes these findings as a sign that role models serve as a "social vaccine," helping "inoculate" underrepresented individuals against the negative effects of stereotypes on their sense of their own capacity. Notably, the mentors were not even faculty—they were only a few years ahead of their mentees. But they made it possible for the women to imagine their future selves being successful on this path.[38]

In the summer of 2020, MIT's mechanical engineering department assembled a new working group to address the underrepresentation of racial minorities in the department. Research suggests that Black women students are more likely to feel a sense of belonging if they have Black role models, and that Black women engineering students feel a greater sense of belonging even learning that a Black STEM faculty member exists at an institution.[39] As of June 2020, only 4 of 112 mechanical engineering faculty at MIT are Black.

Asegun Henry, one of these four, cites the power of role models in his own path. Growing up in Florida, he never considered that he could be a professor. He pictured corduroy jackets with leather elbows, he told me; he had never encountered a professor like himself. Then, as a student at Florida A&M, a historically Black university, he began doing research with a Black engineering professor. One day he walked past the professor's office and, through the open door, saw him working. He was in a T-shirt and sneakers, with his feet on the desk, listening to the rapper Rakim. In that moment, he realized, "I can *do* this job. I can be a professor, and I don't have to change who I am." Seeing was believing. "That imagery is critical," Henry said. "It's hard to ask someone to become something they've never seen before."[40]

The powerful effect of role models is seen elsewhere. The first female chess grandmaster in the world was crowned in the country of Georgia in 1978. The game has been a traditional part of Georgian women's lives (according to folklore, chess sets were an important part of a bride's dowry), and more chess champions followed. Today, of the top one hundred women chess players in the world, six are from this tiny country: 125 times what its population would predict.[41]

Crucially, real-world role models don't just inoculate targets of prejudice against adversity, they also erode that adversity by changing *others'* perceptions. In the Indian state of West Bengal, a 1993 law stipulated that the leadership of a third of all local villages' councils be reserved for women. Because the villages were chosen randomly, this new system served as a remarkable real-world experiment to test the impact of gender quotas. Economists who studied the effect found that the introduction of these female role models in politics by itself raised the educational aspirations of girls in the villages. It made them more likely to want to postpone marriage beyond age eighteen and to find a job that required an education. But it also raised their parents' aspirations for their schooling. Girls spent less time on household chores, and fathers had greater wishes for their daughter to become a village leader. Before the policy, boys attended more school and had higher literacy than girls; afterward, this gap disappeared. In some cases, it reversed.[42]

When I asked Peko Hosoi about the influence of role models, she used a math analogy. Sometimes in math, in order to proceed with a task, you have to first show that a mathematical object exists. To do this, you need to create an "existence proof," an argument that this object *can* exist. An actual example is one good argument that it can. That's what students need, Hosoi says—not necessarily a person on whom to pattern their lives, but an existence proof: evidence that a particular kind of life is possible.[43]

IN MARCH 2018, ON A week when the heavens had dumped two feet of snow on Boston, I stood at the edge of a cavernous foyer at MIT called Lobby 7. I hadn't been there in over a decade. A young woman crouched at a nearby table, tapping urgently on a laptop covered with NASA stickers. Students spilled down the Infinite Corridor, a 250-meter hallway that runs through the campus's main buildings. Twice a year, the sun hits the entrance on Mass Ave and shines from one length to the other, shooting an electric arc of light down the gray marble floor.

Halfway down the corridor, in another large, light-filled plaza, students were hawking tickets to events on campus. This week, the Shakespeare Ensemble was promoting *Queen Lear*, a gender-swapped version of the original, and students at the far end of the lobby were distributing Whoopie Pies. It was Pi Day, after all: 3.14. Talking to students about their lives and their studies, it became clear that the place had changed. And women explained that the sheer volume of women already majoring in mechanical engineering signaled that this was a field where they could thrive.

One student said that she became interested in the field after joining the solar car team freshman year, which was made up entirely of women. As in Dasgupta's study, her mentors were older students. Another student described how she and some female classmates developed a product that uses optical character recognition to convert text into three-dimensional Braille. She was so accustomed to working with women that she was surprised to find, while developing her patent, that the patent world is mostly men—it hadn't occurred to her. A senior proudly described leading a project in the capstone class for the mechanical engineering major. Her team developed a medical device to combat Parkinson's symptoms—a wristband that vibrates to redirect nerve impulses—which allowed an actual Parkinson's patient to use his hands without shaking. A student of Ethiopian and Chinese descent said that in the department she felt "really supported."

Some students I spoke with noted that men seem to participate more in class, asking questions and volunteering answers. But one woman explained that when she doesn't ask questions in class, it's because she finds office hours more conducive for dealing with complexity. "I like to discuss," she said. "If the question is simple enough to be answered in lecture, I can probably just look it up or figure it out on my own."[44]

And they lit up when talking about their female professors. "You're interacting with them every day," said a recent graduate. "You see how brilliant they are." One woman's face broke into a wide grin when she told me about her thermal fluids professor. "I *love* her," she said. "She's so passionate, she's such a nerd. I called my mom and said, 'I have a woman as a thermo prof and she's *awesome*.'"

Her comment pointed to the way this structural change set in motion a chain reaction whose effects are ongoing. The thermo prof, Betar Gallant, had herself been an undergraduate at MIT after that first wave of women professors arrived. She hadn't been a stereotypical engineer as a child, tinkering with toasters or computers, and even considered becoming a novelist. When her engineer dad died while she was still in high school, she looked into the field as a way to feel closer to him. In college she explored studying mechanical engineering. Then one day in lab, her professor turned to her and made an offhand comment about her own research. It was Yang Shao-Horn, one of the first women hired after the policy change.

Right after class, Gallant emailed her to ask if she had any openings in her lab. Gallant ended up writing her thesis on that research, which launched her toward her PhD. Now she specializes in ultralight, energy-dense batteries designed to be an order of magnitude more powerful than today's lithium-ion batteries. She traces her path in large part to that one offhand comment. "It was rare that a professor saw me," she says. "So when one saw me, it was meaningful. For someone to have a conversation with me—it means you're there and you're worthy of a conversation."[45]

Worthy of a conversation. In that moment, I remember my old classmate Djuna Copley-Woods, how her professor told her to drop the class. I remember the hot flush of shame I felt when my professor deflected a comment about academics with a flirtation. I was seen as a body. Copley-Woods wasn't seen at all. In a culture that wasn't meant for us, how delicate, how breakable, this causal chain that leads any one of us anywhere.

A FEW DAYS LATER, I toured one of MIT's mechanical engineering labs, where blue and orange plumes swirled on LED screens and cubicles were filled with robotic fish and synthetic otter fur. I sat with two grad students—Alice from Texas, Samar from Saudi Arabia. On Samar's laptop, we watched a time-lapse video of an experiment she had conducted of two liquids interacting. Interacting fluids are everywhere, from the cream in coffee unfurling like smoke to supernova explosions; fluids moved over each other at the dawn of the universe. But the shapes we watched on the screen had never been captured on video before, and existing physics does not predict them. There was a real mathematical law at work, but no one in the world knows what it is. Cracking that alone is hard enough, without additional isolation.

"It sounds so idealistic to be a pioneer," Alice said. "But who wants to be a pioneer? I don't. It's nice to be someone who comes in after the pioneers. Like, 'It's fine in here! Jump in!'"[46]

The moment Alice told me she doesn't want to be a pioneer was a lightning moment. It's a seemingly simple point: being the first or only person of your identity in a department, or a field, or an organization, is not a role everyone wants or is suited to. We cheer pioneers, but the path is lonely and alienating. Pioneers are constantly confronted with their "otherness." They must contend with others' stereotyping, discomfort, or outright aggression. To survive, they must be able to maintain a sense of well-being in an environment that is harsher to them than it is to others.

What this means is that in addition to having all of the formal skills to do the job, the first woman, or the first Black person, or the first Native American person, or the first Latino person must have an additional unstated set of skills and characteristics that have nothing to do with the explicit job description. For these individuals to function in so many organizations whose cultures are exclusionary, there is a set of shadow requirements—call them "pioneer requirements."

In engineering, for instance, the skills required for the job might be top-notch technical acumen, creativity, ability to work well in teams, and good communication. But the pioneers in engineering must have all of these skills and *also* be able to withstand solitude, *also* be unaffected by aggressive or disparaging comments, *also* be able to navigate a culture that may be downright hostile. To survive in unwelcoming environments, pioneers are required to contribute without necessarily feeling like they belong. Job skills and pioneer skills might even oppose one another: research might require teamwork, but a pioneer might have to thrive in solitude. And pioneers have to fulfill the shadow requirements while performing at the same level as those who are free of them.

Djuna Copley-Woods went on to become a professional mechanical engineer, recently designing cameras for the 2020 NASA Mars rover. But at her jobs, she has typically been the only woman in her department. She was able to survive, she said, because she was "weird," oblivious to social norms. She had a unibrow, she said; once, a colleague even brought her stylish clothes and gently urged her to wear them. But this obliviousness allowed her to thrive in hostile environments. She sometimes laughs now when she describes some of the sexist comments she's had to listen to throughout her career. People are shocked that she can find them funny, she says. But if she hadn't, she points out, she never would have stayed at those jobs. It was an extra requirement. Other women departed the organizations, one by one. "A lot of people left," she said.[47]

This is yet another danger of homogenous organizations, a deep and little appreciated one: they artificially *shrink* the pool of candidates from underrepresented backgrounds. They require those candidates to possess not only the stated requirements but the shadow traits, too. The number of people from any group who possess both is exceedingly small. One complaint about affirmative action programs is that they artificially boost the number of individuals from underrepresented groups in a given school, organization, or company. But homogenous environments artificially boost the number from the majority culture: the skill set they need is much smaller.

Structural change like that of MIT's mechanical engineering department doesn't just alter one diversity metric: it sets in motion a chain reaction. Changing the faculty makeup increased the number of available role models, which invited more women students into the department, which in turn removed the pioneer requirement for students like Alice. The way structural change unleashes downstream benefits can be seen in other contexts, too. A 1994 program called Moving to Opportunity was an effort to weaken the housing segregation that is a product of past and present exclusion and discrimination. In this program, hundreds of low-income, predominantly Hispanic and Black families in cities like Baltimore, Los Angeles, and Chicago were randomly assigned housing vouchers to move to more affluent areas. The children who grew up in these families earned 31 percent higher incomes as adults. They attended college at higher rates and were less likely to become single parents. They were also less likely to later live in impoverished neighborhoods them- selves. It was yet another top-down intervention that unspooled lasting change.

In West Bengal, we see the same unfolding of consequences. Here, structural change expanded women's political participation even beyond the immediate quotas. Two election cycles after the law reserv- ing village leadership positions for women went into effect, the num- ber of women in other elected positions doubled. After years of seeing women leading effectively, villagers were now also more likely to elect women to positions that were not reserved for women. By the study's publication in 2009, about 40 percent of the village council leaders were women.[48]

The presence of women leaders in West Bengal also changed the policy decisions that were made. While a person's social identity does not necessarily predict their behavior or policy positions, economists who studied the impact of the quota system in West Bengal found that women leaders did in fact make different political decisions. Women vil- lage leaders invested more in roads, repairing health centers, and sani- tation. They also prioritized drinking water and irrigation. In doing so, they better represented the wishes of the village than did leaders in vil- lages that did not have quotas.[49]

Any field that is dominated by a limited range of human experi- ence will find itself hampered by limited access to human ingenuity. Mechanical engineering is no exception. Asegun Henry described to me how his experiences shape his approach to engineering. Because he does not see himself in his field's lineage, he feels less reverence toward its traditions and less afraid to question them. And while he notices that his colleagues favor elegant solutions, fighting uphill battles has predisposed him to a "brute force" method of solving problems. These qualities recently enabled him to disprove the universality of a widely accepted model for how heat moves through solids.[50]

People from groups that have historically been excluded may also feel less allegiance to traditional disciplinary boundaries. Data are still minimal and it is important not to overgeneralize, but research suggests that women conduct more interdisciplinary research, and they draw on a greater number of fields when doing so. A study of U.K. researchers, for instance, found that 21 percent of women cited seven or more fields in their work, compared to only 8 percent of men.[51] In a complex world, being able to slip the artificial boundaries of traditional academic disciplines is increasingly important. One MIT mechanical engineering professor, Lydia Bourouiba, works at the nexus of mechanical engineering, epidemiology, and medicine, studying the mechanics of disease transmission. Her merging of fields demonstrated that the World Health Organization's social distance guidelines surrounding COVID-19 relied on an outdated understanding of fluids. The old model assumes virus-containing droplets can be divided into two sizes: large and small. In fact, like gender, they exist on a continuum. And, Bourouiba shows, they travel in a turbulent gas cloud. A coronavirus sneeze can cruise more than twenty-six feet.[52]

Diversifying a field does not by itself sponge away bias or scrub an environment of its consequential harms. In fact, a recent randomized study found that even in fields where women are well represented, like veterinary medicine, people still recommend lower salaries for women; those who feel gender bias is no longer a problem are the most likely to do so.[53] The MIT students I met noted they still see biases in how women faculty are treated; male students question them more aggressively than they do the male professors, and they are described as "tough" for exhibiting behavior that in men is accepted as normal.

Despite the fact that women make up more than half the mechanical engineering majors, gender bias persists among students. A longtime mechanical engineering professor told me he still sees a pattern in some undergraduate classes in which a quiet woman voices an opinion, and the men reflexively discount what she has to say. Often, the professor noted, she is equally or more skilled. This has happened so many times, he told me, that he now points out the potential for this dynamic early in the semester and warns students away from it.[54]

Bias lingers among faculty, too. One particular year, seven of ten top mechanical engineering undergraduates were women, and the professor heard rumblings among faculty that they must have taken easy classes. He investigated. They had earned the best grades in the hardest classes the department offered.[55] But assumptions remained.

Yang Shao-Horn—the battery engineer who arrived at MIT alongside Peko Hosoi and helped bring Betar Gallant into the field—points out that while the undergraduate statistics are remarkable, many inequalities have been kicked higher up the ladder. She credits Mary Boyce with shaping the department and making it strong. But Boyce is notably no longer at MIT: she now heads the Columbia University School of Engineering and did not respond to multiple requests for comment. When Susan Hockfield was MIT's president, she brought Shao-Horn onto prominent committees. But Hockfield is no longer president.

Today, Shao-Horn notices that her technical skill is unquestioned, but her judgment and leadership are still not as trusted as those of her male colleagues. These are the characteristics that would allow her to have more influence within the university. She would like to see more change further up the chain—more women deans, another woman president. When events feature young women scientists and engineers, Shao-Horn points out, the people who decide who goes onstage are still all men. Women should be making these decisions, too. By itself, she says, "Populating a stage with younger women is not the solution. It's an expression of power."[56]

Indeed, increasing diversity is only one step toward fixing a biased environment. It doesn't guarantee fairness or provide the fuel for long-term success. Ensuring that people thrive and stay in the fields that need them—and that they are elevated to their level of competence—requires more than one top-down structural change. It requires a change of culture.

The Architecture of Inclusion

When Uché Blackstock was growing up in the Crown Heights neighborhood of Brooklyn, New York, she thought most doctors were Black. There was her mother, a nephrologist at Kings County Hospital Center. Her pediatricians were Black, too, as were mother's colleagues, the members of the Provident Medical Society of Black physicians and the Susan Smith McKinney Steward Medical Society for Black women physicians. Blackstock and her twin sister would tag along to meetings, where their mother was often presiding. In elementary school, they'd watch her speak; by high school, they were taking notes. Joining the field of medicine seemed attainable, even natural. By the time she was in college, Blackstock knew she wanted to be a physician, too.

Blackstock enrolled in Harvard Medical School like her mother had decades before. She sat in lectures in Bornstein Hall under oil paintings of the school's past luminaries, whose race and sex were vivid symbols of the field's exclusionary history and the fact that she would have been barred in earlier times. During her emergency medicine residency, she worked hard to build friendships with nurses, who, she noticed, would sometimes act friendly and deferential to the male residents but curt toward her.

Later, as an attending physician and faculty member at a university, Black-stock cared for patients who assumed she was there to change their sheets or wheel them to a CT scan. Sometimes, after she had sat at a patient's bedside and discussed their care in detail, they would complain to the staff that they had never seen a doctor. Other doctors would mentally demote her, too: sometimes she'd be on the phone coordinating patient care and the male physician on the other end would give her orders—*Get these five items at the bedside for me*—in the way he would speak to an orderly. Afterward, he would apologize, once he learned he'd been speaking to a physician.

But she loved emergency medicine—it felt to her like medicine in its most quintessential form. She loved the pace, the variety, caring for a child with an ear infection one moment and a man with a collapsed lung the next. She pioneered a curriculum for teaching ultrasound—a technology she thinks of as "the next stethoscope"—so that the medical students would learn its application not just for prenatal testing but to assess fluid around the heart or bleeding in the belly. She became a star of the emergency department. Students she mentored wrote her notes that said, "Meeting you has been one of my favorite things about this school" and went on to join the faculty at other institutions.

So she kept quiet about the things that gave her pause, like the fact that young White men she had just trained seemed to lap up promotions with uncanny frequency, ascending to such roles as associate medical director. These were positions of real power, in which they set the practices and policies for the department, yet they'd only just finished their training. The process seemed opaque, the candidates handpicked. Blackstock wasn't alone with her concerns—other faculty and students of color whispered their concerns about the fairness of promotions. They also noted how subjective grading was, and how this hurt students of color. But they worried that speaking up would label them troublemakers. Blackstock kept her head down, too. *Do your work, don't make a scene.* That's what her parents had taught her. "For a long time," she told me, "I never really said anything."

Then, finally, she was given an opportunity to speak: she was asked to lead recruitment and advancement of women and faculty of color at the school. Blackstock leapt at the chance. She was bubbling with ideas. "I *am* that role," she said. But she soon found herself stymied. After she and other staff facilitated a workshop to educate faculty about microaggressions—daily experiences that exacerbate burnout and attrition—her office was told the term was too inflammatory. When she defended Harvard Medical School's decision to create a less exclusionary environment by removing the Bornstein Hall oil portraits—the same paintings that had weighed on her as a student—word got back to her that some in leadership deemed her actions unacceptable. Mentoring and role models are proven ways to boost marginalized groups' success, but when she proposed new mentoring programs or newsletters highlighting women's contributions, she was told these ideas were too exclusionary of men.

She also found that just as dissent was discouraged, so was open discussion. Blackstock found herself in the position of possessing the unique point of view the school purported to want but being blocked from speaking, acting, or making real decisions. Her ideas were not solicited; when she shared them they were rejected. Her role, it appeared, was to be a symbol of diversity while making no actual changes. When she realized this, she felt sick to her stomach.

The irony was not lost on her: her position was housed in the Office of Diversity Affairs. Her role existed for the purpose of attracting, retaining, and advancing people who fit her description: a woman of color on the medical faculty. But three Black women, and one Latino man, had left the office in the space of five years. Now it lost yet another. After almost ten years, Blackstock decided to resign. The deans asked to meet with her. She shared her concerns with them: about the climate, about fear of retaliation, about feeling censored. "This is the way we do things," she was told. "You have to fit into the culture." And with that, Blackstock quit the faculty and left academic medicine.[1]

BLACKSTOCK'S EXPERIENCE OF FRUSTRATION AND alienation is sadly familiar. One study of five medical schools across the United States found that faculty from underrepresented backgrounds reported isolation, invisibility, discrimination and stereotype threat, disrespect for their academic interests, and the stress of always being "the one" to point out inequities.* Up to 44 percent of doctors who are racial minorities leave academic medicine, as do up to 47 percent of women, while the departure rate for White doctors and men hovers in the 30 percent range. This pattern is repeated across fields: in 2020, an American Bar Association survey found that as high as 70 percent of women of color have considered or have left the profession, often feeling undervalued. In 2017, seven journalists of color left the *Houston Chronicle* newsroom, prompting colleagues who remained to circulate a petition pleading with management to improve its practices. Around the same time, at Minnesota Public Radio, seven of ten departures were people of color. An estimated 40 percent of women engineers leave the profession. Among various demographic groups, women of color report feeling the least included at work.[2]

"When I say these are daily experiences, I really mean that," said Ajilli Hardy, a midcareer engineer who has worked in energy, aviation, and other "large, slow-moving industries." Hardy, a Black woman with a PhD in mechanical engineering from MIT, said that the problems were multilevel: routine biases both subtle and overt were compounded by a lack of institutional response and by colleagues who saw unfair treatment and did not speak up. Making sure her contributions were appropriately credited and valued was another draining task. At some jobs, she could find "just enough" colleagues to make the days bearable. "Just enough means you can get through the day-to-day," she said. But she's also been in companies where she could not find *any*. And in those jobs, she had to make heartbreaking decisions: even when she loved the actual work, the culture made it impossible to stay. Bias can make it difficult for people to simply do their jobs.[3]

Dozens of interviewees from marginalized backgrounds echoed

---

* For these faculty, many of whom may not have come from wealthy families and thus incurred disproportionate medical school debt, the lower-paying field of academic medicine was also seen as a financial sacrifice.

the sentiment and the story: individuals full of talent, passion, energy, and ideas, raring to contribute and lead, but blocked and drained until they have no choice but to leave. The experiences include the everyday biases of being bypassed for promotions, talked over in meetings, uncredited for contributions, or having one's judgment and expertise mistrusted. There is also the social isolation of not being invited to participate in projects or being excluded from social gatherings. These experiences knit together into a dense web of exclusion. The experience is made all the more bitter when the organization's professed commitment to diversity is exposed as a façade. Sometimes this is revealed when leaders fail to intervene in cases of egregious bias, harassment, or bullying. Sometimes it's revealed, as it was for Blackstock, when that commitment evaporates at the precise moment when new approaches could usher in actual change.

To create less biased environments, it's not enough to simply increase the diversity of the group—to add women or any underrepresented group and hope for the best. If the people who increase a group's diversity feel devalued and unwelcome, diversity is a battle half-won. When organizations fail people from marginalized groups by showing them in ways subtle and overt that they are not valued, they recruit talent only to hemorrhage it.

While there is growing awareness that creating an inclusive environment requires active, meaningful efforts, what "inclusion" actually *means* is less clear. Without a single universal definition, it has come to stand for everything from having meaningful connections to others, to participating in decision-making, to having access to insider information. Some researchers posit that in inclusive environments, people have a voice and feel that they belong. Human resource studies professor Lisa Nishii, drawing on two decades of research, proposes that an inclusive setting has three features: fair and unbiased practices, a welcoming attitude of and respect for people's "whole selves," and the desire to seek different perspectives. In much research, inclusion is assessed by asking questions. Do you feel welcome? Do your ideas matter? Do you belong?[4]

Companies often assess inclusion through subjective reporting—whether people feel welcome, valued, and heard. But as business professor

Robin Ely points out, it's possible to feel valued and respected and included without actually *being* included.[5] Before Blackstock took on her leadership role, for instance, she did feel generally welcome, she told me. Her medical work and teaching were respected. Her colleagues and deans were friendly. But that was before she tried to bring her unique point of view to bear on how the organization worked. When she wanted to use her voice to effect change and make a larger impact—when she wanted not just to feel included but to have actual influence—the limits of her inclusion became unmistakable.

Women of color in particular report coming up against such a barrier. As Yang Shao-Horn, the MIT mechanical engineering professor, noted, her scientific acumen is now accepted, but her judgment about larger strategic issues, like hiring or planning, is not. Doualy Xaykaothao, a journalist of Hmong origin, felt encouraged in her field as a very young reporter. But as she sought greater influence, she saw that supervisors were more comfortable with her in a limited role in which she was not making any real decisions. "You just get to play," she said. "And that's not where any of us want to be—those of us who are serious, those of us who are truly dedicated to whatever work that we're doing."[6]

"Inclusion has come to mean a pablum, feel-good thing," says Ely, making sure everyone's birthday is celebrated, sitting in a circle and sharing thoughts. "But whose thoughts are acted upon?" she asks. "What happens when a person's point of view conflicts with the status quo?" Feeling welcome, having a sense of belonging—these are important. So is sharing thoughts. Yet, says Ely, "we can listen to everyone's voices, but still do things as they've always been done."

"I don't want a birthday card," Xaykaothao told me. "I want to genuinely make an impact."[7]

The real question shouldn't only be whether people feel welcome, but whether they have influence. This can be measured by looking at the actual distribution of decision-making. In a typical company, organizational charts are often starkly revealing: who the organization values can be seen in the apportioning of power.

The challenge, then, is how is to accomplish deep, structural inclusion. How are different perspectives incorporated at the very highest levels of influence? For an answer, we can turn to a frustrated Italian lawyer at a failing French law firm.

IN 2004, GIANMARCO MONSELLATO WAS a star partner at Taj, a French firm focused on the intricacies of tax law. At the time, the firm was second tier and shrinking. That year, Monsellato was approached to become CEO. He accepted. The lead player of a losing soccer team was now the coach. He realized something immediately: he needed a better team.

But there was a constraint. Monsellato wasn't able to bring in additional resources—he couldn't raise money or muster a horde of new employees. Somehow, he had to fix it using only what he had. Here's your team, he was told. Now make it win.

As Monsellato reviewed his options, he remembered something he had witnessed years earlier. A lawyer was in the process of being evaluated for a promotion and bonus. She had been on maternity leave for half the year and had worked full-time for the other half. The committee reviewing her performance assessed her accomplishments as if she had worked full-time and so calculated that she had achieved 50 percent of the objectives set for her. The committee was also evaluating a man who had worked full-time; he had achieved 75 percent of his goals. The committee concluded that the man's achievement surpassed the woman's. To Monsellato, she had actually achieved 100 percent of her objectives if one adjusted for the fact that she'd only worked half the year. The 75 percent performer got the promotion and a bonus; the 100 percenter did not.

A good lawyer, Monsellato saw, was being penalized for no other reason, he said, than that "we were bad at math." People with high potential weren't being given opportunities. Specifically, he saw that the environment was biased against women who had children. "We were telling women that you can be successful here as long as you don't have children," he said. "We would never say that to fathers. We expect them to have families."

This approach was troubling for multiple reasons. Obviously, the company was losing out on talent. Moreover, as research shows, this inability to evaluate people fairly created an environment that dissuaded people from putting forth their best effort: perceived unfairness eats away at people's commitment to their workplace.[8] And the law is a profession that requires a tremendous personal investment on the part of its workers—what Monsellato calls a "high commitment field." Unfairness, he realized, would repel legal talent, and it should. It would be absurd to expect people to be committed to an organization that does not treat them fairly. Fairness, Monsellato saw, was a business decision.

So one of the first things Monsellato did when he became CEO was to try to expunge unfairness from the company's practices. He cleaned up the promotion process, making sure people were evaluated relative to the time they were at their jobs. If a woman was gone for half the year because of maternity leave, she was evaluated against a 50 percent benchmark. This alone had an immediate and dramatic impact; once women's contributions were being valued more fairly, they suddenly began getting promoted.

Monsellato also demanded that promotions be based on objective criteria and measurable data. Before this change, managers had sometimes used subjective judgments. They might say to a committee, "I think this person isn't as committed" or "I think this person isn't doing as well." Now managers using anything other than objective criteria would be questioned. If promotions were uneven between women and men, the manager had to explain why; if Monsellato found the reason unconvincing, he would cancel the promotions and restart the process from the beginning. Monsellato also reviewed pay: if employees were receiving unequal pay for the same job, managers had to either explain or correct it. When there were discrepancies, "We always ended up increasing the salary," Monsellato told me, "because there were no explanations except BS."

In addition to these structural changes, he made adjustments to the day-to-day interactions that shape an office culture. When men at the company made sexist remarks, he would take them aside, privately, and say, *Those jokes may have seemed acceptable when you were in school. But we're not in school anymore. Cut it out.* He made sure special assignments were accessible to everyone, not just a favored group, and that major decisions were only made with women present. Realizing that women were functionally blocked from full participation if they had children, he ensured everyone could work part-time if needed, and part-timers could still become partners. Once promotions were based on measurable results and not subjective impressions of who spent the most time in the office, it didn't matter if people left early to pick up children, came in late, or worked from home.

Finally, Monsellato created routes to leadership for women not just by actively promoting them for leadership but also minimizing risk. Some women worried that shifting away from day-to-day legal work would mean losing their technical expertise. And if they failed at this new leadership role, they might be out of a job. They were right: it was risky. If they had had to work harder to achieve their current position, stepping into a new, uncertain job *was* more risky.

So Monsellato instituted a policy: people could return to their old job if the new one was not a good fit. He handed his own clients over to women lawyers in the firm but eased the transition by telling them that if, in six months, they were not happy with their new lawyers, he would take back the account. None came back to him. After six months, he reports, they all said to him, "She's better than you."

Notably absent from the changes were any programs designed to change women's behavior—their negotiating skills, their networking, their clothing, their demeanor, or their assertiveness.

"There is nothing wrong with women," he said. "The problem is the environment."[9]

IN THE YEARS AFTER MONSELLATO'S changes, the proportion of women equity partners at the company rose. In U.S law firms, women make up only 20 percent of equity partners; at Taj, they grew to 50 percent. The partner ranks included women with children. The executive positions were also half filled with women. And women made up fully 50 percent of the top ten earners at the company. The company leapt into the first tier of French law firms. Over the twelve years of Monsellato's tenure, revenue increased 70 percent.[10]

Part of the reason the client base grew, Monsellato said, was that the company's employees were happier and clients liked working with happy people. Increased trust allowed people at work to deal with one another more openly. One woman in the firm approached her boss and said, "Look, my husband and I are planning to have another child. Within the next six months I'll be pregnant. Within the next year I'll be on maternity leave. What can we do to plan?" Rather than hide her pregnancy in fear of reprisal, this lawyer was able to prepare for her short-term leave.

Men benefited from this increased trust, too. One man at the firm who was struggling with issues at home explained his situation to his boss. She told him to take the time he needed to sort out his difficulties, and his team pitched in to cover for him while he was gone. When he returned six months later, they welcomed him back.

These shifts toward inclusion and equality extended beyond gender. In a country rife with anti-Muslim prejudice, Taj also became one of the first big French law firms to promote a Muslim lawyer to equity partner. Malik Douaoui was born in France to Algerian parents, Muslims who belonged to an ethnic minority indigenous to the mountainous Kabylie region. He grew up speaking Kabyle, a Berber language, at home in the Paris banlieues. His father worked in a factory; his mother never learned to read or write. "No one refers to the fact that my name is Malik and not Vincent," he told me. "You're assessed on what you can do."[11]

BUT WHY DID MONSELLATO SUCCEED? One reason is that he used the right strategies. When sociologists Frank Dobbin and Alexandra Kalev set out to discover which interventions actually shift the apportioning of power and change the proportion of people of different races, ethnicities, and genders in leadership positions, they analyzed thirty years of equal employment data from more than eight hundred U.S. companies. They also interviewed hundreds of employees.

Some common workplace programs, they found, decreased the numbers of women and underrepresented minorities in management.

Instituting performance ratings typically didn't work, possibly because they were ultimately subjective and replicated managers' biases. Job tests aimed at objectively measuring applicants' skills failed because managers gamed the system, overlooking results to select favored applicants. Adding formal procedures for registering complaints often resulted in backlash, in turn decreasing the proportion of women and minorities in management.

But other programs did change the diversity of leadership. Mentoring stood out as particularly effective. Dobbin and Kalev found that when leaders formally mentored employees, the percentage of women of color in leadership, and of Latino and Asian American men, increased between 9 and 24 percent. In some industries, like electronics, mentoring boosted Black men and White women in management by more than 10 percent as well.

Another effective intervention was transparency. When available positions and their criteria were communicated to everyone—not filled in secret—the proportion of White and Latina women in management increased 5 to 7 percent. When paths to promotion were clearly articulated, rather than unspoken and changeable, the percentage of Black men and Asian men and women increased 7 to 10 percent. A third successful strategy was accountability. In fact, when a person or group oversaw these programs, and managers knew they might be questioned about their decisions, diversity increased even more. Five years after the introduction of "diversity task forces," people of color as well as White women in management increased by 12 to 30 percent.[12]

At Taj, Monsellato used all three methods: he went beyond mentoring, specifically promoting women, smoothing the way for them, and reducing the risk inherent in their leap. He ensured clear criteria for promotions. And he created accountability by examining whether managers were being fair and implementing consequences for those who acted in discriminatory ways.

ALONG WITH THE RIGHT STRATEGIES, something else contributed to Monsellato's success, something often absent from conversations about diversity and inclusion. That was the fact that he undertook these changes with a very particular mindset—and a very particular motivation.

In a classic study, business professors Robin Ely and David Thomas sought to understand why some organizations with diversity initiatives functioned well and others did not. Thomas (now president of Morehouse College) had been asked to help a struggling consultancy; in exchange, they agreed to be studied. Ely studied a law firm, and together they examined a bank. And they noticed something striking: among these organizations, which were similar in terms of their level of racial diversity and where people of color held positions of authority, actual employees' experiences ranged widely. In some organizations, among some teams, people of different races felt respected and valued; in others, they felt devalued and mistrusted. The difference, Ely and Thomas discovered, seemed to lie in how each team thought about and responded to diversity—the purpose for seeking it in the first place, the value it brought, and the expectations about what it might achieve.

Some groups focused on the ethical dimension of diversity: promoting it was a way to reduce discrimination and correct for past injustices. At the consulting firm, for instance, the organization saw diversity as the right thing to do. A White manager Ely and Thomas interviewed said that employees of color help the company "live up to our ideals of equality and justice." Here, diversity wasn't expected to change how business was done in any meaningful way.

In other cases, Ely and Thomas found that diversity was viewed as a way to reach customers and open markets. The bank they studied relied on Black employees to serve the banking needs of its predominantly Black local clientele. As a Black manager reported, it was important to customers that bankers understood them on a deep level. If the firm were all White, a White manager explained, "our relationships with the community would be extremely strained." Here, corporate diversity was a means of gaining access to and legitimacy with clients.

Both these approaches produced fraught workplace cultures. At the consultancy, where the ethical imperative predominated, racial differences were downplayed in favor of "color blindness." In reality this meant everyone had to assimilate "to a White cultural standard." The only basis for discussing race was in deciding whether or not someone was being discriminatory. One person interviewed noted that White people were terrified they would be "accused of being a racist, which is almost the worst possible thing that could happen to a White person here, short of dismemberment." Employees from marginalized groups also experienced paternalistic and patronizing attitudes, like being denied honest feedback. On these teams, racial identity, Ely and Thomas wrote, "became a source of apprehension for White people and feelings of powerlessness for many people of color." The irony, wrote Thomas and Ely, was that the company's exclusive focus on fairness meant there was no other way of thinking about differences; disagreements had moral stakes. Conflict was avoided. Mistrust and resentment grew.

In parts of the bank where diversity was pursued for a narrow business interest, Black employees had mixed responses: some felt appreciated and recognized; others felt minimized and stifled, hiding aspects of themselves for the sake of their careers. One Black staffer questioned whether he was seen as someone with deep expertise or simply someone good at "making sure that the paperwork is together and making sure the files are in order." In reality, these parts of the bank were experienced as a two-tiered system: a predominantly White, higher-status department dealing with an affluent national clientele, and a department accorded less prestige in which mostly Black employees served the local retail branches.

There was, however, a third approach to diversity, one that avoided the pitfalls of the others. In this perspective, diversity was seen as necessary because different skills and viewpoints were considered crucial, not just to attract specific customers but for the institution itself to evolve. In a key department of the law firm, for instance, leaders had come to believe that people from different backgrounds and with different life experiences provided essential sources of insight that could influence the firm in far-reaching ways, from strategy to operations. When conflicts arose, they were addressed directly because resolving them was critical for the future of the organization.

Since difference was seen as a vital resource, the firm also worked to fully integrate everyone into the organization. Staff were encouraged to learn from one another's experiences; they were expected to be curious about new perspectives and open to revising their own beliefs and behaviors. The firm itself had to be open to changing the way things had been done in the past. As a result, people from marginalized groups felt heard, not erased. Their viewpoints actually changed the way the firm operated. Differences weren't avoided or downplayed because doing so would waste an opportunity for new information and insight. Employees felt respected and valued for their unique contributions because, in fact, they were.[13]

According to Martin Davidson, a management professor who is often asked to help companies whose diversity efforts have failed, an attitude of learning and growth is what distinguishes organizations that achieve lasting inclusion. Seeing differences as riches and being willing to learn from those differences allows people to see conflict as an opportunity for growth, not a land mine to avoid. "Whenever one steps into a context of diversity," Davidson says, "the working assumption has to be, 'I don't know what's going on here.'" The people most skillful at navigating differences, he's observed in his thirty-year career, are those who are constantly seeking to update their knowledge about others.[14] All situations in which there is diversity have the potential for conflict and confusion; an attitude of learning defuses those tensions and transforms them into useful material.

This approach is also helpful, Davidson says, for people from marginalized groups. "One of the things that's been the most liberating thing for me as a Black person," he told me, "is the working assumption that every White person I encounter isn't just another White person." The notion that there is something to learn about everyone he encounters frees him. There's an important caveat to this approach, however, says social psychologist Evelyn Carter. It only works in a safe environment; if people from marginalized groups are going to let down their defenses, it's crucial that everyone work to create an atmosphere free of harm.[15]

It's true that people *can* work together productively without being open to learning from one another's differences, Davidson says. They can still get things done. But, he points out, these situations are inherently fragile. They won't be sustainable for the long term. They won't benefit from the wealth of insight diversity brings. And they won't last in the face of conflict. They may succumb to what scholar Sara Ahmed describes as diversity "image management," making an institution appear inclusive while actually including very few.[16]

That's precisely what Uché Blackstock experienced. She was able to work within the medical school as long as there was no conflict. She was accepted as long as she didn't rock the boat. But as soon as she suggested new, potentially transformative ideas, as soon as she wanted to use her perspective to propose changes, the fragile situation crumbled. Leaders didn't see her vantage point as a trove of insight and richness. They did not see her as a person from whom they could learn. The organization wanted diverse faces but not diverse minds.

AT TAJ, THE CHANGES MONSELLATO introduced were aimed at making use of all the minds; as in some teams Ely and Thomas studied, Monsellato saw differences as essential to the business. The company had been faltering, he believed, because it did not reflect society. "We were an old gentleman's club from the nineteenth century," he said, a largely White and male assemblage out of step with a changing and interconnected world.

Leaders from other companies would, after hearing about Taj's changes, sometimes pull Monsellato aside and insist their companies were succeeding just fine. "That's exactly what the dinosaurs said," Monsellato responded, "before the comet hit Earth." Most leaders, Monsellato said, want to reduce the challenges they might experience, so they promote clones of themselves. (As we've seen, homophily also plays a role.) But to Monsellato's thinking, people who are *not* like you are the most important to hire. They are the ones who will challenge you the most, and better decisions come from ideas being scrutinized and then defended. "Either you're right, or I'm right because I've convinced you," he said. "The more you're challenged internally, the less you're challenged externally."

Monsellato took pains to point out that he had not diversified Taj out of generosity. "As a man, I said, 'Gender diversity is a key factor of our success.' I don't care about fashion, or to be nice. It was meant for business."[17]

Of course, in a hierarchal organization, there's a limit to the possibilities of sharing power and decision-making. But if an environment is fair, if an organization welcomes and values differences, if influence is not restricted to members of a majority group, all this will go a long way toward creating an inclusive experience. A partner named Sophie Blégent-Delapille succeeded Monsellato as head of the company. Asked about the office climate, she said the changes made work life better on a daily basis, not just for women but for everyone. A fair environment is simply a more pleasant place to work. "In my view, that was 80 percent of his success."[18]

WHAT'S KNOWN AS THE "BUSINESS case for diversity" has been widely touted as a reason to pursue it, but diversity is not a fairy dust that unleashes magic when liberally sprinkled. There is, in fact, no evidence that simply making a staff more diverse is an automatic boon. Some studies have found that diverse groups make better decisions and think in more complex ways (mixed-gender teams, for instance, produce the most highly cited technology patents), but other studies have found that diverse groups perform worse. One review of forty years of research about diversity in organizations found that while diverse groups are often more creative than homogenous groups, they can also, as Martin Davidson suggested, be prone to conflict and miscommunication.[19]

But there is a logic to these mixed results. In her analysis of decades of studies looking at diverse groups, business professor Martha Maznevski found that diverse groups do outperform homogenous teams, but under very specific conditions: if people are able to understand each other and build on each other's ideas. That is, if they are able to learn from one another. Ely and Thomas came to a similar conclusion after analyzing almost five hundred branches of a bank. They, too, found that while diversity alone did not result in better performance, in branches where all employees were encouraged to share their insights and able to feel psychologically safe enough to learn from one another, diversity did improve performance. A sense of safety allowed people to take risks, absorb new information, and grow. When a branch was able to use everyone's talents, the results were impressive: among branches that both White and non-White employees described as supportive, the highest-performing were in fact the most diverse. The link between diversity and performance turns out to be an atmosphere of genuine learning.[20]

And here leaders are critical. "What is required," Ely said, "is culture change, and leaders are the stewards of their organization's culture."[21] Leaders create policies and demonstrate appropriate and acceptable behavior. This in turn influences how people act and interact with one another, influencing their daily experiences and feelings, and ultimately shaping their trajectories.

Removing bias from everyday practices is essential but not sufficient for creating a truly inclusive environment. To foster a climate that includes all, everyday practices must be built on a foundation of learning from and valuing differences. And this environment need not be a workplace. These dynamics play a role in places where people live, worship, and learn. In fact, in a remarkable example of the impact of real inclusion, we can look to an environment most people do not usually associate with comfort and belonging at all: the mathematics classroom.

MATHEMATICIAN FEDERICO ARDILA-MANTILLA GREW UP in Colombia, an indifferent student but gifted in math. He was failing most of his classes at his high school in Bogotá when someone suggested he apply to MIT. He had not heard of the school. To his surprise, he got in, and he went on scholarship. Mathematically, he did well. One of his professors—an acid-tongued theoretician known to compare his audience to a herd of cows—routinely tucked "open" math problems into homework assignments, without telling the students. These had never been solved by *anyone*. Ardila solved one. He went on to receive his bachelor's and PhD in math from MIT.

But his academic experience was also one of isolation. Part of it had to do with his own introversion. (An outgoing mathematician, the joke goes, is someone who looks at your shoes when talking to you instead of their own.) Part of it was cultural. As a Latino he was very much in the minority in the department, and he did not feel comfortable in American mathematical spaces. No one had tried to explicitly exclude him, yet he felt alone. In math, collaborating with others opens up new kinds of learning and thinking. But in his nine years at MIT, he worked with others only twice.

At the time, he didn't clearly see the problem. But later, as a professor, he noticed a pattern. Ardila's women, Black, and Latino students who went on to PhD programs also told stories of isolation and exclusion, of trying to join a study group but finding that no one wanted to work with them. Indeed, research has shown, STEM students from ethnic and racial minorities often feel isolated on university campuses, and women STEM students find themselves routinely denigrated and underestimated, even when outperforming men.[22]

Mathematics as an academic field is notoriously homogenous—mostly White or Asian and male—and though mathematicians are not seen as the epitome of masculinity, the culture is macho and aggressive. "Abusive language," Ardila told me, "is completely normalized." While the elders of the field set this tone, the tradition is carried on by younger professors, too. Andrés Vindas-Meléndez, one of Ardila's former grad students, described an experience he had as an undergraduate math major at UC Berkeley when asking an advisor for a signature. "You're not going to be a mathematician," the advisor had told him. As Vindas-Meléndez was walking out the door, the advisor said, "Don't embarrass yourself. And don't embarrass the department."[23]

To Ardila, now a professor at San Francisco State University, the problem was significant: 60 percent of his students come from ethnic minority groups. Nearly half are first-generation college students. So Ardila decided to do what mathematicians do when faced with a huge conundrum: begin by focusing on a smaller problem. He set out to create, in his own classroom, a new kind of math environment.

First, he had to reimagine what math culture could be. To avoid perpetuating macho aggressiveness and instead to make the classroom a place where students would feel comfortable and supported, he devised a class agreement. Students were asked to commit to taking "an active, patient, and generous role" in their learning and that of their classmates. Achieving the right tone also meant rethinking how he spoke about math. Mathematicians frequently use phrases like *It's obvious* or *It's easy to see*, which can be profoundly discouraging for a student who does not immediately find a concept simple. In math, grappling with extremely difficult problems is part of the learning process. "A challenging experience," says Ardila, "can easily become an alienating one." It's especially important to make sure students are not discouraged during early challenges—what's hard to see now may become easier in time. He struck this typically demoralizing math language from his teaching.

Other changes followed. Ardila observed that only a few students would speak in class, so after he posed a question, he asked to see three hands before calling on anyone. The first hand usually shot up quickly, and sometimes the second. Eventually, a third hand would rise, tentatively. Then he would ask students to share their ideas in reverse order. They eventually caught on, he says, but in the process, they understood that all their voices were welcome and encouraged. Classes that began the semester with only a sliver of vocal participants would end with everyone talking.

"Many students feel pressure to leave their true selves at the door," Ardila said, especially if they are from groups not usually visible in the field. So he found ways to invite them to bring more of themselves to math. He would play music to make the classroom more comfortable. Then he invited students to bring in music of their choice. In one calculus session, he assigned a classic challenge—identifying the optimal shape of a can to maximize its volume and minimize the materials used to make it—and asked people to bring a can of food from home to explore the problem. Some reflected students' cultural backgrounds: cans of refried beans, or coconut milk. Others brought in trendy coconut waters and juice.

From a materials standpoint, the wide, short cans of refried beans were the most efficient, students discovered, while coconut water cans,

which tended to be tall and thin, looked larger but were the least efficient. The exercise prompted a spirited discussion of cultures and foods
and competing values in the marketplace. Ardila realized that he didn't
need to demand that students discuss their identities by, say, writing a
word problem about refried beans. He could simply make a conversation possible, and then listen with curiosity and openness. Slowly, as
students shared, a mathematical community began to form.

This community expanded when Ardila developed a collaboration
between San Francisco State and the elite Universidad de los Andes in
Colombia. He conducted joint classes in English via video. Each group
was impressed with the other—the Los Andes students noting the dedication and work ethic of the SFSU students, while they in turn were
inspired by the advanced math background of the Los Andes class. The
final projects were done in pairs, with the collaborations taking place,
as Ardila says, "in the whole Spanish-English spectrum." Many of the
U.S. students were Latino and had only spoken Spanish with their families; now they were learning to communicate about advanced math in
Spanish, too. The international partnerships, Ardila noted, proved the
most fruitful, another instance of differences being generative in an
atmosphere of genuine learning.

To further solidify this nascent community, Ardila created a math
conference in Colombia, which has grown to include people from
twenty countries, mostly in Latin America. Experts and students work
on problems together, share open problems, cheer each other on, and
even dance salsa together. Here, too, Ardila devised an agreement that
lays out the conference's goals: "to offer a rewarding, challenging, supportive, and fun experience to every participant." All participants are
required to read it aloud and discuss it in small groups, reflecting on
how they themselves will contribute to those goals.[24]

"Math is human," said Andrés Vindas-Meléndez, who now considers
Ardila a mentor, and the conference does not hide this. "Clapping after
solutions, salsa dancing. These are all human experiences." This approach
has had far-reaching consequences. There was no explicit mention of the
LGBTQ community, for instance, but one participant shared that as a
gay man he had never felt so comfortable at a math conference.[25] In that

environment, Vindas-Meléndez said, students feel able to try daunting problems, sometimes beyond their internalized limits.[25]

The same effect was happening in Ardila's classrooms. "We often have ideas about who are the good students and who are the bad students," Ardila told me. In typical school settings, students who can do well on tests or solve problems quickly are labeled the best. Ardila offered other ways to succeed, assigning open-ended problems, which is closer to the actual practice of science. Students who may not have performed well in the past revealed new strengths. "I see students who got low scores on tests," he told me, but when they're deeply and personally involved in the mathematics, "they're able to really show a very different kind of work."

For a final project in Euclidean and non-Euclidean geometry, for instance, one student of Mexican and Indigenous descent wanted to learn how his ancestors did math. The student built a replica of the Chichén Itzá temple of Kukulcán, the Mayan snake god. The temple was designed so that at the equinox, the light and shadow cast by the setting sun appears like a serpent slithering from the top of the stairs to the bright snake head at the bottom. The student uncovered the math needed to re-create the structure, complete with the undulating light of the serpent. The project was, said Ardila, of a noticeably higher caliber than the student had demonstrated before. "When students see themselves reflected in the curriculum it qualitatively changes the kind of work they can do. It's really moving."*[26]

Math, after all, is personal, emotional. "Anybody who does mathematics knows this. I just don't think we have the emotional awareness or vocabulary to talk about this as a community."[27]

---

* The idea that seeing oneself reflected in the curriculum matters has been demonstrated elsewhere. In Tucson, Arizona, a program called MAS—Mexican American Studies—was developed to help promote the academic achievement of Mexican American students in the district. The curriculum included Chicano and Latino history, literature, and social issues relevant to the community. An analysis of 8,400 Tucson students found that MAS students—who had generally entered the program with lower GPAs and test scores than their peers—outperformed their classmates, graduating and passing standardized tests at higher rates. The program was eliminated in 2010.

And the feelings experienced during learning, like those experienced at a workplace, have real consequences for people's trajectories. As Nilanjana Dasgupta found, feelings of belonging closely predict whether students will persist in a field where they are not well represented. Much research suggests that feeling accepted and having a sense of belonging—the hallmarks of inclusion—help people persist through difficulty and boosts their achievement. It also helps them stay motivated to remain in their fields.[28]

In the case of Ardila's students, inclusion has had an astonishing impact. Of the twenty-one students in the first joint math class with the Universidad de los Andes, twenty went on to get graduate degrees in math and related fields. Half of these students were from San Francisco State. Fifteen went on to PhDs in math and related fields, and fourteen are already professors. This would be an astounding number even at an elite university, but at a non-PhD-granting state school like SFSU, it's unprecedented. Many of the students originally had no intention of pursuing math PhDs. Of the two hundred students who have participated since the program's founding, fifty have gone on to get math PhDs. Almost all of the U.S. participants are women and/or from historically underrepresented ethnic minority backgrounds.

"It feels immodest to say," Ardila told me, "but students say being in spaces like this is what made them decide to choose a career in math." Vindas-Meléndez, who had been told by another professor not to embarrass the department, said that before meeting Ardila, his automatic response to a challenging problem was to assume he didn't know how to attack it. Ardila encouraged him to withhold that judgment. "Federico has known me for seven years now. He'll say, 'Okay, this is an abstract problem, let's start with some concrete examples.'" Vindas-Meléndez, too, is completing his PhD in mathematics and heading back to Berkeley for a postdoctoral fellowship this fall. Certainly Ardila is, for Vindas-Meléndez and others, a role model—an "existence proof," as Peko Hosoi put it. But Ardila also created an environment in which thriving was possible. Said Vindas-Meléndez, "Federico was the first person at the PhD level who showed me I could succeed."

* * *

To CREATE CULTURES THAT DON'T systematically exclude people, it's important to be comfortable acknowledging differences. A recent study of nearly seven hundred college students found, in fact, that acknowledging differences affects perceptions of bias and may even help student achievement.[29] The students, assigned to an online chemistry, physics, or math class, were presented with one of two teaching philosophies, or a control. In one, students heard an audio welcome message in which the instructor (a middle-aged man) explained that it was important for them to keep in mind the ways they were similar to one another, and that this would promote collaboration and learning. They also received a syllabus that further explained that the classroom was to be a place where students can flourish, and that keeping similarities in mind would increase empathy and better interactions.

Another set of students encountered a different welcome message asking them to keep in mind their differences; their syllabus asserted that considering differences would foster better interactions. When presented with the "acknowledging differences" philosophy, students of color, including Black, Latino, East Asian, South Asian, Native American, Middle Eastern, and Native Hawaiian students, saw the instructor as less biased than when he advocated for focusing on similarities. White students, by contrast, saw the instructor as *more* biased when he acknowledged differences, and least biased when he presented a "color-blind" philosophy.

Indeed, many of my interviewees described moments when others acknowledged their unique backgrounds as instances of true inclusion. Dominique DeGuzman, a computer programmer in San Francisco who is part of the LGBTQ community, remembered that LGBTQ employees in her company felt especially vulnerable after the Orlando nightclub shooting that targeted the gay community. It was especially important, DeGuzman said, for leaders to acknowledge the event and the pain it engendered.

DeGuzman also noted the need for sensitivity to "invisible minorities," such as people from working-class backgrounds, who might experience the request to work overtime differently than those from wealthier backgrounds.[30] Anyone who feels a heightened sense of job insecurity may be less likely to resist longer hours or weekend work.

Ignoring an individual's identity can be felt as deeply alienating. Leslie Miley was one of a tiny number of Black leaders at a large tech company when he learned the company was hosting a group of prominent Black leaders. He hadn't been invited or even told about the event; he felt shocked and hurt. It would have been a crucial moment for someone in the company to recognize his identity, but they missed it entirely.[31]

Fostering an inclusive environment also requires leaders to set the right tone. Mekka Okereke, an engineering leader who is a Nigerian American, was in a meeting where people were discussing an outgoing email. Someone chimed in with an attempt at humor that the email "should sound like our company sent it, not like Nigerians sent it." The room went silent as others looked at him, unsure of how to react. Okereke took a deep breath and said, "Hi. Mekka here. I run all our email and notifications systems. Too bad, Nigerians are sending it anyway." He defused the situation and made it clear that hurtful comments would not be tolerated. But, he added later, while he doesn't mind using humor to advocate for himself, this was a critical moment in which a teammate could have stepped in to address the remark.[32]

Yolanda Davis, a software manager from Atlanta, described a former manager who was especially effective at fostering inclusion: he created opportunities for real connection, sponsoring game nights and meals for people to share foods from their culture. Trust grew, and as it did, difficult conversations felt safer. Davis recalled her manager asking her about "the nod"—the way one Black person might subtly acknowledge another. "In a world that sees you as invisible, it's an awareness of your presence," she explained to him. "It's a sign of respect for your presence."

Remembering the conversation, Davis recalled, "It was a moment. I explained it and he got it." It was an open, honest conversation, it nurtured understanding, and without a sense of mutual trust and belonging, it would not have happened. This openness is something Davis now works to foster among her own team, learning from her Indian colleague about the obstacles he faced growing up on a farm in South India, for instance, about colorism and the caste system.[33]

Of course, the extent to which people want to bring their identities to work or school can vary from person to person. And even in a culture that welcomes this, it can be difficult to acknowledge differences without making a person feel like an exotic specimen. A recent college graduate told me about a writing class he had taken in which the instructor repeatedly referred to the fact that he, the student, was Korean American, and frequently suggested he write about being Korean American. This excessive attention felt alienating to the student: he did not especially want to write about his identity. He had not even mentioned it in class.

But Ardila's lesson with the food cans is an example of how to go about welcoming difference with sensitivity and care. He didn't force it or demand that people share when they weren't comfortable. Instead, he created a space in which people were able to bring their full identities, signaling that they were welcome. Through language, through class policies, through an environment of respect, curiosity, and mutual encouragement, he communicated that there was room for all to succeed. He did not push people to divulge their experiences but was open to it and listened with respect and attention when they did.

WHAT WOULD HAVE HAPPENED IF Blackstock's deans had believed her perspective was an asset of enormous value and had met her concerns with curiosity and openness? Perhaps they would have said, *Tell us about your singular experience. Tell us what you see. We'd like to understand; we'd like to make it right.* Their neglect of her concerns made it impossible for Blackstock to stay, as did their failure to value her insight. After leaving academia, she began her own organization focused on eliminating racial disparities in health and health care, and has become a nationally sought-after expert. The mind that academic medicine didn't want is now creating a new kind of medical culture, just as Ardila is creating a new kind of math culture, uplifting others as they go.

Unbreaking Culture

Transforming the minds, hearts, and habits of individuals is one way to change bias. Another, as we've seen, is to change processes, structures, and the culture of organizations. The two, of course, are intertwined: individuals create the processes, structures, and organizational culture, and these in turn shape individuals' thoughts and actions. But we are also the product of our larger culture—the broader environment in which we live. Change can additionally begin from this third starting point. In fact, prejudice scholars such as psychologist Glenn Adams have called for a shift in focus away from individual hearts and minds "to changing the sociocultural worlds in which those hearts and minds are immersed."[1] Remaking the actual world acts on bias further "upstream," chipping away at the source of biased assumptions.

But representations of the world are also influential, and these may be easier to change. As we know, the media often portray marginalized groups in stereotyped ways, from the Black men in the trailer for *Straight Outta Compton*, to Mexican immigrants on the news, to the Middle Easterners on *Homeland*, whose representations so incensed the Egyptian-born artists who were hired to add realistic graffiti to the set that they spray-painted, in Arabic, "*Homeland* is a joke, and it didn't make us laugh."[2]

Negative images of stigmatized groups are obviously harmful, but positive representations are just the beginning. Over the past several years, Abdelatif Er-rafiy and Markus Brauer, a French psychologist of North African origin and an American psychologist of German origin, respectively, have looked closely at how media might work to shift discriminatory behavior, with a particular focus on anti-Arab discrimination in France. While replacing negative images with positive ones seems intuitively to be the right step, Er-rafiy and Brauer's studies suggest that there's a better route.

In one set of experiments, the researchers created large, glossy posters showing photographs of people of Arab origin. In French, they labeled each person's name, age, and a key, distinct personality trait. Next to the image of one woman, the poster read, YAMINA, 59 YEARS OLD, OPTIMISTIC. Under another, AICHA, 30 YEARS OLD, STINGY. At the bottom, the poster announced in a large font, NOTRE POINT COMMUN: LA DIVERSITÉ. What makes us the same is that we are all different.

Er-rafiy and Brauer then displayed the posters to hundreds of people in France—in health-care settings, high schools, and universities. Each location was matched with an equivalent setting in which there was a different poster or none at all. One physical therapist's office, for instance, featured the poster in the waiting room; a matched office had no poster. In one set of high schools, the posters hung in classrooms and on doors; in the other set there were none.

The researchers found that people who were exposed to the photo poster—while waiting for appointments or walking to and from class—subsequently acted in a less biased way toward people of Arab origin. As part of the experiment, physical therapy patients who had seen the posters returned for a follow-up weeks after the poster had been removed. A person of Arab origin sat in the waiting room (purportedly a fellow client, but in fact a colleague of the researchers, planted for the purpose of the experiment). These patients sat closer to the person of Arab origin than did other patients who had not seen the poster.

At the high schools, when students were asked to support a petition protesting discrimination against Arabs, the ones who had seen the poster were more likely to sign. Undergraduates exposed to the poster became more willing to volunteer for a group that promoted the rights of Arabs. In another experiment, a woman of Arab descent (another friend of the researchers) spilled her bag in front of the subjects. Only 59 percent of the people in the control group offered to help her. In the group that had seen the poster, 91 percent helped the woman pick up her spilled belongings.[3]

Why would a poster highlighting a spectrum of personalities have this effect? It seemed to have something to do with the idea of difference—not the difference between Arabs and non-Arabs but among the Arabs themselves. Rather than trying to foster positive attitudes by, say, introducing French people to Arab traditions or highlighting positive role models, the poster instead emphasized how varied people are within the group identified as Arab. Stereotypes rest on the idea that all members of a group have traits in common, and studies show that people do indeed show more bias toward individuals in a group if they see them as homogenous. By contrast, the more we perceive that a group is composed of people who are wildly different from one another, the less we tend to stereotype. One reason the posters may have worked is that by insisting that Arabs are not alike, they undermined others' ability to stereotype them. "If members are seen as dissimilar," the authors wrote, "it is nearly impossible to feel toward all of them alike."[4]

The fact that the poster featured both positive and negative characteristics seems particularly important. Another version of the poster—one that included only positive attributes, such as "optimistic," "warm," "honest"—was not as effective.

This approach—emphasizing the differences within a group identity—runs counter to common multicultural awareness campaigns, which typically emphasize what sets a group apart from other groups, featuring, for instance, a group's culture or customs. In any pluralistic society, this is a challenge with serious ramifications: the tension between acknowledging and celebrating group differences, on the one hand, and hardening group boundaries, on the other.

* * *

In 2002, the United Nations Children's Fund approached Sesame Workshop, the company that produces the children's television program *Sesame Street*, to see if it would be willing to create a version of the show for Kosovo. The region, roughly the size of Connecticut, was at the time a U.N. protectorate but considered by some to be a Serbian province; it had recently emerged from a devastating ethnic conflict. That war saw over ten thousand people killed or missing, and nearly one million ethnic Albanians, and as many as one hundred and fifty thousand Serbs, fled, sometimes on tractors and trailers, into neighboring countries. The UNICEF officials thought that *Sesame Street* might be a way to begin to address ethnic hatred. While a reconciliation campaign aimed at Kosovar adults seemed futile, perhaps one focused on preschool children could work. *Sesame Street* had played a role in racial integration in the United States. Maybe it could make headway here, too.[5]

At that point, Serbs and Albanians were geographically segregated: the ethnic groups lived in different areas, and children attended separate schools. Many Kosovar children had never even met their ethnic counterparts. When one Serbian girl was asked, "Why do you go to school separately from Albanian children?" she responded, "Because we are Serbian and they are Albanian." "What's the difference?" The girl paused. "I don't know," she said. "Would you like to meet an Albanian child?" the interviewer asked. "No," she said.[6] She did not know what the differences were, only that they mattered immensely.

Sesame Street agreed and recruited a mix of Albanian and Serbian producers to cocreate the series. Tensions ran so high among the producers that it wasn't clear whether the two groups would even agree to meet in the same room to develop the project. Still, over the following months, the team did collaborate. In the small room where they worked on the program, bars covered the windows to protect them from the unrest in the street. At one point, a violent riot erupted, interrupting the project for months.[7]

One of the producers' first decisions was to have two names for the same program—*Rruga Sesam* in Albanian and *Ulica Sezam* in Serbian.

Each would be dubbed into its corresponding language. A basic component of *Sesame Street*, teaching letters and words, was omitted; Albanian parents would not accept Serbian and the Cyrillic alphabet, and Serbian parents would not accept Albanian and Latin letters. The producers, unable to show the words visually, instead created a "picture dictionary." To teach the word "sunglasses," children wore whimsical sunglasses and spoke the word in different languages. Even putting the program's title on-screen was so politically volatile that the producers ultimately decided just to have children shouting the name out loud.[8]

Because Kosovar children had limited access to preschool education, the program's curriculum had immediate practical goals, like teaching counting and safety. But a primary theme was learning respect and appreciation for the children of different ethnic groups. They had never met in real life, but perhaps they could meet through *Sesame Street*.

The team shot dozens of live-action sequences, in which Serbian and Albanian children were filmed separately, cooking with grandparents, observing holidays, being with family, playing games. These were interspersed with traditional *Sesame Street* features, such as Muppet skits and animations. The point of the live footage was to let the children learn about each other and perhaps even identify with one another. *That child bakes with her grandmother, just like I do. I play with my siblings, just like that child does.*

The producers also wanted the children to learn about the rival group while not focusing all their attention on differences between the two, which could promote stereotyping. Too little understanding of another group leads to fear and hatred; too much emphasis on what makes one "other" can lead to prejudice.

Charlotte Cole, head of international education at Sesame Workshop at the time, said that balancing these two goals was difficult. The team needed to distinguish the groups from one another while also showing the diversity within each ethnic community. They also needed to show the similarities between the two rival groups. They did this by including as many different children in the live-action footage as possible. In one sequence, a group of Serbian, Albanian, and Roma children are filmed separately, while a child sings, "We are kids, and we may differ from one another" and "In this world, everyone comes with something unique." The song continues, "Some are very energetic, some are very quiet." But "we all get angry . . . we all smile the same." The children were portrayed as individuals, and as different in experience but not in essence.[9]

An independent research group was recruited to study the effect of the program. It examined whether the children had learned to count, recognize numbers, and cross the street safely. But it also looked at the crucial question: Had the show shifted the Kosovar children's thoughts about one another? Had it worked as some kind of interpersonal contact, allowing the children to gain more nuanced ideas about the other group, just as the CSP officers in Watts had acquired more complex notions about the communities they served?

More than five hundred five- and six-year-olds were recruited for the study, most of whom had not previously seen the program. (Prior to the study, only 2 percent of Serbian children had watched the show while 23 percent of Albanian children had.) Before they watched, the young viewers were asked how they felt about different children. *Is it okay to play with a child who doesn't speak your language? What about a Roma child—would you approach a child from this group? If an Albanian child asked you for help, would you help him? Imagine a Serbian child needs help. Is that okay?* Half the children then saw at least twelve episodes of the show in their native language. The other half did not.

What the researchers found was that among both Albanian and Serbian children, those who watched *Sesame Street* were more open to others. They were more likely to want to interact with a Roma child and more open to playing with someone who didn't speak their language. Among the Serbian children, the percentage expressing mutual respect rose from 37 to 68 percent. Among Albanian children, the percentage rose from 23 to 33 percent. After the intervention, the children who had watched Sesame Street were also more willing to help a child who belonged to a rival ethnic group. When asked why that was okay, the children gave reasons that suggested they were imagining the other child's inner experience. "He is completely lost," they said. Why else? The children responded:

He does not have anything.

He is alone.

He has lost his way.

Somebody from the street might take him away, somewhere very far.

Regardless of everything one should help a child.

Because some dogs might eat him during the night.

In small ways, they had begun to identify with the enemy's child.[10] *We all get angry, we all smile the same.**

REPRESENTATIONS, THROUGH TV SHOWS AND films, can change how one sees "the other." But representations can do something more: they can also change how a group sees itself. To understand how it works, we can turn to a remarkable experiment that took place in the aftermath of discrimination at its most horrific and devastating.

During the period of the Rwandan genocide in 1994, an estimated eight hundred thousand people—about 75 percent of the Tutsi minority population—were killed by individuals aligned with the Hutu-led government. The genocide was carried out by both state-sponsored groups and ordinary civilians.

The roots of the conflict were complex. Hutu-Tutsi divisions had been exacerbated by European colonists, and while there is debate about what Hutu and Tutsi categories meant in precolonial times, in general they corresponded more to occupation, roles, and status than ethnicity, and people could move from one group to another. Colonial powers, elevating Tutsi as a proxy elite, cultivated rigid ethnic boundaries and propagated stereotypes (the Tutsi were "proud" and "polite," the Hutu were "expansive, noisy, and cheerful"). In the 1930s, the Belgian colonial power fixed the divide

---

* In keeping with the concept of "contact theory," the process of producing the program also helped the adult producers. They had come together to cooperate toward a shared goal, a better future for their children. In doing so, the producers grew more tolerant of one another.[11]

by imposing ethnic identity cards, relying on oral histories to determine ethnicity. Legend has it that they even assigned ethnicities to people based on the number of cattle they owned, giving Tutsi status to those with more than ten cows, though this has been disproven.[12]

A 1959 Hutu uprising brought a Hutu majority to power at the country's independence in 1962, and many Tutsi fled in exile. While violence erupted between the two groups over the following decades, Hutu and Tutsi also intermarried and lived in integrated communities. A peace agreement forged in 1993, settling a three-year war between the Hutu government and a Tutsi refugee army, exploded the next year when the Hutu president was assassinated.

At the time, many Rwandans were listening to a new radio station called RTLM. It had been launched in 1993, officially as a private station though it drew power from generators inside the Hutu president's mansion. It mixed music and comedy with call-in programs and anti-Tutsi commentary, referring to Tutsi as "*inyenzi*"—cockroaches—and describing them repeatedly as a group with traits like a "thirst for power." "You can drive out nature with a pitchfork but she keeps coming back," one announcer said. "This thirst is supernatural."[13]

As the violence began, the station exhorted Hutu to exterminate the "cockroaches," sometimes even explaining where to find targets. As *génocidaires* roamed, one journalist noted, they might carry a machete in one hand and a radio in the other. Researchers who studied RTLM found that the greater access people had to the station, the more they participated in the genocide. Violence also spilled over to villages near those with good radio reception. Radio played such a powerful role in the killings that the International Criminal Tribunal for Rwanda argued that its propaganda was what had made the genocide possible.[14] The station's founder was convicted for crimes of genocide.

Ten years later, a group of Rwandans, along with a Dutch nonprofit and a Jewish American psychologist who was a Holocaust survivor, began to ask whether radio could be used not just as a source of hatred but also as a means of healing. Together, they developed a radio soap opera called *Musekeweya*, or *New Dawn*—a program that explicitly

sought to change Rwandan people's beliefs about violence and reconciliation. Post-genocide laws in Rwanda made it illegal to discuss ethnicity,
so the New Dawn writers told stories about two fictionalized hilltop villages that had been peaceful until they fell under leaders who promoted
one group and stoked bigotry and violence. The violence that followed,
in the soap opera, mirrored what happened in the country. Afterward,
some of the New Dawn characters worked to foster communication
and peace and vocally opposed the leaders' messages. Two young lovers
from the rival villages pursued a romance and built a youth-led peace
coalition. The villages ultimately reconciled, working the land side by
side.[15]

The stories in New Dawn were also interlaced with messages about
the roots of prejudice, showing, for instance, that ordinary people can
become violent. It also communicated the importance of bystander
intervention and that loving relationships across group differences
can help reduce prejudice. The goal of all these stories was to begin to
change Rwandans' beliefs and attitudes about one another.

But that's not what happened. Psychologist Elizabeth Levy Paluck
studied the impact of the soap opera over the course of a year by
gathering Rwandans for monthly listening sessions. (A control group
listened instead to a soap opera about health.) Because Rwandans typically listen to the radio in large groups, the participants did the same,
in meetings of forty people. After the yearlong study, they were asked a
battery of questions about their beliefs. Was intermarriage a good idea?
Was bystander intervention important?

In some ways, the intervention was a bust. After the year was over,
Paluck found that listening to New Dawn had not affected people's beliefs
about the benefits of intermarriage or the responsibility of bystand-
ers. They had heard that intermarriage could promote peace, but they
were not more likely to believe it now. They had heard that bystander
intervention was crucial; their middling agreement had not shifted. But
something else had changed: people's thoughts about what they *ought*
to believe. New Dawn listeners would say things like, "I don't want to
encourage my son or daughter to marry someone from the other ethnic group, but I should." They didn't believe it was important to vocally

oppose a harmful message, but they thought they *should* speak up. The soap opera had signaled that tolerance was now popular. And it conveyed which opinions and behaviors had become most desirable.[15]

New Dawn, it seems, didn't change people's convictions but rather functioned as a signal of norms. As Paluck told me, "Media doesn't tell people what to think, but it tells people what other people think. It's a signal of public consensus."[17]

And public consensus, it turns out, can alter how people act: if people learn that a particular behavior is normal and popular, they engage in it more. In one study, for instance, placards were placed in hotel bathrooms, encouraging people to reuse their towels for environmental reasons. One group received cards that said reusing towels would help save the environment. The other received cards that said 75 percent of hotel guests were reusing their towels to help the environment. Those who learned that a majority of the guests were reusing towels were significantly more likely to do so than those who were only asked to be eco-friendly.[18]

In another study, households were encouraged to conserve energy by using fans instead of air-conditioning. Each household was given one of a variety of encouraging messages: using fans would save money, it was socially responsible, it reduced greenhouse gases, or it was being practiced by a majority of their neighbors. The group that was told conserving energy was popular among their neighbors decreased its energy use the most.[19]

Norms are so powerful, they affect what people do even if their behavior is being discouraged: trying to stop bad behavior by telling people that others are doing the same thing can actually increase it. In one study, researchers at the Petrified Forest National Park tested dissuading visitors from stealing wood by posting signs saying that too much wood was being stolen and this harmed the forest. But these signs made theft increase. Writes psychologist Robert Cialdini, the study's author, "Within the statement 'Many people are doing this undesirable thing' lurks the powerful and undercutting normative message 'Many people *are* doing this.'"[20]

Perhaps it is unsurprising, then, that although the *New Dawn* soap opera messages didn't change people's personal beliefs, it did change their behavior. As part of Paluck's study, each listening group was later given a portable stereo system and recordings of the program. When deliberating about how to share the stereo, those who had listened to *New Dawn* over the previous year were more cooperative with one another than the group that had heard the health program. The *New Dawn* group negotiated more and offered more suggestions for sharing. They also engaged in more open communication. *New Dawn* hadn't changed people's beliefs about cooperation, communication, and tolerance, but it did change how much they actually cooperated and communicated and tolerated one another.

As Paluck pointed out, this contrast between beliefs and behavior mirrored the way in which the ethnic violence erupted. Rwandans told her "the violence came like rain," as if it had materialized without warning. It was not motivated by long-standing personal prejudice but was fostered by cultural messages that made killing accepted and desirable. The radio had communicated that everyone is prejudiced, that killing is normal. Perceived consensus propelled the *génocidaires* to action. "Prejudice is social," Paluck told me. "Once you believe you are with the consensus a lot of things happen."[21]

MEDIA INTERVENTIONS TO REDUCE PREJUDICE, such as *New Dawn* and *Sesame Street* in Kosovo, are usually introduced in the context of extreme, often violent conflict, not unconscious discrimination. But changing representations and shifting norms can affect more common, everyday kinds of bias, too.

In one study, White college students on a college campus in western New York were asked if they would answer a short questionnaire about their attitudes about different social groups. Half were invited to learn the attitudes of others in the vicinity, and told that most people in the region held Black individuals in high regard. The other half were given no normative information. Then, after completing the questionnaire, as each student was walking away, either a Black or White student passed by and dropped a stack of papers. Among those who had been given no information about others' beliefs, only 36 percent stopped to help the Black student. Of those who had been told that most in their region held Black people in high esteem, 86 percent stopped to help.[22]

Psychologist Stacey Sinclair and others suggest that the motivation to be in tune with others can affect even very subtle expressions of prejudice. In their study, participants took an implicit association test that measured anti-Black racial bias. Some were tested by a person wearing a blank T-shirt, others by a person in a T-shirt that said ERACISM. The point was to suggest that the slogan wearer was anti-racist. When students took the test from someone with the slogan, their scores showed less bias. Even ostensibly automatic expressions of prejudice were influenced by their perception of others' beliefs.[23]

By manipulating perceived norms, even in wildly different contexts, it's possible to reshape a person's ideas about how their own group thinks and acts. By changing representations of other groups, it's possible to alter their ideas about how others think and act. But putting these tools to work, in the context of media representations, is complicated by the fact that who "us" and "not us" is depends on who is doing the looking. As Charlotte Cole of *Sesame Street* said, "When I learn about myself, the person from the other group is learning about the other. When I learn about the other, someone from the other group is learning about themselves."[24]

Additionally, notions about who belongs in and out of our own group are also a product of culture. It is culture that teaches us who fits inside and outside these categories. If much of bias results from our propensity to place humans in categories, what if the categories could be changed? What if it were possible to make them less rigid?

The question took me to Stockholm, Sweden, where a preschool director has for more than two decades been part of an unusual experiment, an effort to create a preschool where gender no longer matters.

LOTTA RAJALIN'S EXPERIMENT BEGAN IN 1998, when the Swedish government issued a mandate that preschools must provide equal opportunities to boys and girls. At the time, Rajalin was the director of

Nicolaigarden, a preschool in Stockholm's old town. She and her colleagues suspected that children might be acting unfairly toward one another. It could be that the boys were rejecting girls who wanted to play with them, or the girls were shunning boys who sat with them at the art table. To locate sources of unfairness, the teachers decided to start filming the children's day-to-day interactions, hoping to catch some of this bias in action. As the teachers went about their work, painting with the children, comforting them, resolving disputes, and encouraging naps, they filmed one another. When they gathered to watch the videos, they were eager to spot opportunities to correct the children's behavior. But the behavior problems, to their surprise, did not come from the children at all.

These highly trained, well-meaning teachers—in a country ranked number four in the world for gender equality, a country so progressive it has its own word for gender equality, *jämställdhet*—were themselves treating the boys and girls differently. When children cried, the teachers saw that they comforted the girls more than boys. When they picked the children up, they held the girls close to them, face-to-face, while they kept the boys more distant and facing outward. In video after video, the teachers saw that they tolerated more noise from the boys, more running and rowdiness. They told the girls to be quiet, nice, helpful. While they were impatient with girls' impulsiveness, expecting them to have greater self-control, they did not demonstrate the same expectations for the boys. When the children sat in a circle, the teachers were more welcoming toward contributions from the boys.

"We were shocked," Rajalin told me. "It was a very bad time for the school." To carry out the government's mandate to provide an equal learning environment, she realized, the teachers would have to change themselves.[25]

Over the next several years, the teachers began, step by step, to alter the way they acted. They let boys cry if they wanted to cry and comforted them with the same tenderness they displayed toward the girls. They picked up and held the children in the same way. They stopped shushing loud and rowdy girls, letting them be as rambunctious as the boys. The teachers also started flipping the gender of characters in the stories they read, so a mischievous child would be a girl, and a charming, obedient child would be a boy. They stopped starting the day with "Good morning, boys and girls," instead saying "Good morning, friends." Any time they referred to a particular child, they didn't use "he" or "she" but simply the child's name.

The teachers eventually began using a new word: "*hen.*" "*Hen*" was introduced in the 1960s and added to the Swedish dictionary in 2015 to serve as a gender-neutral pronoun, and the teachers began deploying it for someone whose gender wasn't known, such as a cook coming for a job interview. The goal was to encourage children to stop thinking of some professions as male and others as female. The teachers began using "*hen*" during story time, too: a playful bear or kitten character might simply be "*hen.*"

As the school's approach became more widely known, some in the community reacted with hostility. Graffiti appeared on the building. Critics condemned a dystopian school that was brainwashing the children and eliminating the concepts of "male" and "female." Rajalin has a scrapbook of all the hate mail she received.

When a team of psychologists came to study the school, they asked: Had Nicolaigarden eliminated the concept of gender? Children around the world are sensitive to gender at astonishingly young ages—infants can distinguish women's faces from men's, and by age three, children strongly prefer peers of their own gender.[26] Culture, as developmental psychologist Rebecca Bigler's work demonstrates, plays a large and important role in children's attunement to social categories. With its lack of emphasis on gender, had Nicolaigarden removed the "male" and "female" categories from the children's way of seeing the world?

To find an answer, the researchers showed the students pictures of children and asked them to identify their gender. They compared the answers to those of students attending standard Stockholm preschools. The Nicolaigarden children, it turned out, still perceived boys and girls. They could distinguish among genders as easily as children at other Swedish preschools.

What was different about the children at Nicolaigarden, however, was what they *thought* about boys and girls. Compared to others their age, they were less likely to make assumptions about which toys girls and boys preferred. When presented with an unfamiliar child as a potential playmate, they were more likely to choose a child of another gender. In other words, they were less inclined to rely on gender as a shorthand. They saw boys and girls, but they stereotyped them less.[27]

This hadn't really been part of the plan. The goal had been for teachers to treat children equally and ensure they all had the same opportunities to grow and develop. In the process, Nicolaigarden had accomplished a second, more subtle goal. They changed the way the children saw one another. A future study might explore how these children, compared to others, perceive people who do not fit into a gender binary. As it is, the studies showed that children were able to perceive social categories but do less stereotyping of those categories. There was a way to see difference that did not necessarily lead to discrimination. The teachers hadn't eliminated the concept of gender. But they had begun to change its meaning.

RAJALIN HAS NOW OPENED A second school founded on the same principles as Nicolaigarden; it is called Egalia, which means "equal." The school's goal is antidiscrimination of all sorts—gender, class, sexuality, race, and ethnicity. When I arrived, outdoor playtime was just wrapping up, and three-, four-, and five-year-olds came rushing through the door. Rajalin greeted them with almost impish joy. As we walked through the school, children zigzagged around and bumped into her legs, like June bugs against a windowpane.

Rajalin's face is lined and expressive. She owes her own existence, she explained, to the fact that a country valued children—her parents were sent from Finland to live with Swedish families during the violence of World War II. As a child growing up in Sweden, Rajalin said, she was impulsive and wild. She was constantly told that who she was wasn't acceptable, that she needed to calm down, sit still, be nice. "If you tell a little girl, 'Oh, you're so nice and helpful,' she will want to be helpful. Later she might want to be wild and will feel bad about it, because adults want her to be nice." The Egalia and Nicolaigarden teachers, by contrast, are conscious of the signals they are giving to young children and what effect they might have. Critics of the school complain that it indoctrinates children. "Everyone indoctrinates children," Rajalin said, flatly. "Do you want to be aware of what you're doing or not?"[28]

In some ways, the school's effort to free kids from gendered expectations were apparent immediately. A boy in a glittery My Little Pony shirt ran up to me, announced the pony's name, and then darted away. Girls wore skirts and so did some boys. Boys had very short hair and so did some girls. Dollhouses and construction toys were grouped together as "building supplies." In the woodshop, real saws were lined up below a giant leopard-print tapestry. There were more male teachers than I'd ever seen as a preschool. And grouping children by gender was avoided even at the pronoun level; the teachers used no pronouns to refer to any child. Instead, they only spoke each child's first name.

In other ways, Egalia resembles a typical preschool. At lunch, I crushed myself into a tiny child's chair, as the children passed around bowls of hearty bread and a profoundly bland vegetable soup. I asked a five-year-old, one of the few children who spoke English, who he liked to play with. He listed names, a mix of all genders. Then he paused and took a bite of bread. "Everyone," he said, chewing. "You!" But his favorite activity, he told me, was cutting sheets of paper with scissors.

After lunch, three boys tumbled like puppies inside a playhouse made of real tree branches. Children rambled around a play office furnished with an unplugged keyboard, a coatrack, and an artificial potted plant—a modern toddler workplace. Circle time, led by a quiet teaching assistant named Martin, was about math, but sometimes the focus was feelings or how it's okay to be yourself. Notably, the school gives children a lot of autonomy. While adults supervised the sawing in the workshop, children were free to skip over the child-size tools and choose even the toothiest, full-size saw.

The school, it seemed, isn't really "gender neutral" at all. It's simply not gender *corrective*. No child is dissuaded from a particular gender role. There were girls in pink dresses and boys playing with trucks. There were loud children and quiet children of all genders. This didn't seem to matter to anyone. A pink tutu was hanging on a hook for anyone to try. The goal was to avoid actively gendering the children and instead simply letting them be who they already are.

And this, the teachers told me, is really difficult. Ana Rodriguez Garcia, a teacher who grew up in Spain, said that when she started at Egalia, she noticed how deeply ingrained her own expectations about gender were—and how automatic were her habitual responses. Even what seemed like a harmless compliment, such as saying to a girl, "What a pretty dress," was rooted in an assumption, she realized. The statement assumes that being pretty is the girl's goal. And it communicates that prettiness is important. This has profound consequences, because, as Rajalin urges teachers to see, children will do more of what adults praise and reward.

For Garcia, it took time to unravel and change her instant reactions. But now she responds differently. "If a child says, 'Look at my dress,'" she said, "I'll say, 'Oh, is it comfortable?' instead of giving affirmation. Or 'I see you have a dress on. Tell me about it.' Why should 'pretty' be the only thing they hear?"[29]

Her assumptions were so instinctive, Garcia told me, that she didn't realize how much she relied on these default patterns until they were no longer available. When she encountered a child with a name that wasn't identifiably male or female, she could feel her brain crave that information, as though it were a necessary precursor for engaging with that child. "It wasn't relevant information," she said, "but my brain wanted to know."

Garcia's comment reminded me of an experience I had during my early twenties, when the idea of gender existing on a spectrum was not mainstream in the United States. Meeting someone at a conference who I couldn't categorize as a man or a woman, I stammered awkwardly through our conversation. My discomfort, I later realized, came from the fact that I could not draw on my preset modes of interacting. I hadn't realized that I had one way of engaging with women and another way with men, until neither way was available to me.

Over time, Garcia felt that craving diminish, but it's not entirely gone. "A lot of people think fighting stereotypes is just taking away cars and dolls, but you have to work on yourself and constantly fight against your own inclinations," she explained. She still feels the pressure of deeply entrenched ideas. The boy in the My Little Pony T-shirt, for instance, likes to roughhouse but also loves wearing skirts, and Garcia finds herself struggling with the fact that both characteristics are embodied in a single person. Another teacher said that she actually prefers not to know a child's sex. Without the information, she has no ingrained tendencies to resist. An assistant teacher now notices that teachers at his own daughter's school say, "Hey little girls, come on little boys." So did he, before coming to work at Egalia. "I've changed how I talk to kids in the five years I've been here," he said. "Now I say 'friends.'"[30]

The attempt to avoid using labels extends beyond gender, too. Garcia told me that in Spain, teachers quickly tag children: the bossy kid, the violent kid, the one who needs extra help. But labeling a child as "bossy" ascribes to the child a fixed and inherent trait. At Egalia, teachers look for contextual reasons for a child's behavior, and then inquire what the child needs in that moment. And, as has become increasingly popular in many child-rearing contexts, they strive to avoid commenting on a child's talents. This, too, is a form of limiting children, so exquisitely attuned are they to adult affirmation and approval. When a child eagerly shares a painting, it is tempting to say, "You're a great artist! You should be an artist." Instead, the teachers ask, "Why did you paint that apple or that tree? Did you enjoy painting it?"

Perhaps it's not surprising that these schools are emerging in Sweden, given the culture's concern with children and their development. It was among the first countries to ratify the United Nations Convention on the Rights of the Child—a set of fifty-four articles that affirm children's rights to express opinions, be listened to, even have a private life. (The United States has not yet ratified it.) As I walked past the sorbet- and ice-cream-colored buildings in Stockholm's old town, I noticed that paths were designated with an illustration of an adult holding a child's hand. Stairs were bordered with miniature ramps to facilitate pushing a baby stroller up and down the stairs. Some of these amenities emerged in the twentieth century as a way to strengthen the country by making fertility more appealing; laws and policies continue to ensure that caring for children is compatible with work. (There's even a single world in Swedish—*vabba*—that means "staying home to care for a sick child.") But the emphasis on children as autonomous selves is notable. Sweden was the first country in the world to make corporal punishment against children a crime. The law states that "children are to be treated with respect for their person and individuality."[31]

What struck me about Rajalin's schools was the way their approaches to gender fit into a larger philosophy of childhood: that children are people who matter, people who should have agency over their lives. They aren't miniature, slightly ridiculous version of adults—they're humans whose integrity should be respected. One teacher explained to me that when a child says something that makes her laugh because it's absurd or unintentionally amusing, she tries to hide that she's laughing. If the child meant to be serious, laughter is hurtful. "This is also a form of indoctrination—that you don't matter."

Watching Egalia's teachers engage with the children, I began to see gender equity as just one part of a larger goal: reducing bias against children themselves. Indeed, treating children as inferior and unworthy, subjecting them to diminishment and assaults to their dignity—is also a form of prejudice. In 1972, the psychiatrists Chester Pierce and Gail Allen contended that prejudice against children underlies all other forms of oppression, as its widespread practice "teaches everyone how to be an oppressor." Changing "childism" would create a foundation for altering other subjugation.

Of course, children need, developmentally, adult guidance. But this necessity so often morphs into a kind of possession, with parents seeing their children as assets or extensions of themselves, and every pejorative act communicates to a child their inferiority. Pierce describes a situation of an adult asking a child to fetch something for her as featuring all four aspects of any oppression: the oppressor controls the space, mobility, time, and energy of the oppressed. The child's stress grows as they cede each one of these.[32] In truth, the domination of children by adults is perhaps the very first form of domination any of us experience.

The project of the Swedish schools is, by contrast, to create an environment of children's self-determination. The teachers' careful speech and their efforts to hold back are aimed at the same goal: respecting the child's essential self. If the children are being indoctrinated with anything, it seems, it's a philosophy of their own inviolable integrity.

During my time at the school, I thought about how often I project my own stereotypes and assumptions onto children—the many times I've said, "What a pretty dress!" to a friend's daughter or "You're so strong!" to another's son. If I encounter a boy and a girl, I don't say, "Hello, boy and girl." That would seem obviously absurd. But in front of a group, we say, "Hello, boys and girls" without thinking, in the process cementing the boundaries around those categories. How frequently I've burst out laughing at a child's unintentionally funny comment, only to have the child look at me, bewildered, because they were being serious. With pejorative tones, I have made them feel less valuable. With dismissive gestures, I have treated them as less valuable. I thought about my interactions with adults, too—all the times my own need for sensemaking produced assumptions and projections. You must be angry, you shouldn't feel this way. You should. You must. You are.

RAJALIN AND HER TEACHERS UNDERSTAND that freedom isn't one thing but two. There is freedom *to*, but there is also freedom *from*. To give her students freedom *to*—to grow, to explore, to choose—she must also give them freedom from the tightening net of others' expectations. In a larger sense, it is freedom from bias. The opportunity to create this possibility resides in every interaction, every featherweight moment. Only when adults take these moments seriously can the child emerge into the world as their complicated irreducible self.

The same is true for all of us, in all of our encounters with others, real or imagined, momentous or tiny. It is simpler and more expedient to rely on inherited ideas and myths, statistics and stereotypes, to interpret the people we encounter, including small children. It is our habit and our conditioning. Can I, will I, sacrifice this convenience for the sake of others? Can we, will we, practice this with each other?

At the end of my day at the preschool, I sat in the courtyard and watched children chase each other, screaming, as they waited for their parents. In the center of the courtyard were two benches in the shape of horses, the pale paint mostly rubbed off. A child wearing a police cruiser shirt and a blue skirt shrieked and raced into a playhouse. One tiny child wandered over to me and pushed a book into my hands. It was a Swedish translation of an old Richard Scarry volume, illustrating and identifying everyday objects. *Car. Airplane. Train. Truck.* The book was the right level for both of us, and we had a rudimentary conversation, trading words for animals and foods. What an awesome responsibility, teaching a child how to label the things of the world.

After we finished the book, a teacher introduced us. I learned the child was a girl. Would I have treated the child differently had I known her name? What if she had had other visible differences—would I have immediately begun the work of bias? Her history had barely begun; she was just starting to put the world into categories. The world, too, had just barely begun to drape her with meanings, like a set of garish chains, meanings that may have nothing to do with her but that would confer on her advantages and disadvantages, propelling her forward, dragging her back. But there was a moment in the courtyard, before I learned her name, when we were turning pages together, sounding out the words. The courtyard walls above us framed a vast expanse of sky.

Conclusion

Can we overcome biases that are unconscious, unintentional, or unexamined? I began this project because I wanted to find the answer. I now believe the answer is yes. I saw evidence of change in the stories and studies I present here, in preschool teachers and professors, in cricket players and rural villagers, in police like the Watts patrol officer who said he would never return to the "old style" of policing. I saw it in my own family and friends. I saw it in myself. This kind of transformation—relinquishing false, unexamined ideas and reflexes that have been passed down for hundreds or even thousands of years—requires great effort and, before effort, the will to change. But I also came to see how, in an open mind, the commitment to change becomes sturdier with the addition of knowledge, the way a sapling grows stronger after exposure to sunlight and rain.

Altering one's own habits of thinking is not quick or straightforward, even for the most well-intentioned. It is not a cure-all, either. Reducing individual bias won't end disparities and societal inequities: these are the legacy of historical exclusion, unequal access, extractive economic policies, and other invidious structures built on corrupt foundations.

Only large systemic changes—from the reinvention of public safety and prisons to broad economic repair—can redress such gross and long-standing injustices.

But the role of authentic inner, individual change can't be overstated. Laws and institutions emerge from human hearts, motivations, and consciousness. Policies are created by people, and it is people who interpret, implement, and abide by them. We might dismantle structures or laws, but we are still left with human minds that must imagine their replacement. Moreover, political or social action without internal transformation risks re-creating the oppressive and hierarchical thinking that enabled the original wrongs. To avoid that possibility, it is necessary to unwind harmful and unexamined patterns of thinking, practice seeing one another and ourselves with new eyes, and build cultures to support this transformation. All of this work strengthens the foundation for larger, more systemic repair.

The approaches explored in this book offer starting points. We can begin by noticing our own biased reactions, which are often so habitual they are difficult to see. Once seen, they can be questioned and interrupted. We can practice mindful awareness to help observe these reactions more clearly and better regulate our internal landscape so that bias is less likely to overtake our responses. We can form meaningful, collaborative connections with people unlike ourselves; in doing so, we can increase the complexity with which we see others.

Further, we can build structured decision-making into our institutions and organizations to reduce the role of bias in everyday practices. Through creatively redesigning on-ramps to opportunities in work, school, and beyond, we can expand access for those who have been stigmatized or marginalized. Simultaneously, we can ensure that our organizations value all their members, recognizing as essential wealth the contributions of people who have historically been sidelined. We can be intentional about the messages we convey to children and adults and rethink the media we project to avoid reinforcing harmful assumptions. We can also spread new norms about how we engage with one another so that undermining bias becomes ordinary and natural—a daily practice, a widespread movement.

Underlying all of this is a cellular-level shift: a change of heart. When I began this book, I thought I was writing a work of science. My plan was to read, study, synthesize the best evidence, and share what I found. The journey would be straightforward; it would be scientific and outward-facing—as if, when we study the world, we are not always, in some way, seeking ourselves. Instead, what I discovered broke me open.

As the mechanics of bias and oppression outside and inside myself became apparent, the original plan dissolved. The process became something deeper and more recursive, each new revelation sending me back to examine my own assumptions, and then out into my community with my perceptions adjusted, only to start again. Old, inherited reflexes and mythologies surfaced repeatedly, the way the sea washes pollution ashore in waves. But with practice, I grew more alert and better able to catch the debris.

The most essential part of this journey was making and learning from mistakes. When an article I wrote was criticized for racist paternalism by people I respect, I was shown something about my own thinking I had not seen before. First, I defensively rejected the claim, embodying all of philosopher Marilyn Frye's observations about people she terms "whitely": "We do not like to be challenged. We want to fix things. We want to be helpful. We want to be seen as good, and will do just about anything to restore, center, and highlight our goodness."[1] In time, I was able to settle into an unfamiliar but authentic humility. Once I set aside my attachment to notions of my own rightness, I could take an honest inventory of the fears, errors, lack of skill, and biases that had led me to act in ways that harmed others. This happened again and again as I faced my mental conditioning regarding myriad social identities, including my own.

These moments were deeply uncomfortable, even painful. Navigating the emotional dimension of this path is, I think, an underexplored area of research, one that is crucial to address because it can easily subvert the process of change. Feelings move people to irrational and counterproductive extremes, like the men who, amid the wave of #MeToo revelations, vowed never to take another solo meeting with a woman. Facing the reality of bias can evoke shame at one's sense of personal complicity and set off a cycle of self-flagellation, abnegation, and paralysis. When a person tries and fails—for example, causes unintended harm—this can heighten shame, or spark such embarrassment and regret that a person may withdraw their efforts entirely. In her work with organizations on issues of race, social psychologist Evelyn Carter sees White people disengage so frequently after a failure that she believes the most important element of this work is persisting *after* a misstep. Mindfulness and self-compassion are among the skills that may be of assistance in moving through this emotional terrain.[2]

Another underexplored dimension of uprooting bias is the role of understanding history. Studies exploring the Marley Hypothesis—which holds that a person's perception of racism increases in tandem with their knowledge of the past—suggest that a firm grounding in history can be an engine of change. For me, tracing social prejudices to their inception was like tracking a bloodstream infection back to its originating abscess.

The concept of male superiority, for instance, unfolded as a process in particular cultures at particular times. One strand was the gradual replacement of images of female divinity with male ones—the Greek goddess Demeter becoming Saint Demetra and finally Saint Demetrius—so that, step by step, divine authority itself acquired a male face. Today, those who imagine God as a White man are more likely to believe earthly bosses should be the same.[3] Seeing the emergence of patriarchy as an evolving reality rather than a timeless fact was an illumination, a lightning flash disclosing an entire landscape. For lies repeated often enough do not become truth, they become invisible. Understanding this progression, I could now see how the idea of women's inferiority was ingrained like an axiom into the very bases of my culture, its motifs, language, and symbols.

In a moment of lucidity while descending into madness, the writer Shulamith Firestone asked her sister, "Are you even on my side? Are you on your own side?"[4] Learning the history of patriarchal ideas also helped me perceive ways in which I had not been on my own side, how I had internalized falsehoods and discounted my authority and that of other women. The more I perceived these ideas as a human invention, the looser their grip on me became. Likewise, after learning the history of gender expression, I began to view efforts to force gender or sexual orientation into one of two buckets as barbaric and ahistorical, a denial of the complex reality of human experience, as absurd as my grandfather being punished for writing with his left hand.

Studying the history of racism brought about the same kind of increasing freedom from its grip. Whites are trapped "in a history they do not understand," wrote James Baldwin, "and until they understand it, they cannot be released from it." One moment of comprehension occurred as I read the delusions about race that suffused nineteenth-century medical journals, like the idea that African Americans were diseased, decrepit, and "naturally" headed toward annihilation. When I read a Chicago doctor's exhortation that White society "should help along the process of extinction," the force of understanding was so physical I involuntarily stood up and walked out of the room. Audre Lorde wrote "we were never meant to survive," and this was true, quite literally: the lies of racism and the sociopathic assignation of a group of people to extinction were kneaded into society's consciousness, even by those who vowed to do no harm.[5] My body registered this. Centuries of oppression, repression, segregation, and violence, culminating in present-day bias and discrimination, suddenly all appeared to me as a sort of White psychosis.

This book focuses on the harm done to those on the receiving end of bias, harm that is physical, material, economic, psychological, and spiritual. It pays little attention to the material benefits and other resources that accrue as a result to those who experience the flip side of bias: unearned advantage. It deals even less with the harms hidden inside those advantages. But the ticket, as Baldwin wrote, has a price, and oppression—whether flagrant human rights violations or everyday bias and discrimination—also damages the perpetrator. This is in no way to suggest equivalence between the experiences of society's marginalized people and those who are advantaged: there is no comparison. I only want to say that in blindly perpetrating bias, something is lost for the perpetrator as well, and until the most advantaged take seriously the ways in which they are also harmed, any understanding will be partial, and action may be misguided by a sense of "altruism" rather than rooted in the reality of human interdependence. Half understanding, they may succumb to the same kind of saviorism that has underpinned so much justice work.

What is lost? Acts of bias erode trust, the foundation of authentic relationships, promoting alienation and separation. There are also what have been described as "the ordinary vices of domination," which philosopher Samantha Vice lists as "indifference or callousness, cowardice or dishonesty, the failure of imagination and empathy, or just plain laziness."[6] There is, in the privileged mind, ongoing delusion—habitual, unexamined distortions about who one is, where one comes from, who does and does not deserve safety, comfort, opportunity, and care.

There may even be what has come to be called "moral injury," what philosopher Nancy Sherman describes as the inner conflict that derives from having committed moral transgressions that overwhelm one's own sense of humanity. It is the recognition that one's own self has fallen short of standards "that befit good persons."[7] Perhaps "White fragility" or "male fragility," or any similar instability that emerges under stress, is actually a felt connection to an old moral injury, one that could even have been committed by one's forebears. When people in positions of privilege or advantage cry or defensively protest while engaging with these issues, it is like the response of an animal when one approaches a site of damage on its body—one that has never been acknowledged and has instead been left to fester.

In the course of writing this book, I learned that members of my own family had been enslavers. I have always understood and identified more with my father's family, Jewish immigrants who fled persecution in the late nineteenth and early twentieth centuries. My mother's Christian family, long silent about itself, stretches back in this country to the 1600s. I knew that they had relocated to California in the nineteenth century and worked in construction and in canneries. Studying genealogical records, I learned that some had moved from Arkansas, Missouri, and Virginia, and that at least one line had enslaved people. No one ever mentioned them.

I saw these ancestors for the first time in a photo I found in a cardboard box at my parents' house, stuffed among teaspoons and papers: a pale, stern-faced man, his dark-haired wife in stiff crinoline and an unreadable expression. Their daughter Ada stood behind them: my grandfather's grandmother. I held the photo of these people who had made my life possible. I thought about the harm they had inflicted on others and the way it recoiled back into them. Something inside them broke—a sense, perhaps, of being part of the human family. The breaking ricocheted down the generations, unquantifiable. I remembered the grip of my grandfather's hand as he lay in his hospital bed, disfigured by decades of drink, linking us all in a chain. I set down the photo. I walked outside and lay on the ground. This history was my history, was in me, had made me. The grief I poured into the earth was both understanding and release.

If a person and a family can carry a moral injury, perhaps so can a nation and its people. This country, among many, has not yet faced the depths of its crimes, starting with the atrocities committed against the land's first inhabitants and with slavery. It has not reckoned with the resulting injuries to its citizens or mutilation of its own moral center. This is how history becomes psychology, how the rituals of the past become the reflexes of the present. Without confronting the past, we cannot understand it and we cannot take the actions required to repair it.

Contending with the truth about a nation or a family or one's own habits of thinking requires the same skills: a willingness to confront reality, a commitment to keep looking even when one does not like what one sees, the emotional skills to manage and move through the discomfort necessary to all human growth, and the courage to act. As I examined and confronted my own biases, my relationship with the world began to change. Friendships grew deeper and richer, both among people with whom I shared many social identities and among those with whom I shared few. Difficult conversations became more manageable; as I became more confident I could repair my relations with others, it became easier to risk venturing into the unfamiliar. When someone across a social difference reached out to me in friendship or trust, I rushed in, knowing there was information in that exchange that I needed.

For much of this project I struggled with what felt like a paradox, the fact that emphasizing differences carries the risk of entrenching essentialist stereotypes and increasing prejudice and discrimination, but downplaying them can generate feelings of invisibility and disrespect. In time I came to see that the choice was false and impossible. We are all of these: we are similar, sharing the need for belonging, fresh air, vegetables, and human connection; we have differences, born from our ancestries and bodies, and contexts created by people long dead; we are individuals, as particular as the markings of a human iris. "We have no patterns," wrote Lorde, "for relating across our human differences as equals."[8] The problem is not in seeing difference, but in the values and meanings we attach to it. If we can grapple with our biases deeply enough to see one another in all our facets, perhaps we can begin to create the patterns Lorde imagined.

We might, too, be able to feel our way into another's experience. This act of imagination, notes South African scholar Pumla Gobodo-Madikizela, is a prelude to caring, a prelude, perhaps, to love. Gobodo-Madikizela, who served on her country's Truth and Reconciliation Commission, writes that Descartes' famous notion "I think, therefore I am" reflects a sense of individual existence that is independent of others. In fact, she says, we exist in and through each other: our humanity depends on our ability to bestow humanity on others.[9] This truth surpasses the business case for ending bias; it strengthens the culture case and underpins the justice case. We end bias for the sake of others and for our own.

Who might we become without our illusions and denials? We might become human, and trustworthy. We might all become free.

NOTES

INTRODUCTION

1. Ben Barres, *Autobiography of a Transgender Scientist* (Cambridge, MA: MIT Press, 2018), 48, 56; Ben Barres, interview with Charlie Rose, "Gender Identity," *Charlie Rose*, PBS, June 18, 2015, https://charlierose.com/videos/21056. See also Ben Barres, "A Sense of Discomfort with Myself, Part 1/21," *Web of Stories: Life Stories of Remarkable People*, April 11, 2017, https://www.youtube.com/watch?v=_wMLbuHhZwk&list=PLVV0r6CmEsFz74WZPchYPmr9l5oSBLAAR&index=2&t=0s; and Ben Barres, "Feeling Like a Boy, Part 2/21," *Web of Stories: Life Stories of Remarkable People*, April 11, 2017, https://www.youtube.com/watch?v=icNq2uqsvVw&list=PLVV0r6CmEsFz74WZPchYPmr9l5oSBLAAR&index=2; Keay Davidson, "Transsexual Tackles Sexism in Sciences," *SF Gate*, July 13, 2006, https://www.sfgate.com/science/article/STANFORD-Transsexual-tackles-sexism-in-sciences-2531694.php.
2. Ben Barres, "I Like Myself Now, Part 7/21," *Web of Stories: Life Stories of Remarkable People*, April 11, 2017.
3. Ben Barres, "Does Gender Matter?," *Nature* 442 (2006): 133–36.
4. Barres, *Autobiography of a Transgender Scientist*, 17.
5. Larry Summers, "Remarks at NBER Conference on Diversifying the Science and Engineering Workforce," Harvard University, Cambridge, MA, January 14, 2005; and Barres, "Does Gender Matter?"
6. Ben Barres, interview with the author, August 15, 2014; and Shankar Vedantam, *The Hidden Brain: How Our Unconscious Minds Elect Presidents, Control Markets, Wage Wars, and Save Our Lives* (New York: Spiegel & Grau, 2010), 103.

7. Ben Barres, interview with the author, August 15, 2014.

8. Kristen Schilt, personal correspondence with the author, June 18, 2020.

9. Katherine Milkman, Modupe Akinola, and Dolly Chugh, "What Happens Before? A Field Experiment Exploring How Pay and Representation Differentially Shape Bias on the Pathway into Organizations," *Journal of Applied Psychology* 100, no. 6 (2015): 1678–712; Katherine Milkman, Modupe Akinola, and Dolly Chugh, "Temporal Distance and Discrimination: An Audit Study in Academia," *Psychological Science* 23, no. 7 (2012): 710–17; Hua Sun and Lei Gao, "Lending Practices to Same-Sex Borrowers," *Proceedings of the National Academy of Sciences of the United States of America* 116, no. 19 (2019): 9293–302; Lisa Friedman et al., "An Estimate of Housing Discrimination Against Same-Sex Couples," U.S. Department of Housing and Urban Development Office of Policy and Development Research (2013); Devah Pager, Bruce Western, and Bart Bonikowski, "Discrimination in a Low-Wage Labor Market: A Field Experiment," *American Sociological Review* 74, no. 5 (2009): 777–99. See also Marianne Bertrand and Sendhil Mullainathan, "Are Emily and Greg More Employable than Lakisha and Jamal? A Field Experiment on Labor Market Discrimination," *American Economic Review* 94, no. 4 (2004): 991–1013.

10. Salimah Meghani, Eeeseung Byun, Rollin Gallagher, "Time to Take Stock: A Meta-analysis and Systematic Review of Analgesic Treatment Disparities for Pain in the United States," *Pain Medicine* 13, no. 2 (2012): 150–74; see also Nathalia Jimenez et al., "Perioperative Analgesic Treatment in Latino and Non-Latino Pediatric Patients," *Journal of Healthcare for the Poor and Underserved* 21, no. 1 (2010): 229–36; Erica Kenney et. al, "The Academic Penalty for Gaining Weight: A Longitudinal, Change-in-Change Analysis of BMI and Perceived Academic Ability in Middle School Students," *International Journal of Obesity* 39, no. 9 (2015):1408–13; Lauren Rivera and András Tilcsik, "How Subtle Class Cues Can Backfire on Your Resume," *Harvard Business Review* (December 21, 2016; updated April 4, 2017), https://hbr.org/2016/12/research -how-subtle-class-cues-can-backfire-on-your-resume; Jason Okonofua and Jennifer Eberhardt, "Two Strikes: Race and the Disciplining of Young Students," *Psychological Science* 26, no. 5 (2015): 617–24; Steven Foy and Rashawn Ray, "Skin in the Game: Colorism and the Subtle Operation of Stereotypes in Men's College Basketball," *American Journal of Sociology* 125, no. 3 (2019): 730–85; Tracey Colella et al., "Sex Bias in Referral of Women to Outpatient Cardiac Rehabilitation? A Meta-analysis," *European Journal of Preventive Cardiology* 22, no. 4 (2015): 423–41; Katarina Hamberg, Gunilla Risberg, and Eva Johansson, "Male and Female Physicians Show Different Patterns of Gender Bias: A Paper-Case Study of Management of Irritable Bowel Syndrome," *Scandinavian Journal of Public Health* 32, no. 2 (2004): 144–52; Corinne Moss-Racusin et al., "Science Faculty's Subtle Gender Bias Favors Male Students," *Proceedings of the National Academy of Sciences of the United States of America* 109, no. 41 (2012): 16474–79; and Christine Wennerås and Agnes Wold,

"Nepotism and Sexism in Peer-Review," *Nature* 387 (1997): 341–43. When Wennerås and Wold submitted their study more than twenty years ago, it was originally rejected with the note that it might be interesting "to Scandinavian audiences." More recent studies continue to find patterns of disparate competence ratings.

11. Joseph Wertz et. al. "A Typology of Civilians Shot and Killed by US Police: A Latent Class Analysis of Firearm Legal Intervention Homicide in the 2014–2015 National Violent Death Reporting System," *Journal of Urban Health* 97 (2020): 317–28; and Ali Winston, "Medical Examiner Testifies Eric Garner Died of Asthma Caused by Officer's Chokehold," *New York Times*, May 15, 2019, https://www.nytimes.com/2019/05/15/nyregion/eric-garner-death-daniel-pantaleo-chokehold.html.

12. George Lamming, "The Negro Writer and His World," *Presence Africaine: The First International Conference of Negro Writers and Artists* (Paris: Sorbonne, 1956), 332.

13. Kenneth Clark and Mamie Clark, "Racial Identification and Preference in Negro Children," in *Readings in Social Psychology*, ed. Theodore Newcomb and Eugene Hartley (New York: Henry Holt and Co., 1947): 169–78; interview with Kenneth Clark, conducted by Blackside, Inc. for *Eyes on the Prize: America's Civil Rights Years (1954–1965)*, Washington University Libraries, Film and Media Archive, Henry Hampton Collection (November 4, 1985); Lawrence Nyman, "Documenting History: An Interview with Kenneth Bancroft Clark," *History of Psychology* 13, no. 1, (2010): 74–88.

14. Dawn Lundy Martin, "When a Person Goes Missing," *n+1*, no. 30: Motherland (Winter 2018), https://nplusonemag.com/issue-30/politics/when-a-person-goes-missing/.

15. Jae Yun Kim, Grainne Fitzsimons, and Aaron Kay, "Conflating a Solution with a Cause: The Potential Harmful Effects of Women's Empowerment Messages," *Academy of Management Proceedings* 2018, no. 1 (2018): 125–37.

16. Robert Livingston, Ashleigh Shelby Rosette, and Ella Washington, "Can an Agentic Black Woman Get Ahead? The Impact of Race and Interpersonal Dominance on Perceptions of Female Leaders," *Psychological Science* 23, no. 354 (2012): 354–58; Ashleigh Shelby Rosette et al., "Race Matters for Women Leaders: Intersectional Effects on Agentic Deficiencies and Penalties," *Leadership Quarterly* 26 (2016): 429–45; and Kristen Schilt, *Just One of the Guys?: Transgender Men and the Persistence of Gender Inequality* (Chicago: University of Chicago Press, 2010), 85.

17. Stephanie Fryberg and Arianne Eason, "Making the Invisible Visible: Acts of Commission and Omission," *Current Directions in Psychological Science* 26, no. 6 (2017): 554–59; and Adrienne Rich, "Disloyal to Civilization," in *On Lies, Secrets, and Silence: Selected Prose 1966–1978* (New York: W. W. Norton, 1979), 308.

18. Evelyn Carter, interview with the author, April 16, 2016.

19. Rebecca Hetey and Jennifer Eberhardt "Racial Disparities in Incarceration

Increase Acceptance of Punitive Policies," *Psychological Science* 25 (no. 10), 2014: 1949–54; and Connie Rice, interview with the author, October 10, 2018.

20. Ben Barres, interview with the author, June 26, 2016.

21. Ben Barres, interview with the author, August 15, 2014; Schilt, *Just One of the Guys*, 70–75; and Kristen Schilt, interview with the author, August 16, 2014.

22. Joan Roughgarden, interview with the author, August 20, 2014; and Paula Stone Williams, "I've Lived as a Man and a Woman. Here's What I Learned," *TEDx Talks*, December 19, 2017, https://www.youtube.com/watch?v=lrYx7HaUlMY.

23. Elisabeth Kübler-Ross, *On Death and Dying: What the Dying Have to Teach Doctors, Nurses, Clergy and Their Own Families* (New York: Scribner, 2014), 273; and Claudia Rankine, "The Condition of Black Life Is One of Mourning," *New York Times Magazine*, June 22, 2015.

24. Mark Brinson et al., *Riparian Ecosystems: Their Ecology and Status* (Kearneysville, WV: Eastern Energy and Land Use Team National Water Resources Analysis Group, U.S. Fish and Wildlife Service, 1981), 73.

CHAPTER 1. THE CHASE

1. Patricia Devine, interviews with the author, February 23 and August 23, 2017.

2. Faye Crosby et al., "Recent Unobtrusive Studies of Black and White Discrimination and Prejudice: A Literature Review," *Psychological Bulletin* 87, no. 3 (1980): 546–63. The Swedish economist Gunnar Myrdal called this conflict between American democratic ideals and reality the "American Dilemma": "The American thinks, talks, and acts under the influence of high national and Christian precepts, and, on the other hand . . . consideration of community prestige and conformity; group prejudice against particular persons or types of people; and all sorts of miscellaneous wants, impulses, and habits dominate his outlook." Gunnar Myrdal, *An American Dilemma*, vol. 1, *The Negro Problem and Modern Democracy* (New York: Harper & Row, 1944), xliii.

3. John Duckitt, "Psychology and Prejudice: A Historical Analysis and Integrative Framework," *American Psychologist* 47, no. 10 (1992): 1182–85; John Haller, *Outcasts from Evolution: Scientific Attitudes of Racial Inferiority, 1859–1900* (Urbana: University of Illinois Press, 1971), 6–11, 77; Ann Fabian, *The Skull Collectors: Race, Science, and America's Unburied Dead* (Chicago: University of Chicago Press, 2010), 15; Robert Richards, "The Beautiful Skulls of Schiller and the Georgian Girl: Quantitative and Aesthetic Scaling of the Races, 1770–1850," in *Johann Friedrich Blumenbach: Race and Natural History, 1750-1850*, ed. Nicolaas Rupke and Gerhard Lauer (London: Routledge, 2018), 151; and R. Meade Bache, "Reaction Time with Reference to Race," *Psychological Review* 2, no. 5 (1895): 475–86.

4. Frederick Douglass, "The Claims of the Negro, Ethnologically Considered" (Rochester, New York: Lee, Mann & Co., 1854), 34; Franz Samelson, "From

'Race Psychology' to 'Studies in Prejudice': Some Observations on the Thematic Reversal in Social Psychology," *Journal of the History of the Behavioral Sciences* 14 (1978): 268–69; Vernon Williams Jr., *Rethinking Race: Franz Boas and His Contemporaries* (Lexington: University of Kentucky Press, 1996), 44–45; Thomas Garth, "A Review of Racial Psychology," *Psychological Bulletin* 22, no. 6 (1925): 359; and Emory Bogardus, "Measuring Social Distances," *Journal of Applied Sociology* 9 (1925): 299–308.

5. Samelson, "From 'Race Psychology' to 'Studies in Prejudice,'" 265, 267, 273; Martha McLear, "Sectional Differences as Shown by Academic Ratings and Army Tests," *School and Society* 15 (1922): 676–78; Horace Bond, "Intelligence Tests and Propaganda," *Crisis* 28 (1924): 64; Howard Long, "Race and Mental Tests," *Opportunity* 1, no. 3 (1923): 22–28; Carl Brigham, *A Study of American Intelligence* (Princeton, NJ: Princeton University Press, 1923), 183; Carl Brigham, "Intelligence Tests of Immigrant Groups," *Psychological Review* 37, no. 2 (1930): 158–65; William McDougal, *Is America Safe for Democracy?* (New York: Charles Scribner's Sons, 1921), 185–88; David Milner, *Children and Race: Ten Years On* (London: Ward Lock Educational, 1983), 27; John Duckitt, "Psychology and Prejudice: A Historical Analysis and Integrative Framework," *American Psychologist* 47, no. 10 (1992): 1182–93; David Owen, "Inventing the SAT," Alicia Patterson Foundation, 1985, https://aliciapatterson .org/stories/inventing-sat; and Thomas Garth, "A Review of Race Psychology (1924–1929)," *Psychological Bulletin* 27 (1930): 329–56.

6. Samelson, "From 'Race Psychology' to 'Studies in Prejudice,'" 270.

7. Howard Schuman, Charlotte Steeh, Lawrence Bobo, and Maria Krysan, *Racial Attitudes in America: Trends and Interpretations* (Cambridge, MA: Harvard University Press, 1970), 66; August Meier and Elliott Rudwick, *Black Detroit and the Rise of the UAW* (Oxford: Oxford University Press, 1979), 181–82; Amy Fried, "Race, Politics and American State Capacity: U.S. Government Monitoring of Racial Tensions During World War II" (unpublished manuscript, 2017), 13–15; and Herbert Shapiro, *White Violence and Black Response: From Reconstruction to Montgomery* (Amherst: University of Massachusetts Press, 1988), 312–13. The Detroit Housing Commission, surveying the racism that permeated the city, noted: "Any attempt to change the racial pattern of any area in Detroit will result in violent opposition to the housing project. This could very easily reach a point where war production efforts of this entire community could be endangered."

8. "Lynching and Liberty," *Crisis* (July 1940), 209; and Langston Hughes, *The Collected Poems of Langston Hughes* (New York: Vintage Books, 1995), 281.

9. Quoted in James Whitman, *Hitler's American Model* (Princeton, NJ: Princeton University Press, 2017), 10, 118, 147, Kindle. The memo discussed at this meeting argued for the prevention of public "mixing" of *Volk* and "colored" races: "Protection of racial honor of this kind is already practiced by other *Volker*. It is well-known, for example, that the southern states of North

America maintain the most stringent separation between the White population and coloreds in both public and personal interactions."

10. Fried, "Race, Politics and American State Capacity," 1–5; Amy Fried, "'Negro
Morale,' the Japanese-American Internment, and U.S. Government Opinion
Studies During World War II" (Annual Meeting Paper, American Political Sci-
ence Association, 2011): https://ssrn.com/abstract=1903210; Schuman et al.,
Racial Attitudes in America, 65, 66–76; "Negroes in a Democracy at War," May
27, 1942, RG 208, entry 20, box 11, National Archives, College Park, MD; Ken-
neth Clark, "Morale of the Negro on the Home Front: World Wars I and II,"
Journal of Negro Education 12, no. 3 (1943): 428; Mildred Schwartz, "Trends
in White Attitudes Toward Negroes, Report No. 119," *National Opinion Research
Center* (1967): 1–2; and Lawrence Bobo et al., "The Real Record on Racial Attitudes," *Social Trends in American Life: Findings from the General Social Survey
Since 1972* (Princeton, NJ: Princeton University Press, 2012), 38–83.

11. Bobo et al., "The Real Record on Racial Attitudes," 38–74; and Schuman et al.,
Racial Attitudes in America, 66.

12. George Galster, "Racial Steering in Urban Housing Markets: A Review of the
Audit Evidence," *Review of Black Political Economy* 18, no. 3 (1990): 105–29;
Richard Lowy, "Yuppie Racism: Race Relations in the 1980s," *Journal of Black
Studies* 21, no. 4 (1991): 445–64; John Yinger, "Measuring Racial Discrimination with Fair Housing Audits: Caught in the Act," *American Economic Review*
76, no. 5 (1986): 881–93; Marianne Page, "Racial and Ethnic Discrimination
in Urban Housing Markets: Evidence from a Recent Audit Study," *Journal of
Urban Economics* 38, no. 2 (1995): 183–206; Devah Pager and Hana Shep-
herd, "The Sociology of Discrimination: Racial Discrimination in Employment, Housing, Credit, and Consumer Markets," *Annual Review of Sociology*
34 (2008): 181–209; Margery Austin Turner, Michael Fix, and Raymond
Struyk, *Opportunities Denied, Opportunities Diminished: Racial Discrimination in Hiring* (Washington, DC: The Urban Institute Press, 1991), 63; George
Galster, "Race Discrimination in Housing Markets During the 1980s," *Journal of Planning Education and Research* 9, no. 3 (1990): 165–75; Lisa Belkin,
"Show Me a Hero," [excerpt], *New York Times*, March 14, 1999, book review,
https://archive.nytimes.com/www.nytimes.com/books/first/b/belkin-hero
.html; Reuters, "Shoney's Bias Suit Settled," *New York Times*, January 28, 1993,
http://www.nytimes.com/1993/01/28/business/shoney-s-bias-suit-settled
.html; and Ashley Southall, "Bias Payments Come Too Late for Some Farmers," *New York Times*, May 25, 2010, https://www.nytimes.com/2010/05/26/us
/26farmers.html.

13. Faye Crosby et al., "Recent Unobtrusive Studies of Black and White Discrimination and Prejudice," 546–63; Edward Donnerstein and Marcia Donnerstein,
"The Effect of Attitudinal Similarity on Interracial Aggression," *Journal of Personality* 43, no. 3 (1975): 485–502; Marcia Donnerstein and Edward Donnerstein, "Variables in Interracial Aggression: Exposure to Aggressive Interracial

Interactions," *Journal of Social Psychology* 100, no. 1 (1976): 111; and Edward Donnerstein, et al., "Variables in Interracial Aggression: Anonymity, Expected Retaliation, and a Riot," *Journal of Personality and Social Psychology* 22, no. 2 (1972): 236–45. See also Gary I. Schulman, "Race, Sex, and Violence: A Laboratory Test of the Sexual Threat of the Black Male Hypothesis," *American Journal of Sociology* 79, no. 5 (1974): 1260–77.

14. J. Nicole Shelton, "A Reconceptualization of How We Study Issues of Racial Prejudice," *Personality and Social Psychology Review* 4, no. 4 (2000): 374–90.

15. Patricia Devine, interview with the author, August 23, 2017; Diane Melius, interview with the author, August 31, 2017; and Terry Devine, interview with the author, September 5, 2017.

16. Patricia Devine, "Getting Hooked on Research in Social Psychology: Examples from Eyewitness Identification and Prejudice," in *The Social Psychologists: Research Adventures*, ed. Gary G. Brannigan and Matthew R. Merrens, 161–84 (New York: McGraw-Hill, 1995); Patricia Devine, interview with the author, August 23, 2017; and Roy Malpass, interview with the author, September 1, 2017.

17. Roy Malpass and Patricia Devine, "Eyewitness Identification: Lineup Instructions and the Absence of the Offender," *Journal of Applied Psychology* 66, no. 4 (1981): 482–89.

18. James Jones, *Prejudice and Racism* (New York: McGraw-Hill, 1972); and James Jones, interview with the author, May 17, 2021.

19. Patricia Devine, "A Modern Perspective on the Classic American Dilemma," *Psychological Inquiry* 14, no. 3–4 (2003): 244–50; Devine, "Getting Hooked on Research in Social Psychology"; John Bargh and Paula Pietromonaco, "Automatic Information Processing and Social Perception: The Influence of Trait Information Presented Outside of Conscious Awareness on Impression Formation," *Journal of Personality and Social Psychology* 43, no. 3 (1982): 437; Thomas Srull and Robert Wyer, "The Role of Category Accessibility in the Interpretation of Information About Persons: Some Determinants and Implications," *Journal of Personality and Social Psychology* 37, no. 10 (1979): 1660–72; John Bargh, "Automatic and Conscious Processing of Social Information," in *The Handbook of Social Cognition*, ed. Robert Wyer Jr. and Thomas Srull, 1–43 (Hillsdale, NJ: Lawrence Erlbaum, 1984); and E. Tory Higgins, William Rholes, and Carl Jones, "Category Accessibility and Impression Formation," *Journal of Experimental Social Psychology* 13, no. 2 (1977): 141–54.

20. David Meyer and Roger Schvaneveldt, "Facilitation in Recognizing Pairs of Words: Evidence of a Dependence Between Retrieval Operations," *Journal of Experimental Psychology* 90, no. 2 (1971): 227–34.

21. Patricia Devine, interview with the author, August 23, 2017.

22. This concern about priming was prescient. In recent years, the "reproducibility crisis" in science has led some to argue that priming studies have not been reliably reproduced, such as studies that show that people walk more slowly when primed with the words associated with the elderly, like "Florida" and "bingo."

But as psychologist Daniel Kahneman writes, while priming actions remains to be fully proven, "There is adequate evidence for all the building blocks: semantic priming, significant processing of stimuli that are not consciously perceived, and ideo-motor activation." See Stéphane Doyen et al., "Behavioral Priming: It's All in the Mind, But Whose Mind?," *PloS One* (January 18, 2012), https://journals.plos.org/plosone/article?id=10.1371/journal.pone.0029081; B. Keith Payne, Jazmin Brown-Iannuzzi, and Chris Loersch, "Replicable Effects of Primes on Human Behavior," *Journal of Experimental Psychology: General* 145, no. 10 (2016): 1269–79; and Daniel Kahneman, February 14, 2017, 8:37 p.m., comment on Ulrich Schimmack, Moritz Heene, and Kamini Kesavan, "Reconstruction of Train Wreck: How Priming Research Went Off the Rails," February 2, 2017, https://replicationindex.com/2017/02/02/reconstruction-of -a-train-wreck-how-priming-research-went-of-the-rails/comment-page-1 /#comments.

23. Patricia Devine, "Stereotypes and Prejudice: Their Automatic and Controlled Components," *Journal of Personality and Social Psychology* 56, no. 1 (1989): 5–18; and John McConahay, Betty Hardee, and Valerie Batts, "Has Racism Declined? It Depends upon Who's Asking and What Is Asked," *Journal of Conflict Resolution* 25, no. 4 (1981): 563–79.

24. Devine, "Stereotypes and Prejudice," 5–18.

25. Patricia Devine, interview with the author, August 23, 2017.

26. Richard Shiffrin and Walter Schneider, "Controlled and Automatic Human Information Processing: II. Perceptual Learning, Automatic Attending and a General Theory," *Psychological Review* 84, no. 2 (1977): 127–90; and James Neely, "Semantic Priming and Retrieval from Lexical Memory: Roles of Inhibitionless Spreading Activation and Limited-Capacity Attention," *Journal of Experimental Psychology* 106, no. 3 (1977): 226–54.

27. Patricia Devine, interview with the author, February 23, 2017.

28. Defendant's Motion for New Trial, State of Minnesota v. Hayward [*sic*], 4th Dist. Ct. (no. 26421), 1928; Lena Olive Smith, quoted in Ann Juergens, "Lena Olive Smith: A Minnesota Civil Rights Pioneer," *William Mitchell Law Review* 28, no. 1 (2001): 447–48; Samuel Gaertner and John McLaughlin, "Racial Stereotypes: Associations and Ascriptions of Positive and Negative Characteristics," *Social Psychology Quarterly* 46, no. 1 (1983): 23–30; John Dovidio, Nancy Evans, and Richard Tyler, "Racial Stereotypes: The Contents of Their Cognitive Representations," *Journal of Experimental Social Psychology General* 22, no. 1 (1986): 22–37; Russ Fazio et al., "On the Automatic Activation of Attitudes," *Journal of Personality and Social Psychology* 50, no. 2 (1986): 229–38; and Samuel Gaertner and John Dovidio, "The Aversive Form of Racism," in *Prejudice, Discrimination, and Racism*, ed. Gaertner and Dovidio (Orlando, FL: Academic Press, 1986): 61–89.

29. Patricia Devine, interview with the author, April 11, 2016; and Mahzarin R. Banaji, "The Dark Dark Side of the Mind," Harvard University, last

modified July 3, 2014, http://www.people.fas.harvard.edu/~banaji/research /publications/articles/2011_Banaji_OTH.pdf.

30. Janet Shibley Hyde, "The Gender Similarities Hypothesis," *American Psychologist* 60, no. 6 (2005): 581–92; Erika Hall et al., "MOSAIC: A Model of Stereotyping Through Associated and Intersectional Categories," *Academy of Management Review* 44, no. 3 (2019): 643–72.

31. Tunette Powell, "My Son Has Been Suspended Five Times. He's 3," *Washington Post*, July 24, 2014, https://www.washingtonpost.com/posteverything /wp/2014/07/24/my-son-has-been-suspended-five-times-hes-3/?utm_term= .02bf86ee260e; Melissa Harris-Perry, "The Racial Disparity in School Suspensions," interview with Tunette Powell, MSNBC, August 3, 2014, video, 5:07; and Ira Glass, "Is This Working?," *This American Life*, WBEZ Chicago, October 17, 2014, transcript.

32. Tony Fabelo et al., "Breaking Schools' Rules: A Statewide Study of How School Discipline Relates to Students' Success and Juvenile Justice Involvement," Council of State Governments Justice Center and the Public Policy Research Center (2011); Russell Skiba et al., "The Color of Discipline: Sources of Racial and Gender Disproportionality in School Punishment," *Urban Review* 34, no. 4 (2002): 317–42; Anne Gregory, Russell Skiba, and Pedro Noguera, "The Achievement Gap and the Discipline Gap: Two Sides of the Same Coin?," *Educational Researcher* 39, no. 1 (2010): 59–68; Phillip Atiba Goff et al., "The Essence of Innocence: Consequences of Dehumanizing Black Children," *Interpersonal Relations and Group Processes* 106, no. 4 (2014): 526–45; and Jason Okonofua and Jennifer Eberhardt, "Two Strikes: Race and the Disciplining of Young Students," *Psychological Science* 26, no. 5 (2015): 617–24.

33. Philip Guo, "Silent Technical Privilege," *Slate*, January 15, 2014, https://slate.com /technology/2014/01/programmer-privilege-as-an-asian-male-computer -science-major-everyone-gave-me-the-benefit-of-the-doubt.html; Philip Guo, "People Assumed I Was a Tech Whiz Because I'm Asian," interview with Celeste Headlee, *Tell Me More*, NPR, January 23, 2014, transcript.

34. See Samuel Museus and Peter Kiang, "Deconstructing the Model Minority Myth and How It Contributes to the Invisible Minority Reality in Higher Education Research," *New Directions for Institutional Research* 142 (2009): 5–15; Anemona Hartocollis, "Harvard Rated Asian-American Applicants Lower on Personality Traits, Suit Says," *New York Times*, June 15, 2018, https://www.nytimes.com/2018/06/15/us/harvard-asian-enrollment -applicants.html.

35. Ben Barres, interview with the author, June 23, 2016.

36. Michael Moritz, "I Got to Silicon Valley with No Grand Plan," interviewed by Emily Chang, *Studio 1.0*, Bloomberg TV, December 2, 2015; Shawn Tully and Ani Hadjian, "How to Make $400,000 in Just One Minute," CNN Money, May 27, 1996, https://money.cnn.com/magazines/fortune/fortune_archive

/1996/05/27/212866/index.htm; and Chris Blackhurst, "Michael Moritz: The Billionaire Venture Capitalist That Has Studied Alex Ferguson to Steve Jobs on Leadership," *Evening Standard*, December 18, 2015, https://www.standard .co.uk/business/michael-moritz-the-billionaire-venture-capitalist-that-has -studied-alex-ferguson-to-steve-jobs-on-a3140476.html.

37. Sukhinder Singh Cassidy, "Fields of Study and STEM Degrees on 2015 Midas List," Medium.com, December 6, 2015, https://medium.com/@sukhinder singhcassidy/fields-of-study-of-the-2015-midas-list-da884e5886e0.

38. James Sidanius, interview with the author, February 2, 2021; James Sidanius and Felicia Pratto, *Social Dominance: An Intergroup Theory of Social Hierarchy and Oppression* (New York: Cambridge University Press, 1999); and Patricia Devine, interview with the author, May 28, 2019.

39. Anthony Greenwald and Mahzarin Banaji, "Implicit Social Cognition: Attitudes, Self-Esteem, and Stereotypes," *Psychological Review* 102, no. 1 (1995): 4–27; Anthony Greenwald, Debbie McGhee, and Jordan Schwartz, "Measuring Individual Differences in Implicit Cognition: The Implicit Association Test," *Journal of Personality and Social Psychology* 74, no. 6 (1998): 1464–80; and Mahzarin Banaji and Tony Greenwald, "The Implicit Association Test," *Edge*, February 12, 2008, https://www.edge.org/conversation/the-implicit-association-test.

40. Brian Nosek et al., "Pervasiveness and Correlates of Implicit Attitudes and Stereotypes," *European Review of Social Psychology* 18, no. 1 (2007): 36–88; Brian Nosek, Mahzarin Banaji, and Anthony Greenwald, "Harvesting Implicit Group Attitudes and Beliefs from a Demonstration Web Site," *Group Dynamics: Theory, Research, and Practice* 6, no. 1 (2002): 101–15; and Marlene Schwartz et al., "The Influence of One's Own Body Weight on Implicit and Explicit Anti-fat Bias," *Obesity* 14, no. 3 (2006): 440–47; Tessa Charlesworth and Mahzarin Banaji, "Patterns of Implicit and Explicit Attitudes: 1. Long-Term Change and Stability from 2007–2016," *Psychological Science* 30, no. 2 (2019): 174–92; Bentley Gibson, Philippe Rochat, Erin Tone, and Andrew Baron, "Sources of Implicit and Explicit Intergroup Race Bias Among African-American Children and Young Adults," *PLoS One* 12, no. 9 (2017); and Anna-Kaisa Newheiser and Kristina Olson, "White and Black American Children's Implicit Intergroup Bias," *Journal of Experimental Social Psychology* 48, no. 1 (2012): 264–70. See also Karen Gonsalkorale, Jeffrey Sherman, and Karl Christoph Klauer, "Measures of Implicit Attitudes May Conceal Differences in Implicit Associations: The Case of Antiaging Bias," *Social Psychological and Personality Science* 5, no. 3 (2014): 271–78.

41. Phillip Atiba Goff et al., "Not Yet Human: Implicit Knowledge, Historical Dehumanization, and Contemporary Consequences," *Journal of Personality and Social Psychology* 94, no. 2 (2008): 292–306; Jennifer Eberhardt, *Biased: Uncovering the Hidden Prejudice that Shapes What We See, Think, and Do* (New York: Viking, 2019), 145–49.

42. Brian Nosek, Anthony Greenwald, and Mahzarin Banaji, "The Implicit

Association Test at Age 7: A Methodological and Conceptual Review," in J. A. Bargh, ed., *Automatic Processes in Social Thinking and Behavior* (New York: Psychology Press, 2007), 265–92; Anthony Greenwald, T. Andrew Poehlman, Eric Luis Uhlmann, and Mahzarin Banaji, "Understanding and Using the Implicit Association Test: III. Meta-analysis of Predictive Validity," *Journal of Personality and Social Psychology* 97, no. 1 (2009): 17–41; and Frederick Oswald, Gregory Mitchell, Hart Blanton, James Jaccard, and Philip Tetlock, "Predicting Ethnic and Racial Discrimination: A Meta-analysis of IAT Criterion Studies," *Journal of Personality and Social Psychology* 105 (2013): 171–92.

43. Karen Gonsalkorale, Jeffrey Sherman, and Karl Christoph Klauer, "Aging and Prejudice: Diminished Regulation of Automatic Race Bias Among Older Adults," *Journal of Experimental Psychology* 45, no. 2 (2009): 410–14.

44. Bertram Gawronski, interview with the author, April 18, 2017; and Anne Roefs et al., "The Environment Influences Whether High-Fat Foods Are Associated with Palatable or with Unhealthy," *Behavior Research and Therapy* 44, no. 5 (2006): 715–36.

45. Gordon Moskowitz and Peizhong Li, "Egalitarian Goals Trigger Stereotype Inhibition: A Proactive Form of Stereotype Control," *Journal of Experimental Social Psychology* 47, no. 1 (2011): 103–16; Kai Sassenberg and Gordon Moskowitz, "Don't Stereotype, Think Different! Overcoming Automatic Stereotype Activation by Mindset Priming," *Journal of Experimental Social Psychology* 41, no. 5 (2005): 506–14; Kai Sassenberg et al., "Priming Creativity as a Strategy to Increase Creative Performance by Facilitating the Activation and Use of Remote Associations," *Journal of Experimental Social Psychology* 68 (2017): 128–38; Margo Monteith, interview with the author, September 6, 2017; Brian Keith Payne, Heidi Vuletich, and Kristjen Lundberg, "The Bias of Crowds: How Implicit Bias Bridges Personal and Systemic Prejudice," *Psychological Inquiry* 28, no. 4 (2017): 233–48; and Tessa Charlesworth and Mahzarin Banaji, "Patterns of Implicit and Explicit Attitudes: 1. Long-term Change and Stability from 2007–2016," *Psychological Science* 30, no. 2 (2019): 174–92.

46. Russ Fazio, "Multiple Processes by Which Attitudes Guide Behavior: The MODE Model as an Integrative Framework," *Advances in Experimental Social Psychology* 23 (1990): 75–109; Russ Fazio and Tamara Towles-Schwen, "The MODE Model of Attitude-Behavior Processes," in Shelly Chaiken and Yaacov Trope, eds., *Dual-Process Theories in Social Psychology* (New York: Guilford Press, 1999), 97–116; and Ad van Knippenberg, Ap Dijksterhuis, and Diane Vermeulen, "Judgment and Memory of a Criminal Act: The Effects of Stereotypes and Cognitive Load," *European Journal of Social Psychology* 29 (1999): 191–201.

47. Jeffrey Sherman, Bertram Gawronski, and Yaacov Trope, eds., *Dual-Process Theories of the Social Mind* (New York: Guilford Press, 2014), 12–16. Additional information provided by Jeffrey Sherman, interview with the author, January 24, 2018.

48. Shelton, "A Reconceptualization of How We Study Issues of Racial Prejudice."
49. Margo Monteith, "Self-Regulation of Prejudiced Responses: Implications for Progress in Prejudice-Reduction Efforts," *Journal of Personality and Social Psychology* 65, no. 3 (1993): 469–85.

CHAPTER 2. INSIDE THE BIASED BRAIN

1. Annalee Newitz, "Facebook's Ad Platform Now Guesses at Your Race Based on Your Behavior," *Ars Technica*, March 18, 2016, http://arstechnica.com /information-technology/2016/03/facebooks-ad-platform-now-guesses-at -your-race-based-on-your-behavior/.
2. Newitz, "Facebook's Ad Platform."
3. Newitz, "Facebook's Ad Platform."
4. Dave McNary, "'Straight Outta Compton' Tops $200 Million in Worldwide Box Office," *Variety*, November 2, 2015, https://variety.com/2015/film/news /straight-outta-compton-200-million-box-office-1201631627/.
5. John Dovidio, personal correspondence with the author, March 31, 2021; and Will Cox, personal correspondence with the author, March 12, 2021.
6. Wendy Berry Mendes et al., "Threatened by the Unexpected: Physiological Responses During Social Interactions with Expectancy-Violating Partners," *Journal of Personality and Social Psychology* 92, no. 4 (2007): 698–716.
7. Xizhou Xie, Patricia Devine, and Will Cox, "Learning in the Absence of Evidence: Untested Assumptions Perpetuate Stereotyping," poster presented at the Society for Personality and Social Psychology Conference, New Orleans, 2020; and Travis Dixon, "Teaching You to Love Fear: Television News and Racial Stereotypes in a Punishing Democracy," in S. J. Hartnett, ed., *Challenging the Prison Industrial Complex: Activism, Arts, and Educational Alternatives* (Chicago: University of Illinois Press, 2011), 106–23.
8. Rebecca Bigler, interview with the author, December 13, 2017.
9. Lin Bian, Sarah-Jane Leslie, and Andrei Cimpian, "Gender Stereotypes About Intellectual Ability Emerge Early and Influence Children's Interests," *Science* 355, no. 6323 (2017): 389–91; Lacey J. Hilliard and Lynn Liben, "Differing Levels of Gender Salience in Preschool Classrooms: Effects on Children's Gender Attitudes and Intergroup Bias," *Child Development* 81, no. 6 (2010): 1787–98; Miao Qian et al., "Implicit Racial Biases in Preschool Children and Adults from Asia and Africa," *Child Development* 87, no. 1 (2016): 285–96; Andrew Baron and Mahzarin Banaji, "The Development of Implicit Attitudes. Evidence of Race Evaluations from Ages 6 and 10 and Adulthood," *Psychological Science* 17, no. 1 (2006): 53–58; Tara Mandalaywala et al., "The Nature and Consequences of Essentialist Beliefs About Race in Early Childhood," *Child Development* 90, no. 4 (2019): 437–53; and Danielle Perszyk et al., "Bias at the Intersection of Race and Gender: Evidence from Preschool-Aged Children," *Developmental Science* 22, no. 3 (2019):e12788.

10. Rebecca Bigler, interview with the author, December 13, 2017.

11. Sandra Lipsitz Bem, "Gender Schema Theory and Its Implications for Child Development: Raising Gender-Aschematic Children in a Gender-Schematic Society," *Signs: Journal of Women in Culture and Society* 8, no. 4 (1983): 598–616; Rebecca Bigler, "The Role of Classification Skill in Moderating Environmental Influences on Children's Gender Stereotyping: A Study of the Functional Use of Gender in the Classroom," *Child Development* 66, no. 4 (1995): 1072–87; and Rebecca Bigler and Lynn Liben, "Developmental Intergroup Theory: Explaining and Reducing Children's Social Stereotyping and Prejudice," *Current Directions in Psychological Science* 16, no. 3 (2007): 162–66.

12. Rebecca Bigler et al. "Social Categorization and the Formation of Intergroup Attitudes in Children," *Child Development* 68, no. 3 (1997): 530–43; and Rebecca Bigler, interview with the author, December 13, 2017.

13. James Baldwin, "Letter from a Region in My Mind," *New Yorker*, November 17, 1962, https://www.newyorker.com/magazine/1962/11/17/letter-from-a-region-in-my-mind.

14. Rebecca Bigler et al., "When Groups Are Not Created Equal: Effects of Group Status on the Formation of Intergroup Attitudes in Children," *Child Development* 72, no. 4 (2001): 1151–62.

15. Bigler has, unsurprisingly, more than once run afoul of the university committees that ensure experiments are ethical. Ethics boards look askance at experiments involving children. Bigler tends to prevail, arguing that what she does in her experiments is no different from how kids experience everyday life—they're subjected to relentless insistence on categories.

16. Graham Vaughan, Henry Tajfel, and Jennifer Williams, "Bias in Reward Allocation in an Intergroup and Interpersonal Context," *Social Psychology Quarterly* 44, no. 1 (1981): 37–42.

17. Kirsty Smyth et al., "Development of Essentialist Thinking About Religion Categories in Northern Ireland (and the United States)," *Developmental Psychology* 53, no. 3 (2017): 475–96.

18. Brock Bastian and Nick Haslam, "Psychological Essentialism and Stereotype Endorsement," *Journal of Experimental Social Psychology* 42, no. 2 (2006): 1799–813; Kristin Pauker, Nalini Ambady, and Evan Apfelbaum, "Race Salience and Essentialist Thinking in Racial Stereotype Development," *Child Development* 81, no. 6 (2010): 1799–813; and Tara Mandalaywala, David Amodio, and Marjorie Rhodes, "Essentialism Promotes Racial Prejudice by Increasing Endorsement of Social Hierarchies," *Social Psychological and Personality Science* 9, no. 4 (2018): 461–69.

19. Amy Loughman and Nick Haslam, "Neuroscientific Explanations and the Stigma of Mental Disorder: A Meta-analytic Study," *Cognitive Research: Principles and Implications* 3, no. 1 (2018): 43.

20. Travis Dixon and Charlotte Williams, "The Changing Misrepresentation of Race and Crime on Network and Cable News," *Journal of Communication* 65,

no. 1 (2015): 24–39; "Murder: Race, Ethnicity, and Sex of Victim by Race, Ethnicity, and Sex of Offender, 2018 [single victim/single offender]," 2018 Crime in the United States Report, Criminal Justice Information Services Division, the Federal Bureau of Investigation (2018), https://ucr.fbi.gov/crime-in-the-u.s/2018/crime-in-the-u.s.-2018/tables/expanded-homicide-data-table-6.xls; and George Yancy, *Look, a White! Philosophical Essays on Whiteness* (Philadelphia: Temple University Press, 2012), 33.

21. Amy Krosch and David Amodio, "Economic Scarcity Alters the Perception of Race," *Proceedings of the National Academy of Sciences* 111 (2014): 9079–84; Kerry Kawakami, David Amodio, and Kurt Hugenberg, "Intergroup Perception and Cognition: An Integrative Framework for Understanding Causes and Consequences of Social Categorization," *Advances in Experimental Social Psychology* 55 (2017): 1–59; Jon Maner et al., "Functional Projection: How Fundamental Social Motives Can Bias Interpersonal Perception," *Journal of Personality and Social Psychology* 88, no. 1 (2005): 63–78; and Jennifer Eberhardt et al., "Seeing Black: Race, Crime, and Visual Processing," *Journal of Personality and Social Psychology* 87, no. 6 (2004): 876–93. See also Jennifer Eberhardt and Phillip Atiba Goff, "Seeing Race," in *Social Psychology of Prejudice: Historical and Contemporary Issues*, ed. Christian Crandall and Mark Schaller, 219–32 (Lawrence, KS: Lewinian Press, 2004).

22. Sabine Lang, "Various Kinds of Two-Spirit People: Gender Variance and Homosexuality in Native American Communities," in *Two-Spirit People: Native American Gender Identity, Sexuality, and Spirituality*, ed. Sue-Ellen Jacobs, Wesley Thomas, and Sabine Lang (Urbana: University of Illinois Press, 1997), 100–115; and Sharyn Graham Davies, *Challenging Gender Norms: Five Genders Among Bugis in Indonesia* (Belmont, CA: Cengage Learning, 2006), 1–30.

23. Stuart Tyson Smith, "Nubian and Egyptian Ethnicity," in *A Companion to Ethnicity in the Ancient Mediterranean* (Hoboken, NJ: John Wiley & Sons, 2014), 194–212.

24. David Brion Davis, "Looking at Slavery from Broader Perspectives," *American Historical Review* 105, no. 2 (2000): 452, 461; David Brion Davis, "A Big Business," *New York Review of Books*, June 11, 1998, https://www.nybooks.com/articles/1998/06/11/a-big-business/; Ibram X. Kendi, *Stamped from the Beginning: The Definitive History of Racist Ideas in America* (New York: Bold Type Books, 2017), 24; Gomes de Zurara, *The Chronicle of the Discovery and Conquest of Guinea*, vol. 1, trans. Charles Beazley and Edgar Prestage (reprint, New York: Burt Franklin, 1963), 82; and David Brion Davis, "The Culmination of Racial Polarities and Prejudice," *Journal of the Early Republic* 19, no. 4 (1999): 757–75.

25. François Bernier, quoted in David Baum, *The Rise and Fall of the Caucasian Race: A Political History of Racial Identity* (New York: New York University Press, 2006), 52; Daniel Segal, "'The European': Allegories of Racial Purity," *Anthropology*

Today 7, no. 5 (1991): 7–9; and Winthrop Jordan, *White over Black: American Attitudes Toward the Negro, 1550–1812*, 2nd ed. (Chapel Hill: University of North Carolina Press, 2012), 95. See also *Records of the Colony of Rhode Island and Providence Plantations, in New England*, vol. 1, *1636–1663*, ed. John Russell Bartlett (Providence, RI: A. C. Greene and Brothers, 1856), 243.

26. Baum, *The Rise and Fall of the Caucasian Race*, 76, 84–85, 88.

27. Matthew Frye Jacobson, *Whiteness of a Different Color: European Immigrants and the Alchemy of Race* (Cambridge, MA: Harvard University Press, 1999), 6–8; Baum, *The Rise and Fall of the Caucasian Race*, 48, 144.

28. Rebecca Bigler, interview with the author, December 13, 2017.

29. Allison Thomason, "Her Share of the Profits: Women, Agency, and Textile Production at Kültepe/Kanesh in the Early Second Millennium BC," in *Textile Production and Consumption in the Ancient Near East. Archaeology, Epigraphy and Iconography*, ed. Mary-Louise Nosch, Henriette Koefoed, and Eva Andersson Strand (Oxford: Oxbow, 2013), 93–112.

30. Robert W. Venables, "The Clearings and the Woods: The Haudenosaunee (Iroquois) Landscape: Gendered and Balanced," in *Archaeology and Preservation of Gendered Landscapes*, ed. Sherene Baugher and Suzanne M. Spencer-Wood (New York: Springer, 2010), 21–55; and Doug George-Kanentiio, *Iroquois Culture and Commentary* (Ann Arbor, MI: Clear Light Publishers, 2000), 55. See also Judith Brown, "Economic Organization and the Position of Women among the Iroquois," *Ethnohistory* 17 (1971): 153.

31. Sir William Johnson, *Papers of Sir William Johnson*, vol. 3 (New York: University of the State of New York, 1921), 707–8.

32. Johnson, *Papers of Sir William Johnson*, vol. 3, 711.

33. June Namias, *White Captives: Gender and Ethnicity on the American Frontier* (Chapel Hill: University of North Carolina Press, 1993), 88, 284.

34. Stuart Tyson Smith, "Nubian and Egyptian Ethnicity," in *A Companion to Ethnicity in the Ancient Mediterranean* (Hoboken, NJ: John Wiley & Sons, 2014), 194–212; Renée Friedman, "Nubians at Hierakonpolis: Week 1: Maiherpra," *Archaeology Magazine*, April 2009, https://interactive.archaeology .org/hierakonpolis/field/maiherpa.html; and "Special Exhibit: The Tomb of Maihipre" and "View the Treasures of Maihirpre from KV 46," November 10, 2007, https://web.archive.org/web/20071110225500/http://www.geocities .com/royalmummies/Maihirpre/Maihirpre.htm.

35. Hippocrates, *Hippocrates*, vol. 1, trans. W. H. S. Jones (Cambridge, MA: Harvard University Press, 1923), 46; Marcus Vitruvius, quoted in *The Architecture of M. Vitruvius Pollio in Ten Books*, trans. J. Gwilt (London: Priestly and Weale, 1826), 168; and Flavius Vegetius Renatus, "The Military Institutions of the Romans (De Re Militari)," trans. John Clark, accessed February 26, 2018, http://www .digitalattic.org/home/war/vegetius/.

36. Marcus Vitruvius, quoted in *The Architecture of M. Vitruvius Pollio in Ten Books*, 16; and John Percy Vyvian Dacre Balsdon, *Romans and Aliens* (Chapel Hill:

University of North Carolina Press, 1980), 214; and Florus, *Epitome of Roman History*, trans. Edward Forster (Cambridge, MA: Harvard University Press, 1929), http://penelope.uchicago.edu/Thayer/E/Roman/Texts/Florus/Epitome/1F*.html.

37. Benjamin Isaac, *The Invention of Racism in Classical Antiquity* (Princeton, NJ: Princeton University Press, 2006), 83–85; Marcus Vitruvius, *The Ten Books on Architecture*, 6.1, trans. Joseph Gwilt https://lexundria.com/vitr/6.1/gw; and Pliny, quoted in Rebecca Futo Kennedy, *Race and Ethnicity in the Classical World: An Anthology of Primary Sources in Translation* (Indianapolis: Hackett, 2013), 48. Skin color had other meanings, too. In some cases, light skin was considered effeminate. Greek women ideally had pale skin—a sign that they were able to stay inside—while elite Greek men, exercising outdoors, were supposed to have darker skin. Among lower classes, the skin colors reversed, as men who were blacksmiths worked indoors, and women selling goods in the market worked outdoors. Reversing skin color ideals signaled lower classes. Apuleius, quoted in Kennedy, *Race and Ethnicity in the Classical World*, 49.

38. Margaret Talbot, "The Myth of Whiteness in Classical Sculpture," *New Yorker*, October 22, 2018, https://www.newyorker.com/magazine/2018/10/29/the-myth-of-whiteness-in-classical-sculpture; Mark B. Abbe, "Polychromy of Roman Marble Sculpture," The Met's Heilbrunn Timeline of Art History, April 2007, https://www.metmuseum.org/toah/hd/prms/hd_prms.htm; and Florus, *Epitome of Roman History*.

39. Toni Morrison, *The Origin of Others* (Cambridge, MA: Harvard University Press, 2017), 18–21; and Nell Irvin Painter, "Toni Morrison's Radical Vision of Otherness," *New Republic*, October 11, 2017, https://newrepublic.com/article/144972/toni-morrisons-radical-vision-otherness-history-racism-exclusion-whiteness.

40. Tannis MacBeth Williams, *The Impact of Television: A Natural Experiment Involving Three Communities* (Orlando, FL: Academic Press, 1986), 2, 275–92, 399; and Laurie Rudman, "Sources of Implicit Attitudes," *Current Directions in Psychological Science* 13, no. 2 (2004): 79–82.

41. Nicola Döring, Anne Reif, and Sandra Poeschl, "How Gender Stereotypical Are Selfies? A Content Analysis and Comparison with Magazine Adverts," *Computers in Human Behavior* 55, B (2016): 955–62; and Laurie Rudman, Julie Phelan, and Jessica Heppen, "Developmental Sources of Implicit Attitudes," *Journal of Personality and Social Psychology Bulletin* 33, no. 12 (2007): 1700–1713.

42. Mark Schaller, "Sample Size, Aggregation, and Statistical Reasoning in Social Interference," *Journal of Experimental Social Psychology* 28, no. 1 (1992): 65–85; Pedro Bordalo, Katherine Baldiga Coffman, Nicola Gennaioli, and Andrei Shleifer, "Stereotypes," *Quarterly Journal of Economics* 131, no. 4 (2016): 1753–94; and Douglas J. Ahler and Gaurav Sood, "The Parties in Our Heads: Misperceptions About Party Composition and Their Consequences," *Journal of Politics* 80, no. 3 (2018): 964–81.

43. Alex Nowrasteh, "Criminal Immigrants in Texas," *Immigrant Research and Policy Brief*, no. 4 (August 27, 2019), https://www.cato.org/publications/immigration-research-policy-brief/criminal-immigrants-texas-2017-illegal-immigrant; Aaron Chalfin, "Do Mexican Immigrants 'Cause' Crime?," University of Pennsylvania School of Arts and Sciences (accessed March 26, 2021), https://crim.sas.upenn.edu/fact-check/do-mexican-immigrants-cause-crime. See also Christopher Ingraham, "Two Charts Demolish the Notion That Immigrants Here Illegally Commit More Crime," *Washington Post*, June 19, 2018, https://www.washingtonpost.com/news/wonk/wp/2018/06/19/two-charts-demolish-the-notion-that-immigrants-here-illegally-commit-more-crime/; Kristin Butcher and Anne Morrison Piehl, "Crime, Corrections, and California: What Does Immigration Have to Do with It?," *Public Policy Institute of California Population Trends and Profiles* 9, no. 3 (2008): 1-24; Aaron Chalfin, "The Long-Run Effect of Mexican Immigrants on Crime in U.S. Cities: Evidence from Variation in Mexican Fertility Rates," *American Economic Review* 105, no. 5 (2015): 220–25; John MacDonald and Jessica Saunders, "Are Immigrant Youth Less Violent? Specifying the Reasons and Mechanisms," *Annals of the American Academy of Political and Social Science* 641, no. 1 (2012): 125–47; and Erin Kearns et al., "Why Do Some Terrorist Attacks Receive More Media Attention than Others?," *Justice Quarterly* 36, no. 6 (2018): 985–1022, https://www.erinmkearns.com/uploads/2/4/5/5/24559611/kearnsbetuslemieux.2018.jq.mediacoverageterrorism.pdf.

44. Aaron Chalfin and Monica Deza, "Immigration Enforcement, Crime, and Demography," *Criminology and Public Policy* 19, no. 2 (2020): 515–62.

45. Travis Dixon, "Good Guys Are Still Always in White? Positive Change and Continued Misrepresentation of Race and Crime on Local Television News," *Communication Research* 44, no. 6 (2017): 775–92; Travis Dixon and Daniel Linz, "Overrepresentation and Underrepresentation of African Americans and Latinos as Lawbreakers on Television News," *Journal of Communication* 50, no. 2 (2000): 131–54; Robert Entman and Andrew Rojecki, *The Black Image in the White Mind: Media and Race in America* (Chicago: University of Chicago Press, 2000), 78–93; Franklin D. Gilliam and Shanto Iyengar, "Prime Suspects: The Influence of Local Television News on the Viewing Public," *American Journal of Political Science* 44, no. 3 (2000): 560; and Travis Dixon, "Crime News and Racialized Beliefs: Understanding the Relationship Between Local News Viewing and Perceptions of African Americans and Crime," *Journal of Communication* 58, no. 1 (2008): 106–25.

46. Martha Lauzen, "It's a Man's (Celluloid) World: Portrayals of Female Characters in the Top Grossing Films of 2019," Center for the Study of Women in Television and Film, San Diego State University, 2019; D. B. Jones, "Quantitative Analysis of Motion Picture Content," *Public Opinion Quarterly* 6, no. 3 (1942): 411–28; and Hanah Anderson and Matt Daniels, "Film Dialogue from

2,000 Screenplays, Broken Down by Gender and Age," *Pudding*, April 2016, https://pudding.cool/2017/03/film-dialogue/.

47. Jessica Nordell, "Stop Giving Digital Assistants Female Voices," *New Republic*, June 23, 2016; and Adrienne LaFrance, "Why Do So Many Digital Assistants Have Female Names?," *Atlantic*, March 30, 2016, https://www.theatlantic.com/technology/archive/2016/03/why-do-so-many-digital-assistants-have-feminine-names/475884/.

48. Yanna J. Weisberg, Colin G. DeYoung, and Jacob B. Hirsh, "Gender Differences in Personality Across the Ten Aspects of the Big Five," *Frontiers in Psychology* 2 (2011): 178.

49. P. T. Costa and R. R. McCrae, "Normal Personality Assessment in Clinical Practice: The NEO Personality Inventory," *Psychological Assessment* 4, no. 1 (1992): 5–13; Madeline Heilman and Julie Chen, "Same Behavior, Different Consequences: Reactions to Men's and Women's Altruistic Citizenship Behavior," *Journal of Applied Psychology* 90, no. 3 (2005): 431–41; and Kieran Snyder, "The Abrasiveness Trap," *Fortune*, August 26, 2014.

50. Catherine Eckel and Philip Grossman, "Men, Women and Risk Aversion: Experimental Evidence," in *Handbook of Experimental Economics Results* (Amsterdam: Elsevier, 2008), 1061–73; Selwyn Becker and Alice Eagly, "The Heroism of Women and Men," *American Psychologist* 59, no. 3 (2004): 163–78; and Geir Mjøen et al., "Long-Term Risks for Kidney Donors," *Kidney International* 86, no. 1 (2014): 162–67, https://doi.org/10.1038/ki.2013.460.

51. Timothy Judge, Beth Livingston, and Charlice Hurst, "Do Nice Guys—and Gals—Really Finish Last? The Joint Effects of Sex and Agreeableness on Income," *Journal of Personality and Social Psychology* 102, no. 2 (2012): 390–407.

52. "Violence Against Women," World Health Organization Fact Sheet, November 29, 2017; and Linda Gorman, "How Childbearing Affects Women's Wages," *National Bureau of Economic Research* 4 (2011).

53. Amy Chua and Jed Rubenfeld, *The Triple Package: How Three Unlikely Traits Explain the Rise and Fall of Cultural Groups in America* (New York: Penguin Press, 2014).

54. Min Zhou and Jennifer Lee, "Hyper-Selectivity and the Remaking of Culture: Understanding the Asian American Achievement Paradox," *Asian American Journal of Psychology* 8, no. 1 (2017): 7–15; and Alejandro Portes, Rosa Aparicio, and William Haller, *Spanish Legacies: The Coming of Age of the Second Generation* (Berkeley: University of California Press, 2016), 99–101.

55. Andrei Cimpian and Erika Salomon, "The Inherence Heuristic: An Intuitive Means of Making Sense of the World, and a Potential Precursor to Psychological Essentialism," *Behavioral and Brain Sciences* 37, no. 5 (2014): 461–80.

56. George Fredrickson, *The Arrogance of Race* (Middletown, CT: Wesleyan University Press, 1988), 211.

57. Cameron Nickels, *Civil War Humor* (Jackson: University Press of Mississippi),

138–39; "Quashee's Dream of Emancipation," *Frank Leslie's Illustrated Newspaper*, March 1863.

58. Information provided by John Mason, interview with the author, February 2, 2018.

59. David Levering Lewis, interview with the author, January 30, 2018; Fredrickson, *The Arrogance of Race*, 215; "Big Film Shown," *Topeka State Journal*, June 16, 1917; "Birth of a Nation Greatest of Films," *Hattiesburg News*, May 12, 1916; and "Birth of a Nation Is a Product of Great Merit," *Ogden Standard*, March 27, 1916.

60. Meline Toumani, *There Was and There Was Not: A Journey Through Hate and Possibility in Turkey, Armenia, and Beyond* (London: Picador, 2015), 133–34.

61. Inas Deeb et al., "The Development of Social Essentialism: The Case of Israeli Children's Inferences About Jews and Arabs," *Child Development* 81, no. 3 (2010): 757–77.

CHAPTER 3. HOW MUCH DOES EVERYDAY BIAS MATTER?

1. Nellie Bowles and Liz Gannes, "All-Male Ski Trip and No Women at Al Gore Dinner: Kleiner's Chien Takes the Stand in Pao Lawsuit," Vox.com, February 25, 2015, https://www.vox.com/2015/2/25/11559418/all-male-ski-trip-and -no-women-at-al-gore-dinner-kleiners-chien-takes; Pao vs. Kleiner Perkins Caufield & Byers LLC, A136090 (Cal. Ct. App. June 26, 2013).

2. Mark Sullivan, "Ellen Pao Wanted Kleiner Perkins to Invest in Twitter in 2008. Kleiner Perkins Passed," Venturebeat.com, March 24, 2015, https:// venturebeat.com/2015/03/24/ellen-pao-wanted-kleiner-perkins-to-invest -in-twitter-in-2008-kleiner-perkins-passed/; Nellie Bowles and Liz Gannes, "Performances Review Rewrites and Pao's 'Genetic Makeup': Pao vs. Kleiner Perkins Trial Day 4," Vox.com, February 27, 2015, https://www.vox.com/2015 /2/27/11559518/a-performance-review-that-changed-dramatically-and-paos -400000; and Nellie Bowles and Liz Gannes, "Kleiner's Matt Murphy on Why Pao Failed as a VC," Vox.com, March 19, 2015, https://www.vox.com/2015/3 /19/11560516/kleiners-matt-murphy-on-why-pao-failed-as-a-vc.

3. Liz Gannes and Nellie Bowles, "FAQ: What Happens Now in the Ellen Pao/ Kleiner Perkins Trial?," Vox.com, March 26, 2015, https://www.vox.com/2015 /3/26/11560718/faq-what-happens-now-in-the-ellen-paokleiner-perkins-trial; Vauhini Vara, "The Ellen Pao Trial: What Do We Mean by 'Discrimination'?," *New Yorker*, March 14, 2015, https://www.newyorker.com/business/currency/the-ellen -pao-trial-what-do-we-mean-by-discrimination; and juror statement, retrieved on November 13, 2017, https://www.youtube.com/watch?v=uenVcbi4tIc.

4. Wal-Mart v. Dukes et al., 564 U.S. 338 (2011); Reed Abelson, "6 Women Sue Wal-Mart, Charging Job and Promotion Bias," *New York Times*, June 20, 2001, https:// www.nytimes.com/2001/06/20/business/6-women-sue-wal-mart-charging-job -and-promotion-bias.html; Washington Post Editors, "Rundown: Wal-Mart Sex Discrimination Suit Goes to Supreme Court," *Washington Post*, June 20, 2011,

https://www.washingtonpost.com/blogs/political-economy/post/rundown-wal
-mart-sex-discrimination-suit-goes-to-supreme-court/2011/03/25/AFrOw0nB
_blog.html; Lisa Featherstone, *Selling Women Short: The Landmark Battle for Work-*
ers' Rights at Wal-Mart (New York: Basic Books, 2005), 2–3; and Joan C. Williams
et al., "You Can't Change What You Can't See: Interrupting Racial and Gender
Bias in the Legal Profession" Commission on Women in the Profession, American Bar Association, 2018, https://www.mcca.com/wp-content/uploads/2018/09/
You-Cant-Change-What-You-Cant-See-Executive-Summary.pdf.

5. Wal-Mart v. Dukes et al., 564 U.S. 338 (2011); and David Yeager et al., "Loss
   of Institutional Trust Among Racial and Ethnic Minority Adolescents: A Consequence of Procedural Injustice and a Cause of Life-Span Outcomes," *Child
   Development* 88, no. 3 (2017): 1033.

6. Jason Okonofua and Jennifer Eberhardt, "Two Strikes: Race and the Disci-
   plining of Young Students," *Psychological Science* 26, no. 5 (2015): 617–24; and
   Jason A. Okonofua, Gregory M. Walton, and Jennifer L. Eberhardt, "A Vicious
   Cycle: A Social-Psychological Account of Extreme Racial Disparities in School
   Discipline," *Perspectives on Psychological Science* 11, no. 3 (2016): 381–98. See
   also Dylan Glover, Amanda Pallais, and William Pariente, "Discrimination as
   a Self-Fulfilling Prophecy: Evidence from French Grocery Stores," *Quarterly
   Journal of Economics* 132, no. 3 (2017): 1219–60.

7. Kathryn Monahan et al., "From the School Yard to the Squad Car: School
   Discipline, Truancy, and Arrest," *Journal of Youth and Adolescence* 43, no. 7
   (2014): 1110–22.

8. Emily Singer, "Ants Build Complex Structures with a Few Simple Rules," *Quanta
   Magazine*, April 9, 2014, https://www.quantamagazine.org/ants-build-complex
   -structures-with-a-few-simple-rules-20140409/; David Green, *Of Ants and
   Men: The Unexpected Side Effects of Complexity in Society* (Berlin: Springer-
   Verlag, 2014); and Danie Strömbom and Audrey Dussutour, "Self-Organized
   Traffic via Priority Rules in Leaf-Cutting Ants," *PLOS Computational Biology*
   14, no. 10 (2018): e1006523.

9. Richard Martell, David Lane, and Cynthia Emrich, "Male-Female Differences:
   A Computer Simulation," *American Psychological Association* 51, no. 2 (1996):
   157; and Joan Williams and Rachel Dempsey, *What Works for Women at Work*
   (New York: New York University Press, 2014).

10. Thomas Schelling, "Some Fun, Thirty-Five Years Ago," *Handbook of Computa-
    tional Economics* 2 (2006): 1640–44.

11. Thomas Schelling, "Dynamic Models of Segregation," *Journal of Mathematical
    Sociology* 1, no. 1 (1971): 143–86. See also Junfu Zhang, "Tipping and Resi-
    dential Segregation: A Unified Schelling Model," *Journal of Regional Science*
    51, no. 1 (2011): 167–93.

12. Kimberlé Crenshaw, "A Black Feminist Critique of Antidiscrimination Doctrine,
    Feminist Theory and Antiracist Politics," in *Feminist Legal Theory: Foundations*,
    ed. D. K. Weisbert (Philadelphia: Temple University Press, 1993), 383–95.

13. Williams et al., "You Can't Change What You Can't See"; Robert Livingston, Ashleigh Shelby Rosette, and Ella Washington, "Can an Agentic Black Woman Get Ahead? The Impact of Race and Interpersonal Dominance on Perceptions of Female Leaders," *Psychological Science* 23, no. 354 (2012): 354–58; Erin Cooley et al., "Bias at the Intersection of Identity: Conflicting Social Stereotypes of Gender and Race Augment the Perceived Femininity and Interpersonal Warmth of Smiling Black Women," *Journal of Experimental Social Psychology* 74: 43–49; and Jennifer Berdahl and Celia Moore, "Workplace Harassment: Double Jeopardy for Minority Women," *Journal of Applied Psychology* 9 (2006): 426–36.

14. Robert Livingston and Ashleigh Shelby Rosette, "Stigmatization, Subordination, or Marginalization? The Complexity of Social Disadvantage across Gender and Race," *Inclusive Leadership: Transforming Diverse Lives, Workplaces, and Societies*, ed. Bernardo Ferdman, Jeanine Prime, and Ronald Riggio (New York: Routledge, 2021).

15. Erika Hall et al., "MOSAIC: A Model of Stereotyping Through Associated and Intersectional Categories," *Academy of Management Review* 44, no. 3 (2019): 643–72; Phillip Atiba Goff et al., "'Ain't I a Woman?': Towards an Intersectional Approach to Person Perception and Sub-Category-Based Harms," *Sex Roles* 59 (2008): 392–403; and Jennifer Berdahl and Ji-A Min, "Prescriptive Stereotypes and Workplace Consequences for East Asians in North America," *Cultural Diversity and Ethnic Minority Psychology* 18: 141–52.

16. Emilio Castilla, "Accounting for the Gap: A Firm Study Manipulating Organizational Accountability and Transparency in Pay Decisions," *Organization Science* 26, no. 2 (2015): 311–33; Corinne Moss-Racusin et al., "Science Faculty's Subtle Gender Biases Favor Male Students," *Proceedings of the National Academy of Sciences* 109, no. 41 (2012): 16474–79; and Heather Sarsons, "Interpreting Signals in the Labor Market: Evidence from Medical Referrals" (working paper, Harvard University, Cambridge, MA, 2017). See also Emilio Castilla, "Gender, Race, and Meritocracy in Organizational Careers," *American Journal of Sociology* 113, no. 6 (2008): 1479–526.

17. Mark Egan, Gregor Matvos, and Amit Seru, "When Harry Fired Sally: The Double Standard in Punishing Misconduct" (Harvard Business School Finance Working Paper No. 19-047, March 13, 2017, updated October 17, 2018), https://papers.ssrn.com/abstract=2931940; Ashleigh Shelby Rosette, Robert Livingston, "Failure Is Not an Option for Black Women: Effects of Organizational Performance on Leaders with Single Versus Dual-Subordinate Identities," *Journal of Experimental Social Psychology* 48 (no. 5) 2012: 1162–67.

18. Madeline Heilman and Michelle Haynes, "No Credit Where Credit Is Due: Attributional Rationalization of Women's Success in Male-Female Teams," *Journal of Applied Psychology* 90, no. 5 (2005): 905–16; Heather Sarsons, "Recognition for Group Work: Gender Differences in Academia," *American*

Economic Review 107, no. 5 (2017): 141–45; and Björk, "The Invisible Woman: A Conversation with Björk," interviewed by Jessica Hopper, pitchfork.com, January 21, 2015, https://pitchfork.com/features/interview/9582-the-invisible -woman-a-conversation-with-bjork/.

19. Monica Biernat, M. J. Tocci, and Joan Williams, "The Language of Performance Evaluations: Gender-Based Shifts in Content and Consistency of Judgment," *Social Psychological and Personal Science* 3, no. 2 (2012): 186–92; Joan Williams, Katherine Phillips, and Erika Hall, "Tools for Change: Boosting the Retention of Women in the STEM Pipeline," *Journal of Research in Gender Studies* 6 (2016): 1–75; Trae Vassallo et al., "Elephant in the Valley," distributed by Women in Tech (2017), https://www.elephantinthevalley.com; and Ellen Pao, *Reset: My Fight for Inclusion and Lasting Change* (New York: Random House, 2017), 110–14.

20. American Bar Association Commission on Women in the Profession, prepared by Janet E. Gans Epner, *Visible Invisibility: Women of Color in Law Firms* (Chicago: American Bar Association, 2006); Joan Williams and Rachel Dempsey, *What Works for Women at Work* (New York: New York University Press, 2014), 106; Maureen Dowd, "The Women of Hollywood Speak Out," *New York Times Magazine*, November 20, 2015; and Stacey L. Smith et al., *Inclusion in the Director's Chair: Gender, Age & Race of Directors Across 1,200 Top Films from 2007 to 2018*, USC Annenberg Inclusion Initiative, January 2019.

21. Paul Sackett, Cathy DuBois, and Ann Wiggins Noe, "Tokenism in Performance Evaluation: The Effects of Work Group Representation on Male-Female and White-Black Differences in Performance Ratings," *Journal of Applied Psychology* 76, no. 2 (1991): 263–67; Monique Lortie-Lussier and Natalie Rinfret, "The Proportion of Women Managers: Where Is the Critical Mass?," *Journal of Applied Social Psychology* 32, no. 9 (2002): 1974–91; and Asya Pazy and Israela Oron, "Sex Proportion and Performance Evaluation Among High-Ranking Military Officers," *Journal of Organizational Behavior* 22, no. 6 (2001): 689–702.

22. Williams et al., "You Can't Change What You Can't See"; and Liane Jackson, "Minority Women Are Disappearing from Big Law—and Here's Why," *ABA Journal*, March 1, 2016, https://www.abajournal.com/magazine/article/minority _women_are_disappearing_from_biglaw_and_heres_why.

23. Toni Schmader, "Stereotype Threat Deconstructed," *Current Directions in Psychological Science* 19, no. 1 (2010): 14–18; Kristen Jones et al., "Not So Subtle: A Meta-analytic Investigation of the Correlates of Subtle and Overt Discrimination," *Journal of Management* 42, no. 6 (2013): 1588–613; Jessica Salvatore and J. Nicole Shelton, "Cognitive Costs of Exposure to Racial Prejudice," *Psychological Science* 18, no. 9 (2007): 810–15; Sarah Singletary, "The Differential Impact of Formal and Interpersonal Discrimination on Job Performance" (PhD diss., Rice University, Houston, TX, 2009); Benoit Dardenne, Muriel Dumont, and Thierry Bollier, "Insidious Dangers of Benevolent Sexism: Consequences for Women's Performance," *Journal of Personal Psychology* 93, no. 5 (2007): 764–79; and Sarah

Singletary and Mikki Hebl, "Compensatory Strategies for Reducing Interpersonal Discrimination: The Effectiveness of Acknowledgments, Increased Positivity, and Individuating Information," *Journal of Applied Psychology* 94, no. 3 (2009): 797–805. See also Theresa Vesico et al., "Power and the Creation of Patronizing Environments: The Stereotype-Based Behaviors of the Powerful and Their Effects on Female Performance in Masculine Domains," *Journal of Personality and Social Psychology* 88, no. 4 (2005): 658–72.

24. Robert Rosenthal and Lenore Jacobson, "Pygmalion in the Classroom," *Urban Review* 3, no. 1 (1968): 16–20; and Mitchell Leslie, "The Vexing Legacy of Lewis Terman," *Stanford Magazine*, July/August 2000, https://stanfordmag.org/contents/the-vexing-legacy-of-lewis-terman.

25. Daan Van Knippenberg and Michaéla Schippers, "Work Group Diversity," *Annual Review of Psychology* 58 (2007): 515–41; Elizabeth Mannix and Margaret Neale, "What Differences Make a Difference? The Promise and Reality of Diverse Teams in Organizations," *Psychological Science in the Public Interest* 6, no. 2 (2005): 31–55; Marie-Élène Roberge and Rolf van Dick, "Recognizing the Benefits of Diversity: When and How Does Diversity Increase Group Performance?," *Human Resource Management Review* 20, no. 4 (2010): 295–308; Alina Lungeanu and Noshir Contractor, "The Effects of Diversity and Network Ties on Innovations: The Emergence of a New Scientific Field," *American Behavioral Scientist* 59, no. 5 (2015): 548–64; Evan Apfelbaum, Katherine Phillips, and Jennifer Richeson, "Rethinking the Baseline in Diversity Research: Should We Be Explaining the Effects of Homogeneity?," *Perspectives on Psychological Science* 9, no. 3 (2014): 235–44; and Samuel Sommers, "On Racial Diversity and Group Decision Making: Identifying Multiple Effects of Racial Composition on Jury Deliberations," *Journal of Personal and Social Psychology* 90, no. 4 (2006): 597–612.

26. Brendon Larson, *Metaphors for Environmental Sustainability: Redefining Our Relationship with Nature* (New Haven, CT: Yale University Press, 2014), 83.

27. Arline Geronimus, "The Weathering Hypothesis and the Health of African-American Women and Infants: Evidence and Speculations," *Ethnicity & Disease* 2, no. 3 (1992): 207–21; Phillip Bickler, John Feiner, and John Severinghaus, "Effects of Skin Pigmentation on Pulse Oximeter Accuracy at Low Saturation," *Anesthesiology* 102, no. 4 2005: 715–19; and Edwin Nieblas-Bedolla et al., "Changing How Race Is Portrayed in Medical Education: Recommendations from Medical Students," *Academic Medicine* 95, no. 12 (2020): 1802–6.

CHAPTER 4. BREAKING THE HABIT

1. John Ridley Stroop, "Studies of Interference in Serial Verbal Reactions," *Journal of Experimental Psychology* 18, no. 6 (1935): 643–62. Details about the Prejudice Habit-Breaking Workshop gathered in person, February 23, 2017.

2. David Miller, "Tech Companies Spend Big Money on Bias Training—but

It Hasn't Improved Diversity Numbers," *Conversation*, July 10, 2015, http://theconversation.com/tech-companies-spend-big-money-on-bias-training-but-it-hasnt-improved-diversity-numbers-44411.

3. Katerina Bezrukova et al., "A Meta-analytical Integration of over 40 Years of Research on Diversity Training Evaluation," *Psychological Bulletin* 142, no. 11 (2016): 1227–74.

4. Frank Dobbin and Alexandra Kalev, "Why Diversity Programs Fail," *Harvard Business Review*, July 1, 2016, https://hbr.org/2016/07/why-diversity-programs-fail. See also Alexandra Kalev, Frank Dobbin, and Erin Kelly, "Best Practices or Best Guesses? Assessing the Efficacy of Corporate Affirmative Action and Diversity Policies," *American Sociological Review* 71, no. 4 (2006): 589–617.

5. Frank Dobbin, Daniel Schrage, and Alexandra Kalev, "Rage Against the Iron Cage: The Varied Effects of Bureaucratic Personnel Reforms on Diversity," *American Sociological Review* 80, no. 5 (2015): 1014–44; Tessa Dover, Brenda Major, and Cheryl Kaiser, "Members of High-Status Groups Are Threatened by Pro-Diversity Organizational Messages," *Journal of Experimental Social Psychology* 62 (2016): 58–67; and Tiffany Brannon et al., "From Backlash to Inclusion for All: Instituting Diversity Efforts to Maximize Benefits Across Group Lines," *Social Issues and Policy Review* 12, no. 1 (2018): 57–90.

6. Kalev, Dobbin, and Kelly, "Best Practices or Best Guesses?," 589–617; and Alfiee Breland-Noble, Twitter thread, August 11, 2020, 1:23 p.m., https://twitter.com/dralfiee/status/1293251668759523337.

7. Elizabeth Paluck, "Interventions Aimed at the Reduction of Prejudice and Conflict," *Oxford Handbook of Intergroup Conflict*, ed. Linda Tropp (Oxford: Oxford University Press, 2012), 179–92.

8. Michael Moritz, "I Got to Silicon Valley with No Grand Plan," interviewed by Emily Chang, *Studio 1.0*, Bloomberg Television, December 2, 2015.

9. Eric Uhlmann and Geoffrey Cohen, "'I Think It, Therefore It's True': Effects of Self-Perceived Objectivity on Hiring Discrimination," *Organizational Behavior and Human Decision Processes* 104, no. 2 (2007): 207–23; and Christopher Begeny et al., "In Some Professions, Women Have Become Well Represented, yet Gender Bias Persists—Perpetuated by Those Who Think It Is Not Happening," *Science Advances* 6, no. 26 (2020): 7814.

10. Victoria Plaut, Kecia Thomas, and Matt Goren, "Is Multiculturalism or Color Blindness Better for Minorities?," *Psychological Science* 20, no. 4 (2009): 444–45. See also Jacqueline Yi, Nathan R. Todd, and Yara Mekawi, "Racial Colorblindness and Confidence in and Likelihood of Action to Address Prejudice," *American Journal of Community Psychology* 65, no. 3–4 (2020): 407–22.

11. David Amodio, "Coordinated Roles of Motivation and Perception in the Regulation of Intergroup Responses: Frontal Cortical Asymmetry Effects on the P2 Event-Related Potential and Behavior," *Journal of Cognitive Neuroscience* 22, no. 11 (2010): 2609–17. See also Ashley Martin and Katherine Phillips, "What 'Blindness' to Gender Differences Helps Women See and Do: Implications for

Confidence, Agency, and Action in Male-Dominated Environments," *Organizational Behavior and Human Decision Processes* 142 (2017): 28–44.

12. John Darley and Paget Gross, "A Hypothesis-Confirming Bias in Labeling Effects," *Journal of Personality and Social Psychology* 44, no. 1 (1983): 20–33.

13. Will Cox, interview with the author, February 23, 2017.

14. Sun Joo-Grace Ahn, Amanda Minh Tran Le, and Jeremy Bailenson, "The Effect of Embodied Experiences on Self-Other Merging, Attitude, and Helping Behavior," *Media Psychology* 16, no. 1 (2013): 7–38.

15. Jason Okonofua, David Paunesku, and Gregory Walton, "Brief intervention cuts suspension rates in half," *Proceedings of the National Academy of Sciences* 113, no. 19 (2016): 5221-26; Jason Okonofua et al., "A Scalable Empathic Supervision Intervention to Mitigate Recidivism from Probation and Parole," *Proceedings of the National Academy of Sciences* 118, no. 14 (2021): e2018036118.

16. Patricia Devine et al., "A Gender Bias Habit-Breaking Intervention Led to Increased Hiring of Female Faculty in STEMM Departments," *Journal of Experimental Social Psychology* 73 (2017): 211–15; and Molly Carnes et al., "The Effect of an Intervention to Break the Gender Bias Habit of Faculty at One Institution: A Cluster Randomized, Controlled Trial," *Academic Medicine* 90, no. 2 (2015): 221–30.

17. Patricia Devine et al., "Long-Term Reduction in Implicit Race Bias: A Prejudice Habit-Breaking Intervention," *Journal of Experimental Social Psychology* 48, no. 6 (2012): 1267–78; and Patricia Devine, "Breaking the Prejudice Habit" (workshop, University of Wisconsin, Madison, February 23, 2017).

18. Bezrukova et al., "A Meta-analytical Integration of over 40 Years of Research on Diversity Training Evaluation."

19. Patricia Devine, interview with the author, February 23, 2017.

20. Paul Regier and David Redish, "Contingency Management and Deliberative Decision-Making Processes," *Frontiers in Psychiatry* 6, no. 76 (2015): 1–13.

21. A. David Redish, *The Mind Within the Brain* (Oxford: Oxford University Press, 2013), 38; and Paul Regier, "Decision Making Gone Awry: Dorsal Striatum, Decision-Making, and Addiction" (PhD diss., University of Minnesota, 2015), 1–132.

22. David Redish, interview with the author, July 26, 2017.

23. Milton Rokeach, *The Nature of Human Values* (New York: Free Press, 1973), 215–34.

24. Patrick Forscher, interview with the author, February 24, 2017.

25. Milton Rokeach, "Long-Range Experimental Modification of Values, Attitudes, and Behavior," *American Psychologist* 26, no. 5 (1971): 453–59; and Okonofua et al., "A Scalable Empathic Supervision Intervention to Mitigate Recidivism from Probation and Parole."

26. Will Cox, interview with the author, February 23, 2017.

27. Evelyn Carter, interview with the author, July 23, 2020.

28. Patrick Forscher, "The Individually-Targeted Habit-Breaking Intervention

and Group-Level Change," Thesis Commons, August 4, 2017; and Elizabeth Levy Paluck, "Peer Pressure Against Prejudice: A High School Field Experiment Examining Social Network Change," *Journal of Experimental Social Psychology* 47, no. 2 (2011): 350–58.

29. Edward Chang et al., "The Mixed Effects of Online Diversity Training," *PNAS* 116, no. 16 (2019): 7778–83.

30. Jessica Nelson, Glenn Adams, and Phia Salter, "The Marley Hypothesis: Denial of Racism Reflects Ignorance of History," *Psychological Science* 24, no. 2 (2013): 213–18; Courtney Bonam et al., "Ignoring History, Denying Racism: Mounting Evidence for the Marley Hypothesis and Epistemologies of Ignorance," *Social Psychological and Personality Science* 10, no. 2 (2018): 257–65.

CHAPTER 5. THE MIND, THE HEART, THE MOMENT

1. Tracy Mumford, "An Inventory of Philando Castile's Car: Life, Interrupted," *MPR News*, June 21, 2017, https://www.mprnews.org/story/2017/06/21/an-inventory-of-philando-castiles-car-life-interrupted; and Libor Jany and Anthony Lonetree, "Quiet, Unassuming Philando Castile Was 'Like Mr. Rogers with Dreadlocks,'" *Minneapolis Star Tribune*, July 8, 2016, https://www.startribune.com/quiet-unassuming-philando-castile-was-like-mr-rogers-with-dreadlocks/385892971/.

2. Andy Mannix, "Police Audio: Officer Stopped Philando Castile on Robbery Suspicion," *Minneapolis Star Tribune,* July 12, 2016; Tracy Mumford, Riham Feshir, and Jon Collins, *74 Seconds: The Trial of Officer Jeronimo Yanez,* "Episode 10: Jeronimo Yanez Takes the Stand," *MPR News*, June 16, 2017, https://live.mprnews.org/Event/The_trial_of_officer_Jeronimo_Yanez/969433148.

3. Minnesota Statutes, 2019. 624.714, "Carrying of Weapons Without Permit; Penalties," https://www.revisor.mn.gov/statutes/?id=624.714; and BETnetworks, "Philando Castile's Girlfriend, Diamond Reynolds, Recounts the Tragic Day," youtube.com, April 7, 2019, https://www.youtube.com/watch?v=ynWj07m-ZYc.

4. "Yanez Case Trial Documents," Ramsey County Attorney's Office, Ramsey County, MN, https://www.ramseycounty.us/your-government/leadership/county-attorneys-office/news-updates/case-updates/state-v-jeronimo-yanez/yanez-case-trial-documents.

5. Brianna Provenzano, "This Emotional Letter Written by Philando Castile's Former Student Shows How Loved He Was," July 9, 2016, https://www.mic.com/articles/148291/this-emotional-letter-written-by-philando-castile-s-former-student-shows-how-loved-he-was.

6. Chao Xiong, "Jury Takes Case Against Jeronimo Yanez in Shooting of Philando Castile," *Minneapolis Star-Tribune*, June 13, 2017; and Steve Karnowski, "Attorney: Officer 'Did What He Had to Do' in Driver Shooting," Associated Press, June 12, 2017.

7. "Yanez Case Trial Documents."

8. "Use-of-Force Experts Analyze Castile Shooting Video," *MPR News*, June 21, 2017, https://www.mprnews.org/story/2017/06/21/use-of-force-experts-analyze-castile-yanez-shooting-video; Simon Baldwin et al., "Stress-Activity Mapping: Physiological Responses During General Duty Police Encounters," *Frontiers in Psychology* 10 (2019): 2216; and Mumford, Feshir, and Collins, *74 Seconds*, "Episode 10: Jeronimo Yanez Takes the Stand."

9. Mitch Smith, "Minnesota Officer Acquitted in Killing of Philando Castile," *New York Times*, June 16, 2017, https://www.nytimes.com/2017/06/16/us/police-shooting-trial-philando-castile.html.

10. Connie Rice, interview by Charles Monroe-Kane, "Fearing the Other," *To the Best of Our Knowledge*, Wisconsin Public Radio, February 5, 2015; Connie Rice et al., *Rampart Reconsidered: The Search for Real Reform Seven Years Later*, Los Angeles Police Department, 2006, http://lapd-assets.lapdonline.org/assets/pdf/Rampart%20Reconsidered-Full%20Report.pdf; and interview transcripts provided by Connie Rice to author.

11. Connie Rice, interview by Charles Monroe-Kane, "Fearing the Other."

12. Jennifer Eberhardt et al., "Seeing Black: Race, Crime, and Visual Processing," *Journal of Personality and Social Psychology* 87, no. 6 (2004): 876–93.

13. John Paul Wilson, Kurt Hugenberg, and Nicholas Rule, "Racial Bias in Judgments of Physical Size and Formidability: From Size to Threat," *Journal of Personality and Social Psychology* 113, no. 1 (2017): 59–80; and Phillip Atiba Goff et al., "The Essence of Innocence," *Journal of Personality and Social Psychology* 106, no. 4 (2014): 526–45.

14. Wison, Hugenberg, and Rule, "Racial Bias in Judgments of Physical Size and Formidability."

15. Jaclyn Ronquillo et al., "The Effects of Skin Tone on Race-Related Amygdala Activity: An fMRI Investigation," *Social Cognitive and Affective Neuroscience* 2, no. 1 (2007): 39–44; Cara Talaska, Susan Fiske, and Shelly Chaiken, "Legitimating Racial Discrimination: Emotions, Not Beliefs, Best Predict Discrimination in a Meta-analysis," *Social Justice Research* 21, no. 3 (2008): 263–396; Robert Abelson et al., "Affective and Semantic Components in Political Person Perception," *Journal of Personality and Social Psychology* 42, no. 4 (1982): 619–30; and Emile Bruneau, Mina Cikara, and Rebecca Saxe, "Parochial Empathy Predicts Reduced Altruism and the Endorsement of Passive Harm," *Social Psychological and Personality Science* 8, no. 8 (2017): 934–42.

16. Osagie Obasogie and Zachary Newman, "The Endogenous Fourth Amendment: An Empirical Assessment of How Police Understandings of Excessive Force Become Constitutional Law," *Cornell Law Review* 104 (2018): 1281.

17. Dean Knox, Will Lowe, Jonathan Mummolo, "Administrative Records Mask Racially Biased Policing," *American Political Science Review*, 114, no. 3 (2020), 619–37; and Eyder Peralta and Cheryl Corley, "The Driving Life and Death of Philando Castile," *NPR News*, July 15, 2016.

18. Stephen Holmes et al., "Individual and Situational Determinants of Police Force: An Examination of Threat Presentation," *American Journal of Criminal Justice* 23, no. 1 (1998): 83–106; Justin Nix et al., "A Bird's Eye View of Civilians Killed by Police in 2015," *Criminology & Public Policy* 16, no. 1 (2017): 309–40; Roland Fryer Jr., "An Empirical Analysis of Racial Differences in Police Use of Force," *Journal of Political Economy* 127, no. 3 (published online, 2019); Frank Edwards, Hedwig Lee, and Michael Esposito, "Risk of Being Killed by Police Use of Force in the United States by Age, Race–Ethnicity, and Sex," *Proceedings of the National Academy of Sciences* 116, no. 34 (2019): 16793–98; Vesla Weaver, Gwen Prowse, and Spencer Piston, "Too Much Knowledge, Too Little Power: An Assessment of Political Knowledge in Highly Policed Communities," *Journal of Politics* 81, no. 3 (2019): 1153–66; and Paul Butler, *Chokehold: Policing Black Men* (New York: New Press, 2017), 42. See also Cody Ross, "A Multi-Level Bayesian Analysis of Racial Bias in Police Shootings at the County-Level in the United States, 2011–2014," *PLoS ONE* 10, no. 11 (2015).

19. Frank Edwards, Michael Esposito, Hedwig Lee, "Risk of Police-Involved Death by Race/Ethnicity and Place, United States, 2012–2018," *American Journal of Public Health* 108, no. 9 (2018): 1241–48; and Elle Lett et al., "Racial Inequity in Fatal US Police Shootings, 2015–2020," *Journal of Epidemiology and Community Health,* EPub ahead of print (2020): 1–4.

20. Sally Hadden, *Slave Patrols: Law and Violence in Virginia and the Carolinas* (Cambridge, MA: Harvard University Press, 2001), 6, 23; *The Statutes at Large of South Carolina,* vol. 7, ed. David J. McCord (Columbia, SC: A. S. Johnston, 1840), https://www.carolana.com/SC/Legislators/Documents/The_Statutes_at_Large_of_South_Carolina_Volume_VII_David_J_McCord_1840.pdf; and Jennifer Eberhardt et al., "Looking Deathworthy: Perceived Stereotypicality of Black Defendants Predicts Capital-Sentencing Outcomes," *Psychological Science* 17, no. 5 (2006): 383–86.

21. Cynthia Najdowski, Bette Bottoms, and Phillip Atiba Goff, "Stereotype Threat and Racial Differences in Citizens' Experiences of Police Encounters," *Law and Human Behavior* 39, no. 5 (2015): 463–77; and Kimberly Kahn et al. "Misinterpreting Danger? Stereotype Threat, Pre-attack Indicators, and Police-Citizen Interactions," *Journal of Police Criminal Psychology* 33 (2018): 45–54.

22. Steve Featherstone, "Professor Carnage," *New Republic,* April 17, 2017, https://newrepublic.com/article/141675/professor-carnage-dave-grossman-police-warrior-philosophy; Dave Grossman and Jim Glennon, *The Bulletproof Warrior* (Elmhurst, IL: Calibre Press, 2003); Brian Schatz, "'Are You Prepared to Kill Somebody?' A Day with One of America's Most Popular Police Trainers," *Mother Jones,* March/April 2017, https://www.motherjones.com/politics/2017/02/dave-grossman-training-police-militarization/; Jennifer Bjorhus, "Officer Who Shot Castile Attended 'Bulletproof Warrior' Training," *Minneapolis Star Tribune,* July 14, 2016; and Response to FOIA request number 2-21018910058, National Personnel Records Center, October 5, 2017.

23. "Cause of Law Enforcement Deaths, 2010-2019," National Law Officers' Memorial Fund, updated October 14, 2020, https://nleomf.org/facts-figures/causes-of-law-enforcement-deaths; "Contacts Between Police and the Public, 2015," Bureau of Justice Statistics, October 11, 2018, https://www.bjs.gov/index.cfm?ty=pbdetail&iid =6406.

24. Mumford, Feshir, and Collins, *74 Seconds*, "Episode 16: The Dashcam Video," *MPR News*, June 16, 2017, https://www.mprnews.org/story/2017/06/21/use-of -force-experts-analyze-castile-yanez-shooting-video.

25. Phillip Atiba Goff, "Dr. Phillip Atiba Goff Addresses Racism and Police Violence," Solve—MIT, October 1, 2020, https://www.youtube.com/watch?v =GPEpz9h9nAk; Briana Bierschbach, "Minnesota Lawmakers Pass Sweeping Package of Police Accountability Measures," *Minneapolis Star Tribune*, July 21, 2020, https://www.startribune.com/state-lawmakers-strike-deal-on -police-reform-proposals/571833891/.

26. David Amodio, "The Neuroscience of Prejudice and Stereotyping," *Nature Reviews Neuroscience* 15, no. 10 (2014): 670–82; Amodio, "Intergroup Anxiety Effects on the Control of Racial Stereotypes: A Psychoneuroendocrine Analysis," *Journal of Experimental Social Psychology* 45 (2009): 60–67; and David Amodio, personal correspondence with the author, November 5, 2018.

27. Zak Stone, "To Protect, Serve, and Meditate," *MEL Magazine*, January 21, 2016.

28. Richard Goerling, interview with the author, December 22, 2016; Mindful.org, "Richard Goerling: Mindful Policing," March 16, 2017, https://www.youtube .com/watch?v=JXqjrWcBdJ4&feature=emb_logo; and Bill Radke and Allie Ferguson, "Can Yoga and Meditation Save America's Police Officers?," March 27, 2016, https://kuow.org/stories/can-yoga-and-meditation-save-america-s -police-officers.

29. Maureen O'Hagan, "To Pause and Protect," *Mindful*, April 26, 2016; Richard Goerling, interview with the author, December 22, 2016.

30. Jon Kabat-Zinn, *Coming to Our Senses: Healing Ourselves and the World Through Mindfulness* (New York: Hachette Books, 2005); Mindful.org, "Jon Kabat-Zinn: Me Me Me," January 11, 2017, https://www.mindful.org/jon-kabat-zinn -defining-mindfulness/; and interview with Rhonda Magee by David Treleaven, Trauma-Sensitive Mindfulness Podcast, accessed August 30, 2020.

31. Richard Goerling, interview with the author, December 22, 2016.

32. Shantha Rajaratnam et al., "Sleep Disorders, Health, and Safety in Police Officers," *JAMA* 306, no. 23 (2011): 2567–78; Zhen Wang et al., "A Prospective Study of Predictors of Depression Symptoms in Police," *Psychiatry Research* 175, no. 3 (2010): 211–16; Parveen Joseph et al., "Police Work and Subclinical Atherosclerosis," *Journal of Occupational and Environmental Medicine* 51, no. 6 (2009): 700–707; Tara Hartley et al., "Health Disparities in Police Officers: Comparisons to the U.S. General Population," *International Journal of Emergency Mental Health* 13, no. 4 (2013): 211–20; Katelyn Jetelina et al., "Prevalence of Mental

Illness and Mental Health Care Use Among Police Officers," *JAMA Network Open* 3, no. 10 (2020): e2019658; "Census of Fatal Occupational Injuries, 2018," Bureau of Labor Statistics, https://www.bls.gov/news.release/cfoi.toc.htm; Joel Shannon, "At Least 228 Police Officers Died by Suicide in 2019, Blue H.E.L.P. Says," *USA Today*, January 2, 2020; and Richard Goerling, interview with the author, February 1, 2018.

33. Avdi Avdija, "Police Use of Force: An Analysis of Factors That Affect Police Officer's Decision to Use Force on Suspects," *International Research Journal of Social Sciences* 2, no. 9 (2013): 1–6; Greg Ridgeway, "Officer Risk Factors Associated with Police Shootings: A Matched Case-Control Study," *Statistics and Public Policy* 3, no. 1 (2016): 1–6; and Samuel Carton et al., "Identifying Police Officers at Risk of Adverse Events," *Proceedings of the 22nd ACM SIGKDD International Conference on Knowledge Discovery and Data Mining* (August 2016): 67–76.

34. Cristina Queiros et al., "Burnout as Predictor of Aggressivity Among Police Officers," *European Journal of Policing Studies* 1, no. 2 (2013): 110–35; Robyn Gershon, Susan Lin, and Xianbin Li, "Work Stress in Aging Police Officers," *Journal of Occupational and Environmental Medicine* 44, no. 2 (2002): 160–67; Arne Nieuwenhuys, Rouwen Cañal-Bruland, and Raôul Oudejans, "Effects of Threat on Police Officers' Shooting Behavior: Anxiety, Action Specificity, and Affective Influences on Perception," *Applied Cognitive Psychology* 26, no. 4 (2012): 608–15; Shantha Rajaratnam et al., "Sleep Disorders, Health, and Safety in Police Officers," *JAMA* 306, no. 23 (2011): 2567–78; Katelyn Jetelina et al., "Cumulative, High-Stress Calls Impacting Adverse Events among Law Enforcement and the Public," *BMC Public Health* 20, no. 1137 (2020); Carton et al., "Identifying Police Officers at Risk of Adverse Events"; and Nicolien Kop and Martin Euwema, "Occupational Stress and the Use of Force by Dutch Police Officers," *Criminal Justice and Behavior* 28, no. 5 (2001): 631-52.

35. Bruce S. McEwen and John H. Morrison, "The Brain on Stress: Vulnerability and Plasticity of the Prefrontal Cortex over the Life Course," *Neuron* 79, no. 1 (2013): 16–29; Bruce McEwen and P. J. Gianaros, "Stress- and Allostatis-Induced Brain Plasticity," *Annual Review of Medicine* 62 (2011): 431–45; Linda Mah, Claudia Szabuniewicz, and Alexandra Fiocco, "Can Anxiety Damage the Brain?," *Current Opinion in Psychiatry* 29, no. 1 (2016): 56–63; Sofi da Cunha-Bang, Patrick Fisher, Liv Hjordt, et al., "Violent Offenders Respond to Provocations with High Amygdala and Striatal Reactivity," *Social Cognitive and Affective Neuroscience* 12, no. 5 (2017): 802–10, doi:10.1093/scan/nsx006; Kerry Ressler, "Amygdala Activity, Fear, and Anxiety: Modulation by Stress," *Biological Psychiatry* 67, no. 12 (2010): 1117–19; and Larry Siever, "Neurobiology of Aggression and Violence," *American Journal of Psychiatry* 165, no. 4 (2008): 429–42.

36. Brian Keith Payne, "Conceptualizing Control in Social Cognition: How Executive Functioning Modulates the Expression of Automatic Stereotyping,"

Journal of Personality and Social Psychology 89, no. 4 (2005): 488–503; Olesya Govorun and Brian Keith Payne, "Ego-Depletion and Prejudice: Separating Automatic and Controlled Components," *Social Cognition* 24, no. 2 (2006): 111–13; C. Neil Macrae, Alan Milne, and Galen Bodenhausen, "Stereotyping as Energy-Saving Device: A Peek Inside the Cognitive Toolbox," *Journal of Personality and Social Psychology* 66, no. 1 (1994): 37–47; Daniël H. J. Wigboldus et al., "Capacity and Comprehension: Spontaneous Stereotyping Under Cognitive Load," *Social Cognition* 22, no. 3 (2004): 292–309; and Mah, Szabuniewicz, and Fiocco, "Can Anxiety Damage the Brain?," 56–63.

37. Meghan Hunt, *Working to Close the Gap: How Work Stress and Fatigue Affect Racial Disparities in Traffic Stops by Oakland Police*, Master's Policy Report, Mills College (2017); and Debbie Ma et al., "When Fatigue Turns Deadly: The Association Between Fatigue and Racial Bias in the Decision to Shoot," *Basic and Applied Social Psychology* 35, no. 6 (2013): 515–24. See also Balbir Singh et al., "When Practice Fails to Reduce Racial Bias in the Decision to Shoot: The Case of Cognitive Load," *Social Cognition* 38, no. 6 (2020): 555–70.

38. Mumford, Feshir, and Collins, *74 Seconds*, "Episode 2: The Officer," *MPR News*, May 23, 2017; and Jeronimo Yanez interview, Bureau of Criminal Apprehension, July 7, 2016.

39. Connie Rice, interview with the author, October 10, 2018; Richard Goerling, interview with the author, December 22, 2016.

40. Cheri Maples, "Mindfulness and the Police | Dharma Talk by Cheri Maples," *Vulture Peak Gathering*, June 15, 2016, https://www.youtube.com/watch?v=NZ4jrd9IIh0; Barbara Casey, "Fierce-Faced Bodhisattva: A Policewoman's Story," *The Mindfulness Bell*, 32 (2002–2003), 15–21; and Joan Duncan Oliver, "She's Got the Beat," *Tricycle: The Buddhist Review*, Winter 2009, https://tricycle.org/magazine/shes-got-beat/.

41. Maples, "Mindfulness and the Police."

42. Daniel Goleman and Richard Davidson, *Altered Traits: Science Reveals How Meditation Changes Your Mind, Brain, and Body* (New York: Avery, 2017), 98–99, 121, 144–45, 163, 189–90, 207; and Yoona Kang, June Gruber, and Jeremy Gray, "Mindfulness: Deautomatization of Cognitive and Emotional Life," in *The Wiley Blackwell Handbook of Mindfulness*, ed. Amanda Le, Christelle Ngnoumen, and Ellen Langer (Chichester, West Sussex, U.K.: Wiley Blackwell, 2014), 168–85.

43. Rhonda Magee, "The Way of ColorInsight: Understanding Race and Law Effectively Through Mindfulness-Based ColorInsight Practice," *Georgetown Law Journal of Modern Critical Race Perspectives* (University of San Francisco Law Research Paper no. 2015-19, 2016), 1–52; Adam Lueke and Bryan Gibson. "Mindfulness Meditation Reduces Implicit Age and Race Bias: The Role of Reduced Automaticity of Responding," *Social Psychological and Personality Science* 6, no. 3 (2015): 284–91; and Sarah Schimchowitsch and Odile Rohmer, "Can We Reduce Our Implicit Prejudice Toward Persons with Disability? The Challenge of Meditation," *International Journal of Disability, Development and*

Education 63, no 6 (2016): 641–50, https://www.tandfonline.com/doi/abs/10 .1080/1034912X.2016.1156656.

44. Yoona Kang, Jeremy Gray, and John Dovidio, "The Nondiscriminating Heart: Lovingkindness Meditation Training Decreases Implicit Intergroup Bias," *Journal of Experimental Psycholology* 143, no. 3 (2014): 1306–13; and Anālayo, *Satipaṭṭhāna: The Direct Path to Realization* (Cambridge: Windhorse, 2003), 195–96.

45. Helen Weng et al., "Compassion Training Alters Altruism and Neural Responses to Suffering," *Psychological Science* 24, no. 7 (2013): 1171–80.

46. Kang et al., "The Nondiscriminating Heart"; and Yoona Kang and Emily Falk, "Neural Mechanisms of Attitude Change Toward Stigmatized Individuals: Temporoparietal Junction Activity Predicts Bias Reduction," *Mindfulness* 11 (2020): 1378–89. See also Yoona Kang, "Examining Interpersonal Self-Transcendence as a Potential Mechanism Linking Meditation and Social Outcomes," *Current Opinion in Psychology* 28 (2018): 115–19.

47. *Zen Master Dogen: An Introduction with Selected Writings*, trans. Yuho Yokai with Daizen Victoria (New York: Weatherhill, 1976), 39; Fynn-Mathis Trautwein, José Naranjo, and Stefan Schmidt, "Decentering the Self? Reduced Bias in Self- vs. Other-Related Processing in Long-Term Practitioners of Loving-Kindness Meditation," *Frontiers in Psychology* 7 (2016): 1785; and Yoona Kang, interviews with the author, September 6 and 13, 2016.

48. Brian Beekman, interview with the author, April 13, 2018.

49. Michael Christopher et al., "A Pilot Study Evaluating the Effectiveness of a Mindfulness-Based Intervention on Cortisol Awakening Response and Health Outcomes Among Law Enforcement Officers," *Journal of Police and Criminal Psychology* 31 (2016): 15–28; and Michael Christopher et al., "Mindfulness-Based Resilience Training to Reduce Health Risk, Stress Reactivity, and Aggression among Law Enforcement Officers: A Feasibility and Preliminary Efficacy Trial," *Psychiatry Research* 264 (2018): 104–15.

50. Public Records Request, Bend Police Department, September 2, 2020.

51. Interviews with officers at Mindful Badge Resilience Training, Bend, Oregon, April 13–15, 2018.

52. Scott Vincent, interview with the author, April 14, 2018.

53. Eric Russell, interview with the author, April 18, 2018.

54. Phillip Atiba Goff and Hilary Rau, "Predicting Bad Policing: Theorizing Burdensome and Racially Disparate Policing through the Lenses of Social Psychology and Routine Activities," *Annals of the American Academy of Political and Social Science* 687, no. 1 (2020): 67–88.

55. Brian Beekman, interview with the author, April 13, 2018.

56. Maples, "Mindfulness and the Police | Dharma Talk."

57. Charles Mill, *The Racial Contract* (Ithaca, NY: Cornell University Press, 1997), 18.

58. Yoona Kang, interview with the author, January 12, 2018; and nico hase, interview with the author, December 15, 2016.

59. Rhonda Magee, "Taking and Making Refuge in Racial [Whiteness] Awareness

and Racial Justice Work," in *Buddhism and Whiteness: Critical Reflections*, ed. George Yancy and Emily McRae (Lanham, MD: Rowman & Littlefield, 2019), 269; and interview with Rhonda Magee by David Treleaven, Trauma-Sensitive Mindfulness Podcast.

60. Robin DiAngelo, *White Fragility* (New York: Beacon Press, 2018), 103; and Pittman McGehee, Christopher Germer, Kristin Neff, "Core Values in Mindful Self-Compassion," in L. M. Monteiro et al., eds., *Practitioner's Guide to Ethics and Mindfulness-Based Interventions* (Cham, Switzerland: Springer International Publishing, 2017), 279–93.

61. Rhonda Magee, *The Inner Work of Racial Justice* (New York: Tarcher Perigee, 2019), 103–6, 152; Rhonda Magee interview by Dan Harris, 10% Happier podcast, June 4, 2020; Thich Nhat Hanh, *Anger: Wisdom for Cooling the Flames* (New York: Riverhead Books, 2002), 119; and Thich Nhat Hanh and Rachel Neumann, *Calming the Fearful Mind* (Berkeley, CA: Parallax Press, 2005), 12.

62. Yoona Kang et al., "Effects of Self-Transcendence on Neural Responses to Persuasive Messages and Health Behavior Change," *Proceedings of the National Academy of Sciences* 115, no. 40 (2018): 9974–79.

63. Armen Marsoobian, "Acknowledging Intergenerational Moral Responsibility in the Aftermath of Genocide," *Genocide Studies and Prevention: An International Journal* 4, no. 2 (2009): 211–20.

CHAPTER 6. THE WATTS JIGSAW

1. Connie Rice, *Power Concedes Nothing: One Woman's Quest for Social Justice in America, from the Courtroom to the Kill Zones* (New York: Scribner, 2012), 248.

2. Connie Rice, interview with the author, October 10, 2018.

3. Rice, *Power Concedes Nothing*, 7–12, 27–29, 107–10, 116, 183–91, 242–43.

4. Connie Rice, interview with the author, October 10, 2018.

5. Donna Murch, "Crack in Los Angeles: Crisis, Militarization, and Black Response to the Late Twentieth-Century War on Drugs," *Journal of American History* 102, no. 1 (2015): 162–73; Patricia Klein, "LAPD Draws Fire for Ramming Home in Raid," *Los Angeles Times*, February 8, 1985, https://www.latimes.com/archives/la-xpm-1985-02-08-mn-4676-story.html; Langford v. Superior Court, 43 Cal. (1987); and John Mitchell, "The Raid That Still Haunts L.A.," *Los Angeles Times*, March 14, 2001, https://www.latimes.com/archives/la-xpm-2001-mar-14-mn-37553-story.html.

6. Connie Rice, interview with the author, October 19, 2018.

7. Jill Leovy, *Ghettoside: A True Story of Murder in America* (New York: Spiegel and Grau, 2015), 9.

8. Anjuli Sastry and Karen Grigsby Bates, "When LA Erupted in Anger: A Look Back at the Rodney King Riots," NPR, April 26, 2017, https://www.npr.org/2017/04/26/524744989/when-la-erupted-in-anger-a-look-back-at-the-rodney; Lou Cannon, "When Thin Blue Line Retreated, L.A. Riot Went Out

of Control," *Washington Post*, May 10, 1992, https://www.washingtonpost.com /archive/politics/1992/05/10/when-thin-blue-line-retreated-la-riot-went-out -of-control/2ccf3e5c-c03b-4d82-bce1-0ea43be30cd3; Patt Morrison, "Daryl Gates: The Star of His Own Police Show," *Los Angeles Times*, April 18, 2010; Rice, *Power Concedes Nothing*, 113; and Jim Newton, "Change in Black and White in L.A.," *Los Angeles Times*, March 6, 2011, http://articles.latimes.com /2011/mar/06/opinion/la-oe-newton-rodney-king-20110306.

9. Joe Domanick, *Blue: The LAPD and the Battle to Redeem American Policing* (New York: Simon and Schuster, 2015), 153–205.

10. "Coast Police Chief Accused of Racism," *New York Times*, May 13, 1982.

11. Connie Rice, interview with the author, October 19, 2018.

12. Connie Rice, interview with the author, October 10, 2018.

13. Connie Rice, interview with the author, October 10, 2018.

14. Connie Rice, interview with the author, October 10, 2018.

15. Gordon Allport, *The Nature of Prejudice: 25th Anniversary Edition* (Reading, MA: Addison-Wesley, 1979), xv, 261–82; Thomas Pettigrew and Linda Tropp, "Allport's Intergroup Contact Hypothesis: Its History and Influence," in *On the Nature of Prejudice: Fifty Years After Allport*, Jack Dovidio, Peter Glick, and Laurie Rudman, eds. (Malden, MA: Blackwell, 2005), 262–77; and Anthony Greenwald., "Under What Conditions Does Intergroup Contact Improve Intergroup Harmony?," in *The Scientist and the Humanist: A Festschrift in Honor of Elliot Aronson*, ed. Marti Hope Gonzales, Carol Tavris, and Joshua Aronson (New York: Psychology Press, 2010), 267–81.

16. Colonel Raymond Ansel, "From Segregation to Desegregation: Blacks in the U.S. Army 1703–1954," U.S. Army War College, December 4, 1989, https://apps .dtic.mil/dtic/tr/fulltext/u2/a219730.pdf; Ulysses Lee, *The Employment of Negro Troops* (Washington, DC: Office of the Chief of Military History, United States Army, 1966); David Colley, *Blood for Dignity* (New York: St. Martin's Press, 2003), 189–94; Imelda Patoglun-An and Jeffrey Clair, "An Experimental Study of Attitudes Toward Homosexuals," *Deviant Behavior* 7, no. 2 (1986): 121–35; and B. J. DiTullio, "The Effect of Employing Trainable Mentally Retarded (TMR) Students as Workers Within the Philadelphia Public School System: Attitudes of Supervisors and Non-handicapped Co-workers Towards the Retarded as a Result of Contact" (PhD diss., Temple University, 1982).

17. John Dixon, Kevin Durrheim, and Colin Tredoux, "Intergroup Contact and Attitudes Toward the Principle and Practice of Racial Equality," *Psychological Science* 18, no. 10 (2007): 867–72; John Dixon et al., "A Paradox of Integration? Interracial Contact, Prejudice Reduction, and Perceptions of Racial Discrimination," *Journal of Social Issues* 66, no. 2 (2010): 402–16; and Jennifer Richeson and Samuel Sommers, "Toward a Social Psychology of Race and Race Relations for the Twenty-First Century," *Annual Review of Psychology* 67 (2016): 439–63.

18. J. Nicole Shelton et al., "Expecting to Be the Target of Prejudice: Implications for Interethnic Interactions," *Personality and Social Psychology Bulletin*

31, no. 9 (2005): 1189–202; and Elizabeth Paluck, Seth Green, and Donald Green, "The Contact Hypothesis Re-evaluated," *Behavioural Public Policy* 3 (no. 2): 129–58.

19. Rice, *Power Concedes Nothing*, 217–24.

20. Rice, *Power Concedes Nothing*, 283.

21. Jorja Leap et al., "Evaluation of the LAPD Community Safety Partnership" (UCLA, 2020), http://www.lapdpolicecom.lacity.org/051220/CSP%20Evalu ation%20Report_2020_FINAL.pdf; and Rice, *Power Concedes Nothing*, 277.

22. Rice, *Power Concedes Nothing*, 306.

23. Leighton Woodhouse, "50 Years After the Riots, Watts Projects and LAPD Learn to Co-Exist," Gawker.com, August 11, 2015, https://gawker.com/50 -years-after-the-riots-watts-projects-and-lapd-learn-1723326136; Nina Revoyr, "How Watts and the LAPD Make Peace," *Los Angeles Times*, June 6, 2015, http://www.latimes.com/opinion/op-ed/la-oe-revoyr-lessons-from-watts -gang-task-force-20150607-story.html; and Phil Tingirides, interview with the author, November 21, 2018.

24. Constance Rice and Susan Lee, "Relationship-Based Policing: Achieving Safety in Watts," A Report for the President's Task Force on 21st Century Policing, Urban Peace Institute (2007), https://static1.squarespace.com/static /55b673c0e4b0cf84699bdffb/t/5a1890acec212d9bd3b8f52d/1511559341778 /President%27s+Task+Force+CSP+Policy+Brief+FINAL+02-27-15updated .pdf; and Susan Lee, interview with the author, November 12, 2018.

25. Susan Lee, interview with the author, November 12, 2018.

26. Connie Rice, interview with the author, October 10, 2018.

27. Emada Tingirides, interview with the author, November 29, 2018.

28. Susan Lee, interview with the author, November 12, 2018; and Connie Rice, interview with the author, October 10, 2018.

29. Susan Lee, interview with the author, October 29, 2018.

30. Andre Christian, interview with the author, January 7, 2019; and Susan Lee, interview with the author, October 29, 2018.

31. Susan Lee, interview with the author, November 12, 2018; and Connie Rice, interview with the author, October 10, 2018.

32. K. B. Turner, David Giacopassi, and Margaret Vandiver, "Ignoring the Past: Coverage of Slavery and Slave Patrols in Criminal Justice Texts," *Journal of Criminal Justice Education* 17, no. 1 (2006): 181–95; H. M. Henry, "The Police Control of the Slave in South Carolina," (PhD diss., Vanderbilt University, 1913); Sally Hadden, *Slave Patrols: Law and Violence in Virginia and the Carolinas* (Cambridge, MA: Harvard University Press, 2001), 24, 114, 123; Philip Reichel, "Southern Slave Patrols as a Transitional Police Type," *American Journal of Police* 7, no. 2 (1988): 51–77; and "Melina Abdullah: It's a Mistake to Equate What Happens to Property with What Happens to Black Lives," *To the Point*, KCRW Los Angeles Public Radio, June 4, 2020, https://www.kcrw.com/news/shows/to-the-point /police-racism-white-supremacy-trump-protests; and Rice interview.

33. Melvyn Hayward, interview with the author, January 4, 2019.

34. Connie Rice, interview with the author, October 10, 2018.

35. Susan Lee, interview with the author, October 29, 2018.

36. Susan Lee, interview with the author, November 12, 2018.

37. Woodhouse, "50 Years After the Riots, Watts Projects and LAPD Learn to Co-Exist."

38. Emada Tingirides, interview with the author, November 29, 2018; Phil Tingirides, interview with the author, November 24, 2018; and Sam Kuhn and Stephen Lurie, *Reconciliation Between Police and Communities: Case Studies and Lessons Learned* (New York: John Jay College, 2018).

39. Jorja Leap, *Project Fatherhood: A Story of Courage and Healing in One of America's Toughest Communities* (Boston: Beacon Press, 2015), 116–24.

40. Andre Christian, interview with the author, January 7, 2019; and Phil Tingirides, interview with the author, November 21, 2018.

41. Rice, *Power Concedes Nothing*, 334; Aaron Hagstrom, "The LAPD Community Safety Partnership: An Experiment in Policing" (master's thesis, USC, 2014), http://digitallibrary.usc.edu/cdm/ref/collection/p15799coll3/id/414752.

42. Phil Tingirides, interview with the author, November 21, 2018; Lupita Valdovino, interview with the author, August 19, 2020; and Jeff Joyce, interview with the author, February 5, 2019.

43. Sandy Banks, "Young Players on the Watts Bears Are Part of a Larger Team Effort," *Los Angeles Times*, September 16, 2013, https://www.latimes.com/local/la-xpm-2013-sep-16-la-me-0917-banks-lapd-football-20130917-story.html; Hagstrom, "The LAPD Community Safety Partnership"; and Emada Tingirides, interview with the author, November 29, 2018.

44. Nina Revoyr, "How Watts and the LAPD Make Peace."

45. Andre Christian, interview with the author, January 7, 2019; Connie Rice, interview with the author, October 10, 2018; Phil Tingirides, interview with the author, February 1, 2019; and Emada Tingirides, interview with the author, November 29, 2018.

46. "The Los Angeles Community Safety Partnership: 2019 Assessment," Urban Institute, March 2020; and Phil Tingirides, interview with the author, February 1, 2019.

47. "The Los Angeles Community Safety Partnership: 2019 Assessment," Urban Institute, March 2020; Sam Kuhn and Stephen Lurie, *Reconciliation Between Police and Communities: Case Studies and Lessons Learned* (New York: John Jay College, 2018); "The Homicide Report," *Los Angeles Times*, http://homicide.latimes.com, accessed February 21, 2021; Phil Tingirides, interview with the author, February 1, 2019; Susan Lee, interview with the author, November 12, 2018; Tracy Meares and Tom Tyler, "Policing: A Model for the Twenty-First Century," in *Policing the Black Man: Arrest, Prosecution, and Imprisonment*, ed. Angela Davis (New York: Knopf, 2018), 162–74 ; and Tracey Meares, "Policing and Procedural Justice: Shaping

Citizens' Identities to Increase Democratic Participation," *Northwestern University Law Review* 111, no. 6 (2017): 1525–36; *Homicide Report 2017*, LAPD Police Commission, http://assets.lapdonline.org/assets/pdf/2017 -homi-report-final.pdf.

48. Tracey Meares, "Norms, Legitimacy, and Law Enforcement," *Oregon Law Review* 79 (2000): 391–416; Andrew Papachristos, Tracy Meares, and Jeffrey Fagan, "Why Do Criminals Obey the Law? The Influence of Legitimacy and Social Networks on Active Gun Offenders," *Journal of Criminal Law and Criminology* 102, no. 2 (2012): 397–440; and Danielle Wallace et al., "Desistance and Legitimacy: The Impact of Offender Notification Meetings on Recidivism Among High Risk Offenders," *Justice Quarterly* 33, no. 7 (2016): 1237–64.

49. Leap et al., "Evaluation of the LAPD Community Safety Partnership," 42, 48–54, 65, 74–75, 93; and Cindy Chang, "LAPD Community Policing Program Has Prevented Crime, and Made Residents Feel Safer, Study Finds," *Los Angeles Times*, May 13, 2020.

50. Elliot Aronson, interview with the author, June 24, 2016; and Matthew Snapp, interview with the author, December 20, 2016.

51. Elliot Aronson et al., *The Jigsaw Classroom* (Ann Arbor, MI: Sage Publications, 1978), 36–67; and Elliot Aronson, interview with Ben Dean, *MentorCoach*, September 24, 2010, https://www.mentorcoach.com/wp-content/uploads/2017/05 /TRANSCRIPT-ELLIOT-ARONSON.pdf.

52. Diane Bridgeman and Elliot Aronson, "Jigsaw Groups and the Desegregated Classroom: In Pursuit of Common Goals," *Personality and Social Psychology Bulletin* 5, no. 4 (1979): 438–46; Elliot Aronson, interview with the author, June 24, 2016.

53. Greenwald, "Under What Conditions Does Intergroup Contact Improve Intergroup Harmony?"; Iain Walker and Mary Crogan, "Academic Performance, Prejudice, and the Jigsaw Classroom: New Pieces to the Puzzle," *Journal of Community Applied Social Psychology* 8, no. 6 (1998): 381–93; and Mary Jenness, *Twelve Negro Americans* (New York: Friendship Press, 1936), 166–69.

54. Elliot Aronson, interview with the author, June 24, 2016.

55. Lewis Hyde, *The Gift: Creativity and the Artist in the Modern World*, 2nd ed. (New York: Random House, 2007), 57, 72, 74–75.

56. Matthew Lowe, "Types of Contact: A Field Experiment on Collaborative and Adversarial Caste Integration," *Behavioral and Experimental Economics eJournal* (CESifo Working Paper No. 8089, 2020); and Salma Mousa, "Building Social Cohesion Between Christians and Muslims Through Soccer in Post-ISIS Iraq," *Science* 369, no. 6505 (2020): 866–70. Interestingly, Indian players assigned to play *against* players from a different caste *also* were more likely to choose a player from another caste as a future teammate.

57. Connie Rice, interview with the author, October 10, 2018.

58. Amos Tversky and Daniel Kahneman, "Judgment Under Uncertainty: Heuristics and Biases," *Science* 185, no. 4157 (1974): 1124–31; and Amos Tversky

and Daniel Kahneman, "Belief in the Law of Small Numbers," *Psychological Bulletin* 76, no. 2 (1971): 105–10.

59. Gary Klein, *Sources of Power: How People Make Decisions* (Cambridge, MA: MIT Press, 1998), 14–15, 42–43.

60. Daniel Kahneman and Gary Klein, "Conditions for Intuitive Expertise: A Failure to Disagree," *American Psychologist* 64, no. 6 (2009): 515–26.

61. Information in this and the following two paragraphs from David Redish, interviews with the author, July 26, 2017, and March 22, 2021; and David Redish, personal correspondence with the author, March 5 and March 8, 2021.

62. Joshua Correll et al., "Stereotypic Vision: How Stereotypes Disambiguate Visual Stimuli," *Journal of Personality and Social Psychology* 108, no. 2 (2015): 219–33.

63. Sandy Sangrigoli et al., "Reversibility of the Other-Race Effect in Facial Recognition During Childhood," *Psychological Science* 16, no. 6 (2005): 440–44.

64. Jim Dwyer, Peter Neufeld, and Barry Scheck, *Actual Innocence: When Justice Goes Wrong and How to Make It Right* (New York: New American Library, 2003); "DNA Exonerations in the United States," *Innocence Project*, accessed December 30, 2020, https://innocenceproject.org/dna-exonerations-in-the-united-states; Taki Flevaris and Ellie Chapman, "Cross-Racial Misidentification: A Call to Action in Washington State and Beyond," *Seattle University Law Review* 38, no. 3 (2015): 861; and Samuel Gross, "What We Think, What We Know, and What We Think We Know About False Convictions," *Ohio State Journal of Criminal Law* 14, no. 2 (2017): 753–86.

65. Beth Crandall and Karen Getchell-Reiter, "Critical Decision Method: A Technique for Eliciting Concrete Assessment Indicators from the Intuition of NICU Nurses," *Advances in Nursing Science* 16, no. 1 (1993): 42–51.

66. Ellen Langer, Richard Bashner, and Benzion Chanowitz, "Decreasing Prejudice by Increasing Discrimination," *Journal of Personality and Social Psychology* 49, no. 1 (1985): 113–20.

67. Amam Saleh et al., "Deaths of People with Mental Illness During Interactions with Law Enforcement," *International Journal of Law and Psychiatry* 58 (2018): 110–16; Kelly Smith, "Minnesota Could Mandate Officer Training for Mental-Illness Calls," *Minneapolis Star Tribune*, April 2, 2017, http://www.startribune.com/minnesota-could-mandate-officer-training-for-mental-illness-calls/417872373/; and "Re-Engineering Training on Police Use of Force," Police Executive Research Forum (2015).

68. Michael Compton et al., "Brief Reports: Crisis Intervention Team Training: Changes in Knowledge, Attitudes, and Stigma Related to Schizophrenia," *Psychiatric Services* 57, no. 8 (2006): 1199–202, https://ps.psychiatryonline.org/doi/full/10.1176/ps.2006.57.8.1199.

69. Michael Compton et al., "The Police-Based Crisis Intervention Team (CIT) Model: II. Effects on Level of Force and Resolution, Referral, and Arrest," *Psychiatric Services* 65, no. 4 (2014): 523–29; Jennifer Skeem and Lynne Bibeau,

"How Does Violence Potential Relate to Crisis Intervention Team Responses to Emergencies?," *Psychiatric Services* 59, no. 2 (2008): 201–4; Amy Watson and Anjali Fulambarker, "The Crisis Intervention Team Model of Police Response to Mental Health Crises: A Primer for Mental Health Practitioners," *Best Practices in Mental Health* 8, no. 2 (2012): 71; Eric Russell, interview with the author, April 18, 2018; and "'CAHOOTS': How Social Workers and Police Share Responsibilities in Eugene, Oregon," NPR, June 10, 2020, https://www .npr.org/2020/06/10/874339977/cahoots-how-social-workers-and-police -share-responsibilities-in-eugene-oregon.

70. Diane Lefer, "Both Sides of the Street," *Sun*, April 2008, https://www.thesunmagazine.org/issues/388/both-sides-of-the-street.

71. Jorja Leap, interview with the author, October 26, 2018; and David Kennedy, "Pulling Levers: Chronic Offenders, High-Crime Settings, and a Theory of Prevention," *Valparaiso University Law Review* 31, no. 2 (1997): 449.

72. Connie Rice, "Reforming the LAPD," interview with Charles Monroe-Kane, *To the Best of Our Knowledge*, Wisconsin Public Radio, December 21, 2014, https://www.ttbook.org/interview/reforming-lapd.

73. Phil Tingirides, interview with the author, November 21, 2018; and "The Los Angeles Community Safety Partnership: 2019 Assessment," Urban Institute.

74. Lupita Valdovinos, interview with the author, August 19, 2020.

75. Leap et al., "Evaluation of the LAPD Community Safety Partnership"; and "The Los Angeles Community Safety Partnership: 2019 Assessment," Urban Institute.

76. "Will Changes to LAPD's Community Safety Partnership Build Trust with Residents?," *Press Play with Madeline Brand*, KCRW Los Angeles Public Radio, July 28, 2020, https://www.kcrw.com/news/shows/press-play-with -madeleine-brand/la-community-safety-police-covid-19-education/lapd-csp -emada-tingirides-social-work; Lupita Valdovino, interview with the author, August 19, 2020; and Leap et al., "Evaluation of the LAPD Community Safety Partnership."

77. Leap et al., "Evaluation of the LAPD Community Safety Partnership."

78. Jeff Joyce, interview with the author, February 5, 2019.

79. Connie Rice, interview with the author, October 19, 2018.

80. Jeff Joyce, interview with author, February 8, 2021; and Nickerson Gardens resident, interview with author, February 8, 2021.

CHAPTER 7. DESIGNING FOR FLAWED HUMANS

1. Chris's story told by Adam Heffelbower, interview with the author, September 15, 2017; Katie Mulloy, interview with the author, September 12, 2017; and Jada Pirius, interview with the author, December 29, 2017.

2. Amanda Clarke et al., "Thirty Years of Disparities Intervention Research: What Are We Doing to Close Racial and Ethnic Gaps in Health Care?," *Medical Care* 51, no. 11 (2013): 1020–26; Kristin M. Lefebvre and Lawrence A. Lavery,

"Disparities in Amputations in Minorities," *Clinical Orthopedics and Related Research* 469, no. 7 (2011): 1941–50; John Van Evrie, *White Supremacy and Negro Subordination* (New York: Van Evrie, Horton, and Co., 1868), 311–13; and John Haller, *Outcasts from Evolution: Scientific Attitudes of Racial Inferiority, 1859–1900* (Urbana: University of Illinois Press, 1971), 40–68.

3. Frederick Hoffman, *Race Traits and Tendencies of the American Negro* (New York: Macmillan for the American Economic Association, 1896), 140–44.

4. Norrisa Haynes, Lisa Cooper, and Michelle Albert, "At the Heart of the Matter: Unmasking and Addressing the Toll of COVID-19 on Diverse Populations," *Circulation* 142, no. 2 (2020): 105–7; Christina Dragon et al., "Transgender Medicare Beneficiaries and Chronic Conditions: Exploring Fee-for-Service Claims Data," *LGBT Health* 4, no. 6 (2017): 404–11; and Sandy James et al., "The Report of the 2015 U.S. Transgender Survey," National Center for Transgender Equality, 2016, https://www.transequality.org/sites/default/files/docs/USTS-Full-Report-FINAL.PDF.

5. Vickie Shavers, Alexis Bakos, and Vanessa Sheppard, "Race, Ethnicity, and Pain Among the U.S. Adult Population," *Journal of Healthcare for the Poor and Underserved* 21, no. 1 (2010): 177–220; Kristian Foden-Vensil, "Emergency Medical Responders Confront Racial Bias," *Kaiser Health News*, January 11, 2019, https://khn.org/news/emergency-medical-responders-confront-racial-bias/; Adil Shah et al., "Analgesic Access for Acute Abdominal Pain in the Emergency Department Among Racial/Ethnic Minority Patients," *Medical Care* 53, no. 12 (2015): 1000–1009; Monika Goyal et al., "Racial Disparities in Pain Management of Children with Appendicitis in Emergency Departments," *JAMA Pediatrics* 169, no. 11 (2015): 996–1002; Sophie Trawalter, Kelly Hoffman, and Adam Waytz, "Racial Bias in Perceptions of Others' Pain," *PLoS One* 7, no. 11 (2012): e48546, and correction, *PLoS One* 11, no. 3 (2016): e0152334; Samuel A. Cartwright, *Report on the Diseases and Physical Peculiarities of the Negro Race*, *New Orleans Medical and Surgical Journal* 7 (1851): 691–715; P. Tidyman, "A Sketch of the Most Remarkable Disease of the Negroes of the Southern States," *Philadelphia Journal of the Medical and Physical Sciences* 3, no. 6 (1826): 306–38; Harriet Washington, *Medical Apartheid: The Dark History of Medical Experimentation on Black Americans from Colonial Times to the Present* (New York: Doubleday, 2008); Haller, *Outcasts from Evolution*, 53–68; Kelly Hoffman et al., "Racial Bias in Pain Assessment and Treatment Recommendations, and False Beliefs About Biological Differences Between Blacks and Whites," *PNAS* 113, no. 16 (2016): 4296–301; Lundy Braun and Barry Saunders, "Avoiding Racial Essentialism in Medical Science Curricula," *AMA Journal of Ethics* 19, no. 6 (2017): 518–27; and Tod Hamilton and Robert Hummer, "Immigration and the Health of U.S. Black Adults: Does Country of Origin Matter?," *Social Science and Medicine* 73, no. 10 (2011): 1551–60. See also David Hyman et al., "Lower Hypertension Prevalence in First-Generation African Immigrants Compared to U.S.-Born African Americans," *Ethnicity & Disease* 10, no. 3 (2000): 343–49.

6. D. Robert Harris, Roxanne Andrews, and Anne Elixhauser, "Racial and Gender Differences in Use of Procedures for Black and White Hospitalized Adults," *Ethnicity and Disease* 7, no. 2 (1997): 91–102; John Canto et al., "Relation of Race and Sex to the Use of Reperfusion Therapy in Medicare Beneficiaries with Acute Myocardial Infarction," *New England Journal of Medicine* 342, no. 15 (2000): 1094–1100; Sameer Arora et al., "Fifteen-Year Trends in Management and Outcomes of Non-ST-Segment-Elevation Myocardial Infarction Among Black and White Patients," *Journal of the American Heart Association* 7, no. 19 (2018): e010203; Khadijah Breathett et al., "African Americans Are Less Likely to Receive Care by a Cardiologist During an Intensive Care Unit Admission for Heart Failure." *Journal of the American College of Cardiology: Heart Failure* 6, no. 5 (2018): 413–20; Rachel Johnson et al., "Patient Race/Ethnicity and Quality of Patient–Physician Communication During Medical Visits," *American Journal of Public Health* 94, no. 12 (2004): 2084–90; Megan Shen et al., "The Effects of Race and Racial Concordance on Patient-Physician Communication: A Systematic Review of the Literature," *Journal of Racial and Ethnic Health Disparities* 5, no. 1 (2018): 117–40; Kristin M. Lefebvre and Lawrence A. Lavery, "Disparities in Amputations in Minorities," *Clinical Orthopedics and Related Research* 469, no. 7 (2011): 1941–50; J. A. Mustapha, "Explaining Racial Disparities in Amputation Rates for the Treatment of Peripheral Artery Disease (PAD) Using Decomposition Methods," *Journal of Racial and Ethnic Health Disparities* 4, no. 5 (2017): 784–95; and Tyler Durazzo, Stanley Frencher, and Richard Gusberg, "Influence of Race on the Management of Lower Extremity Ischemia: Revascularization vs. Amputation," *JAMA Surgery* 148, no. 7 (2013): 617–23.

7. Jodi Katon, "Contributors to Racial Disparities in Minimally Invasive Hysterectomy in the U.S. Department of Veterans Affairs," *Medical Care* 57, no. 12 (2019): 930–36; Lisa Callegari et al., "Associations Between Race/Ethnicity, Uterine Fibroids, and Minimally Invasive Hysterectomy in the VA Healthcare System," *Women's Health Issues* 29, no. 1 (2019): 48–55; Ahsan Arozullah, et al., "Racial Variation in the Use of Laparoscopic Cholecystectomy in the Department of Veterans Affairs Medical System," *Journal of the American College of Surgeons* 188, no. 6 (2000): 604–22; Kristina Schnitzer et al., "Disparities in Care: The Role of Race on the Utilization of Physical Restraints in the Emergency Setting," *Academic Emergency Medicine* 27 (2020): 943–50; Clairmont Griffith et al., "The Effects of Opioid Addiction in the Black Community," *International Journal of Collaborative Research on Internal Medicine & Public Health* 10, no. 2 (2018): 843–50; and Steven Ross Johnson, "The Racial Divide in the Opioid Epidemic," *Modern Healthcare*, February 27, 2016, https://www.modernhealthcare.com/article/20160227/MAGAZINE/302279871/the-racial-divide-in-the-opioid-epidemic. A thorough overview of the evidence of racial and ethnic disparities in health care and medical treatment can be found in H. Jack Geiger, "Racial and Ethnic Disparities in Diagnosis and Treatment: A

Review of the Evidence and a Consideration of the Causes," in *Unequal Treatment: Confronting Racial and Ethnic Disparities in Healthcare*, ed. B. Smedley, A. Stith, and A. Nelson (Washington, DC: National Academies Press, 2003).

8. Francesco Rubino et al., "Joint International Consensus Statement for Ending Stigma of Obesity," *Nature Medicine* 26 (2020): 485–97; Kimberly Gudzune, "Physicians Build Less Rapport with Obese Patients," *Obesity: A Research Journal* 21, no. 10 (2013): 2146–52; James et al., "The Report of the 2015 U.S. Transgender Survey"; and Shabab Ahmed Mirza and Caitlin Rooney, "Discrimination Prevents LGBTQ People from Accessing Health Care," Center for American Progress, January 18, 2018, https://www.americanprogress.org/issues/lgbt/news/2018/01/18/445130/discrimination-prevents-lgbtq-people-accessing-health-care/.

9. Diane Hoffmann and Anita Tarzian, "The Girl Who Cried Pain: A Bias Against Women in the Treatment of Pain," *Journal of Law, Medicine & Ethics* 29 (2001): 13–27, https://papers.ssrn.com/sol3/papers.cfm?abstract_id=383803; Esther Chen et al., "Gender Disparity in Analgesic Treatment of Emergency Department Patients with Acute Abdominal Pain," *Academic Emergency Medicine* 15, no. 5 (2008): 414–18; Hani Jneid et al., "Sex Differences in Medical Care and Early Death After Acute Myocardial Infarction," *Circulation* 118, no. 25 (2008): 2803–10; Tracey Collella et al., "Sex Bias in Referral of Women to Outpatient Cardiac Rehabilitation? A Meta-analysis," *European Journal of Preventive Cardiology* 22, no. 4 (2015): 423–41; Marta Supervía et al., "Cardiac Rehabilitation for Women: A Systematic Review of Barriers and Solutions," *Mayo Clinic Proceedings* 92, no. 4 (2017): 565–77; Cornelia Borkhoff et al., "The Effect of Patients' Sex on Physicians' Recommendations for Total Knee Arthroplasty," *Canadian Medical Association Journal* 178, no. 6 (2008): 681–87; and Robert Fowler et al., "Sex- and Age-Based Differences in the Delivery and Outcomes of Critical Care," *Canadian Medical Association Journal* 177, no. 12 (2007): 1513–19. See also Andreas Valentin, "Gender-Related Differences in Intensive Care: A Multiple-Center Cohort Study of Therapeutic Interventions and Outcome in Critically Ill Patients," *Critical Care Medicine* 31, no. 7 (2003): 1901–7.

10. Lynn Freedman et al., "A 360 Degree Approach to Understanding Disrespect and Abuse of Women During Childbirth: Creating Space for Women and Providers to Define the Challenges," American Public Health Association's 2019 Annual Meeting, November 2–6, 2019; Rob Haskell, "Serena Williams on Motherhood, Marriage, and Making Her Comeback," *Vogue*, January 10, 2018, https://www.vogue.com/article/serena-williams-vogue-cover-interview-february-2018; Susan Mann et al., "What We Can Do About Maternal Mortality—And How to Do It Quickly," *Obstetrical and Gynecological Survey* 75, no. 4 (2020): 217–18; Nina Martin and Renee Montagne, "Black Mothers Keep Dying After Giving Birth. Shalon Irving's Story Explains Why," *All Things Considered*, NPR, December 7, 2017, https://www.npr.org/2017/12/07/568948782/black-mothers-keep-dying-after-giving-birth-shalon-irvings

-story-explains-why?t=1601468947480; Building U.S. Capacity to Review and Prevent Maternal Deaths (program), in "Report from Nine Maternal Mortality Review Committees," CDC Foundation 2018; Kharah Ross et al., "Socioeconomic Status, Preeclampsia Risk and Gestational Length in Black and White Women," *Journal of Racial and Ethnic Health Disparities* 6, no. 6 (2019): 1182–91; David Williams, Naomi Priest, and Norman Anderson, "Understanding Associations Between Race, Socioeconomic Status and Health: Patterns and Prospects," *Health Psychology* 35, no. 4 (2016): 407–11; William Lee Howard, "The Negro as a Distinct Ethnic Factor in Civilization," *Medicine (Detroit) IX* (1903): 420–33; and Donald Chatman, "Endometriosis in the Black Woman," *American Journal of Obstetrics and Gynecology* 125, no. 7 (1976): 987–89.

11. Paul Chodoff, "Hysteria and Women," *American Journal of Psychiatry* 139, no. 5 (1982): 545–51; Maya Dusenbery, *Doing Harm: The Truth About How Bad Medicine and Lazy Science Leave Women Dismissed, Misdiagnosed, and Sick* (New York: HarperOne, 2018), 63–70; and Brian Earp et al., "Gender Bias in Pediatric Pain Assessment," *Journal of Pediatric Psychology* 44, no. 4 (2019): 403–14.

12. Overview from interviews with Steven Epstein, January 19, 2021 and Maya Dusenbery, October 18, 2018; Steven Epstein, *Inclusion: The Politics of Difference in Medical Research* (Chicago: University of Chicago Press, 2009), 60–61; Trisha Flynn, "Female Trouble," *Chicago Tribune*, October 29, 1986, https://www.chicagotribune.com/news/ct-xpm-1986-10-29-8603210488-story.html; and Heather Whitley and Wesley Lindsey, "Sex-Based Differences in Drug Activity," *American Family Physician* 80, no. 11 (2009): 1254–58.

13. Jessica Cerdeña, Marie Plaisime, and Jennifer Tsai, "From Race-Based to Race-Conscious Medicine: How Anti-Racist Uprisings Call us to Act," *Lancet* 396, no. 10257 (2020): 1125–28; Maanvi Singh, "Younger Women Hesitate to Say They Are Having a Heart Attack," NPR, February 25, 2015, http://www.npr.org/sections/health-shots/2015/02/24/388787045/younger-women-hesitate-to-say-theyre-having-a-heart-attack; and Amy V. Ferry et al., "Presenting Symptoms in Men and Women Diagnosed with Myocardial Infarction Using Sex-Specific Criteria," *Journal of the American Heart Association* 8, no. 17 (2019): e012307.

14. Margaret Ann Miller, "Gender-Based Differences in the Toxicity of Pharmaceuticals—the Food and Drug Administration's Perspective," *International Journal of Toxicology* 20, no. 3 (2001): 149–52; Ameeta Parekh et al., "Review Adverse Effects in Women: Implications for Drug Development and Regulatory Policies," *Expert Review of Clinical Pharmacology* 4, no. 4 (2011): 453–66; Heather Whitley and Wesley Lindsey, "Sex-Based Differences in Drug Activity," *American Family Physician* 80, no. 11 (2009): 1254–58; Epstein, *Inclusion*, 233; and Lesley Stahl, "Sex Matters: Drugs Can Affect Sexes Differently," *60 Minutes*, CBS, February 9, 2014, http://www.cbsnews.com/news/sex-matters-drugs-can-affect-sexes-differently/.

15. Carlos Penaloza et al., "Sex of the Cell Dictates Its Response: Differential Gene Expression and Sensitivity to Cell Death Inducing Stress in Male and Female Cells," *Journal of the Federation of American Societies for Experimental Biology* 23, no. 6 (2009): 1869–79; Anatoly Rubtsov et al., "Genetic and Hormonal Factors in Female-Biased Autoimmunity," *Autoimmunity Review* 9, no. 7 (2010): 494–98; and Fariha Angum et al., "The Prevalence of Autoimmune Disorders in Women: A Narrative Review," *Cureus* 12, no. 5 (2020): e8094. See also Anne Fausto-Sterling, "The Five Sexes, Revisited," *Sciences* 40, no. 4 (July 2000): 18–23.

16. Dusenbery, *Doing Harm*, 12; and Gunilla Risberg, Eva Johansson, and Katarina Hamberg, "A Theoretical Model for Analysing Gender Bias in Medicine," *International Journal for Equity in Health* 8, no. 28 (2009).

17. Stephen Vavricka et al., "Celiac Disease Diagnosis Still Significantly Delayed—Doctor's but Not Patients' Delay Responsive for the Increased Total Delay in Women," *Digest of Liver Disorders* 48, no. 10 (2016): 1148–54; Fazlul Karim, "Gender Differences in Delays in Diagnosis and Treatment of Tuberculosis," *Health Policy and Planning* 22, no. 5 (2007): 329–34; Vega Jovani et al., "Understanding How the Diagnostic Delay of Spondyloarthritis Differs Between Women and Men: A Systematic Review and Metaanalysis," *Journal of Rheumatology* 44, no. 2 (2017): 174–83; Maya Dusenbery, "Everybody Was Telling Me There Was Nothing Wrong," BBC, May 29, 2018, http://www.bbc.com/future/story/20180523-how-gender -bias-affects-your-healthcare; Joanne Demmler et al., "Diagnosed Prevalence of Ehlers-Danlos Syndrome and Hypermobility Spectrum Disorder in Wales, UK: A National Electronic Cohort Study and Case–Control Comparison," *BMJ Open* 9, no. 11 (2019); and Nafees Din et al., "Age and Gender Variations in Cancer Diagnostic Intervals in 15 Cancers: Analysis of Data from the UK Clinical Practice Research Datalink," *PLoS One* 10, no. 5 (2015): e0127717.

18. Patricia Jasen, "From the 'Silent Killer' to the 'Whispering Disease': Ovarian Cancer and the Uses of Metaphor," *Medical History* 53, no. 4 (2009): 489–512.

19. Maya Dusenbery, interview with the author, October 18, 2018; and Maya Dusenbery interview with Jenara Nerenberg, "How to Address Gender Inequality in Health Care," *Greater Good*, March 9, 2018.

20. Rodney Hayward, "Counting Deaths from Medical Errors," *JAMA* 288, no. 19 (2002): 2404–5; Jerome Kassirer and Richard Kopelman, "Cognitive Errors in Diagnosis: Instantiation, Classification, and Consequences," *American Journal of Medicine* 86, no. 4 (1989): 433–41; Mark Graber, Nancy Franklin, and Ruthanna Gordon, "Diagnostic Error in Internal Medicine," *Archives of Internal Medicine* 165, no. 13 (2005): 1493–99; Mathieu Nendaz and Arnaud Perrier, "Diagnostic Errors and Flaws in Clinical Reasoning: Mechanisms and Prevention in Practice," *Swiss Medical Weekly* (October 2012): 142; Brad Greenwood, Seth Carnahan, and Laura Huang,

"Patient–Physician Gender Concordance and Increased Mortality Among Female Heart Attack Patients," *PNAS* 115, no. 34 (2018): 8569–74; and Ivuoma N. Onyeador et al., "The Value of Interracial Contact for Reducing Anti-Black Bias Among Non-Black Physicians: A Cognitive Habits and Growth Evaluation (CHANGE) Study Report," *Psychological Science* 31, no. 1 (2020): 18–30.

21. Michele Beckman et al., "Venous Thromboembolism: A Public Health Concern," *American Journal of Preventive Medicine* 38, no. 4 (2010): S495–501; S. Z. Goldhaber, "Preventing Pulmonary Embolism and Deep Vein Thrombosis: A 'Call to Action' for Vascular Medicine Specialists," *Journal of Thrombosis and Haemostasis* 5, no. 8 (2007): 1607–9; and Michael Streiff et al., "Lessons from the Johns Hopkins Multi-Disciplinary Venous Thromboembolism (VTE) Prevention Collaborative," *British Medical Journal* 344 (2012): e3935.

22. Peter Pronovost et al., "An Intervention to Decrease Catheter-Related Bloodstream Infections in the ICU," *New England Journal of Medicine* 355 (2006): 2725–32; and Alex Haynes et al., "A Surgical Safety Checklist to Reduce Morbidity and Mortality in a Global Population," *New England Journal of Medicine* 360 (2009): 491–99.

23. Michael Streiff et al., "The Johns Hopkins Venous Thromboembolism Collaborative: Multidisciplinary Team Approach to Achieve Perfect Prophylaxis," *Journal of Hospital Medicine* 11, no. S2 (2016): S8–S13; Brandyn Lau et al., "Individualized Performance Feedback to Surgical Residents Improves Appropriate Venous Thromboembolism (VTE) Prophylaxis Prescription and Reduces Potentially Preventable VTE: A Prospective Cohort Study," *Annals of Surgery* 264, no. 6 (2016): 1181–87; Elliott Haut et al., "Improved Prophylaxis and Decreased Rates of Preventable Harm with the Use of a Mandatory Computerized Clinical Decision Support Tool for Prophylaxis for Venous Thromboembolism in Trauma," *Archives of Surgery* 147 no. 10 (2012): 901–7; and Amer Zeidan et al., "Impact of a Venous Thromboembolism Prophylaxis 'Smart Order Set': Improved Compliance, Fewer Events," *American Journal of Hematology* 88, no. 7 (2013): 545–49.

24. Brandyn Lau, interview with the author, November 20, 2017.

25. Brandyn Lau et al., "Eliminating Healthcare Disparities via Mandatory Clinical Decision Support: The Venous Thromboembolism (VTE) Example," *Medical Care* 53, no. 1 (2015): 18–24.

26. Richard Thaler, Cass Sunstein, and John Balz, "Choice Architecture" (April 2, 2010), available at SSRN: https://ssrn.com/abstract=1583509 or http://dx.doi.org/10.2139/ssrn.1583509.

27. J. P. Redden et al., "Serving First in Isolation Increases Vegetable Intake Among Elementary Schoolchildren," *PLoS One* 10, no. 4 (2015): e0121283; and Traci Mann, "The Science of Weight Loss," *Inquiring Minds*, April 25, 2015.

28. Joy Buolamwini and Timnit Gebru, "Gender Shades: Intersectional Accuracy

Disparities in Commercial Gender Classification," Proceedings of the 1st Conference on Fairness, Accountability and Transparency, *PMLR* 81 (2018): 77–91.

29. Purva Rawal et al., "Using Decision Support to Address Racial Disparities in Mental Health Service Utilization," *Residential Treatment for Children and Youth* 25, no. 1 (2008): 73–84; Supervía et al., "Cardiac Rehabilitation for Women," 565–77; information about Mayo Clinic internal initiative provided by Sharonne Hayes, interview with the author, December 12, 2018; and Douglas Starr, "Meet the Psychologist Exploring Unconscious Bias and Its Tragic Consequences for Society," *Science*, March 26, 2020.

30. Claudia Goldin and Cecilia Rouse, "Orchestrating Impartiality: The Impact of 'Blind' Auditions on Female Musicians," *American Economic Review* 90, no. 4 (2000): 715–41.

31. Brittany Shammas, "Broward's Gifted Programs Getting More Diverse," *Sun Sentinel*, October 17, 2015.

32. Cynthia Park, interviews with the author, September 6 and 8, 2017.

33. Laura Giuliano, interview with the author, September 2, 2016.

34. Jack Naglieri and Donna Ford, "Addressing Underrepresentation of Gifted Minority Children Using the Naglieri Nonverbal Ability Test (NNAT)," *Gifted Child Quarterly* 47, no. 2 (2003): 155–60.

35. Cynthia Park, interviews with the author, September 6 and 8, 2017; and Donna Turner, interview with the author, November 7, 2017.

36. David Card and Laura Giuliano, "Universal Screening Increases the Representation of Low-Income and Minority Students in Gifted Education," *PNAS* 113, no. 48 (2016): 13678–83.

37. David Card and Laura Giuliano, "Can Tracking Raise the Test Scores of High-Ability Minority Students?," *American Economic Review* 106, no. 10 (2016): 2783–816.

38. David Card and Laura Giuliano, "Universal Screening Increases the Representation of Low-Income and Minority Students in Gifted Education."

39. Stefanie Johnson and Jessica Kirk, "Dual-Anonymization Yields Promising Results for Reducing Gender Bias: A Naturalistic Field Experiment of Applications for Hubble Space Telescope Time," *Astronomical Society of the Pacific* 132, no. 1009 (published online, 2020).

40. Lauren Rivera, "Hiring as Cultural Matching: The Case of Elite Professional Service Firms," *American Sociological Review* 77, no. 6 (2012): 999–1022; and Lauren Rivera, *Pedigree: How Elite Students Get Elite Jobs* (Princeton, NJ: Princeton University Press, 2015), 92–99.

41. Eric Uhlmann and Geoffrey Cohen, "Constructed Criteria: Redefining Merit to Justify Discrimination," *Psychological Science* 16, no. 6 (2005): 474–80.

42. Mahnaz Behroozi et al., "Does Stress Impact Technical Interview Performance?," Proceedings of the 28th ACM Joint Meeting on European Software Engineering Conference (November 2020): 481–92.

43. Jessica Nordell, "How Slack Got Ahead in Diversity," *Atlantic*, April 26, 2018.

44. Oras Alabas et al., "Sex Differences in Treatments, Relative Survival, and Excess Mortality Following Acute Myocardial Infarction: National Cohort Study Using the SWEDEHEART Registry," Journal of the *American Heart Association* 6, no. 12 (2017): e007123; and Sharonne Hayes, interview with the author, December 12, 2018.

45. Roxanne Pelletier et al., "Sex-Related Differences in Access to Care among Patients with Premature Acute Coronary Syndrome," *Canadian Medical Association Journal* 186, no. 7 (2014): 497–504. See also Gabrielle Chiaramonte and Ronald Friend, "Medical Students' and Residents' Gender Bias in the Diagnosis, Treatment, and Interpretation of Coronary Heart Disease Symptoms," *Health Psychology* 25, no. 3 (2006): 255–66.

46. Sharonne Hayes, interview with the author, December 12, 2018.

47. Media company employee, interview with the author, August 11, 2016.

CHAPTER 8. DISMANTLING HOMOGENEITY

1. Lucy Battersby, "Twitter Criticised for Failing to Respond to Caroline Criado-Perez Rape Threats," *Age*, July 29, 2013, https://www.theage.com.au/technology/twitter-criticised-for-failing-to-respond-to-caroline-criadoperez-rape-threats-20130729-2qu8d.html.

2. Charlie Warzel, "'A Honeypot for Assholes': Inside Twitter's 10-Year Failure to Stop Harassment," *Buzzfeed*, August 11, 2016, https://www.buzzfeednews.com/article/charliewarzel/a-honeypot-for-assholes-inside-twitters-10-year-failure-to-s#.vtd6q73YB; Cécile Guerin and Eisha Maharasingam-Shah, "Public Figures, Public Rage," Institute for Strategic Dialogue, 2020, 3, https://www.isdglobal.org/wp-content/uploads/2020/10/Public-Figures-Public-Rage-4.pdf; "Toxic Twitter: a Toxic Place for Women," Amnesty International Report, 2018, https://www.amnesty.org/en/latest/research/2018/03/online-violence-against-women-chapter-1/#topanchor; and Azmina Dhrodia, "We Tracked 25,688 Abusive Tweets Sent to Women MPs—Half Were Directed at Diane Abbott," *New Statesman*, September 5, 2017.

3. Janko Roettgers, "Twitter CEO Admits Company Didn't Fully Grasp Abuse Problem," *Variety*, March 1, 2018, https://variety.com/2018/digital/news/twitter-ceo-abuse-1202714236/; Warzel, "'A Honeypot for Assholes.'"

4. Leslie Miley, interview with the author, September 20, 2016.

5. Jonah Berger and Katherine Milkman, "What Makes Online Content Viral?," *Journal of Marketing Research* 49, no. 2 (2012): 192–205.

6. Leslie Miley, interview with the author, September 20, 2016.

7. Melvin Conway, "How Do Committees Invent?," *Datamation*, April 1968, http://www.melconway.com/research/committees.html.

8. Warzel, "'A Honeypot for Assholes.'"

9. Ev Williams, "Never Underestimate Your First Idea," interview by Reid Hoffman,

Masters of Scale, February 21, 2018, https://mastersofscale.com/wp-content
/uploads/2018/02/ep.-20_-masters-of-scale-with-ev-williams-formatted
-transcript.pdf.

10. Leslie Miley, interview with the author, September 20, 2016.

11. Warzel, "'A Honeypot for Assholes'"; Williams, "Never Underestimate Your
First Idea"; Cécile Guerin and Eisha Maharasingam-Shah, *Public Figures, Public
Rage: Candidate Abuse on Social Media*, ISD Global, October 5, 2020, https://
www.isdglobal.org/isd-publications/public-figures-public-rage-candidate
-abuse-on-social-media/; and Leslie Miley, interview with the author, September 20, 2016.

12. Sirin Kale, "'I Felt Like a Trapped Animal': 6 Women Describe What It's Like
to Be Stalked," *Vice*, July 18, 2018, https://www.vice.com/en/article/8xewn5/
women-describe-stalking-harassment-cases.

13. Craig Silverman, Ryan Mac, and Pranav Dixit, "'I Have Blood on My Hands':
A Whistleblower Says Facebook Ignored Global Political Manipulation," *Buzzfeed*, September 14, 2020, https://www.buzzfeednews.com/article/craigsilverman
/facebook-ignore-political-manipulation-whistleblower-memo.

14. "Labor Force Statistics from the Current Population Survey," Bureau of Labor Statistics (2019), last modified January 22, 2020, https://www.bls.gov/cps/cpsaat11
.htm; Sinduja Rangarajan, "Silicon Valley Diversity Data: Who Released Theirs,
Who Didn't," *Reveal News*, June 25, 2018, https://apps.revealnews.org/silicon-valley
-diversity-list/; and "Fortune 500 List: Which CEOs from Top US Companies Have
MBAs?," U2B Executive Business Education, May 25, 2020, https://u2b.com/2020
/05/25/fortune-500-do-ceos-from-top-us-companies-all-have-mbas/. CEO and
height calculations derived from publicly available demographic information.

15. Maya Beasley, "There Is a Supply of Diverse Workers in Tech, so Why Is Silicon Valley So Lacking in Diversity?," Center for American Progress, March
29, 2017, https://www.americanprogress.org/issues/race/reports/2017/03
/29/429424/supply-diverse-workers-tech-silicon-valley-lacking-diversity/;
Quoctrung Bui and Claire Cain Miller, "Why Tech Degrees Are Not Putting More Blacks and Hispanics into Tech Jobs," *New York Times*, February
25, 2016; Kapor Center, "Understanding the Leaky Tech Pipeline: The Lack of
Racial and Gender Diversity in the Tech Workforce," accessed January 21, 2021,
https://leakytechpipeline.com/pipeline/tech-workforce/; Sinduja Rangarajan,
"Here's the Clearest Picture of Silicon Valley's Diversity Yet: It's Bad. But Some
Companies Are Doing Less Bad," *Reveal News*, June 25, 2018, https://www
.revealnews.org/article/heres-the-clearest-picture-of-silicon-valleys-diversity
-yet/; Jennifer Glass et al., "What's So Special about STEM? A Comparison of
Women's Retention in STEM and Professional Occupations," *Social Forces* 92,
no. 2 (2013): 723–56; and Members Directory, National Academy of Engineering, accessed October 7, 2020, https://www.nae.edu/MemberDirectory.aspx.

16. Anette Hosoi, interview with the author, March 7, 2018.

17. Djuna Copley-Woods, interview with the author, March 18, 2018.

18. MIT alumna, interview with the author, October 4, 2019.

19. Members of the First and Second Committees on Women Faculty in the School of Science, "A Study on the Status of Women Faculty at MIT, 1996–1999," *MIT Faculty Newsletter*, March 1999; and Nancy Hopkins, Lotte Bailyn, Lorna Gibson, and Evelynn Hammonds, "Report of the Committees on the Status of Women Faculty," Massachusetts Institute of Technology, March 2002, https://facultygovernance.mit.edu/sites/default/files/reports/2002-03_Status_of_Women_Faculty-All_Reports.pdf.

20. Hopkins et al., "Report of the Committees on the Status of Women Faculty"; and Heather Antecol, Kelly Bedard, and Jenna Stearns, "Equal but Inequitable: Who Benefits from Gender-Neutral Tenure Clock Stopping Policies?," *American Economic Review* 108, no. 9 (2019): 2420–41.

21. Raleigh McElvery, "3 Questions: Nancy Hopkins on Improving Gender Inequality in Academia," *MIT News*, September 30, 2020, https://news.mit.edu/2020/3-questions-nancy-hopkins-improving-gender-equality-in-academia-0930.

22. Kath Xu, Dawn Wendell, and Andrea S. Walsh, "Getting to Gender Parity in a Top-Tier Mechanical Engineering Department: A Case Study," ASEE Annual Conference & Exposition, Columbus, Ohio, June 2017, https://peer.asee.org/28406; and Hopkins et al., "Report of the Committees on the Status of Women Faculty."

23. Information about MIT history from Doug Hart (February 28, 2018), Kath Xu (February 28, 2018), Dawn Wendell (January 23, 2018), Anette Hosoi (March 7, 2018), and Yang Shao-Horn (March 7, 2018), interviews by the author.

24. Kath Xu, interview with the author, February 28, 2018; and Douglas Hart, interview with the author, February 28, 2018.

25. Thomas Sowell, "Affirmative Action Around the World," *Hoover Digest*, Hoover Institution, Stanford University, October 30, 2004, https://www.hoover.org/research/affirmative-action-around-world; and Richard Sander and Stuart Taylor, Jr., "The Painful Truth About Affirmative Action," *Atlantic*, October 2, 2012.

26. Richard Lempert, David Chambers, and Terry Adams, "Michigan's Minority Graduates in Practice: The River Runs Through Law School," *Law and Social Inquiry* 25, no. 2 (2000): 395–505; James Sterba, "Completing Thomas Sowell's Study of Affirmative Action and Then Drawing Different Conclusions," *Stanford Law Review* 57, no. 2 (2004): 657–93; and William Bowen and Derek Bok, *The Shape of the River* (Princeton, NJ: Princeton University Press, 2000), 259. See also Stacy B. Dale and Alan B. Krueger, "Estimating the Effects of College Characteristics over the Career Using Administrative Earnings Data," *Journal of Human Resources* 49, no. 2 (Spring 2014): 323–58; and Ronald Smothers, "To Raise the Performance of Minorities, a College Increased Its Standards," *New York Times*, June 29, 1994, https://www.nytimes.com/1994/06/29/us/to-raise-the-performance-of-minorities-a-college-increased-its-standards.html.

27. Bowen and Bok, *The Shape of the River*, 202, 263–65; Ashley Hibbett, "The Enigma of the Stigma: A Case Study on the Stigma Arguments Made in Opposition to

Affirmative Action Programs in Higher Education," *Harvard Blackletter Law Journal* 21 (Spring 2005): 75; and Anita Allen, "Was I Entitled or Should I Apologize? Affirmative Action Going Forward," *Journal of Ethics* 15 (2011): 253–63.

28. Davis Hsu, "MIT Remains in Favor of Affirmative Action," *Tech*, October 27, 1995.

29. Madeline Heilman, Caryn Block, and Peter Stathatos, "The Affirmative Action Stigma of Incompetence: Effects of Performance Information Ambiguity," *Academy of Management Journal* 40, no. 3 (1997): 603–25; MIT Diversity Statement, retrieved October 14, 2020, https://mitadmissions.org/policies/#diversity; and Xu, Wendell, and Walsh, "Getting to Gender Parity in a Top-Tier Mechanical Engineering Department."

30. Deirdre Bowen, "Brilliant Disguise: An Empirical Analysis of a Social Experiment Banning Affirmative Action," *Indiana Law Journal* 85, no. 4 (2010): 1197–254; Angela Onwuachi-Willig, Emily Houh, and Mary Campbell, "Cracking the Egg: Which Came First—Stigma or Affirmative Action?," *California Law Review* 96 (2008): 1299–352; and Evelyn Carter, personal correspondence with the author, August 16, 2020.

31. Peter Arcidiacono, Josh Kinsler, and Tyler Ransom, "Legacy and Athlete Preferences at Harvard," National Bureau of Economic Research Working Paper 26316, September 2019, https://www.nber.org/papers/w26316.

32. Kelly Rivoire, "Number of Female Faculty Increases," *Tech* 124, no. 20 (2004), http://tech.mit.edu/V124/N20/20womenfac.20n.html; Xu, Wendell, and Walsh, "Getting to Gender Parity in a Top-Tier Mechanical Engineering Department"; and Doug Hart, interview with the author, February 28, 2018.

33. Xu, Wendell, and Walsh, "Getting to Gender Parity in a Top-Tier Mechanical Engineering Department."

34. Doug Hart, interview with the author, February 28, 2018.

35. Xu, Wendell, and Walsh, "Getting to Gender Parity in a Top-Tier Mechanical Engineering Department."

36. "Enrollment Statistics 2018–2019," MIT Registrar's Office, https://registrar.mit.edu/stats-reports/enrollment-statistics-year/all; Xu, Wendell, and Walsh, "Getting to Gender Parity in a Top-Tier Mechanical Engineering Department"; and Luwen Huang, Elizabeth Qian, and Karen Willcox, "Gender Diversity: The Past Two Decades at the Massachusetts Institute of Technology," MIT Diversity Dashboard, October 2017, https://kiwi.oden.utexas.edu/mit-gender-diversity.php.

37. Anette Hosoi, interview with the author, March 7, 2018; MIT faculty member, interview with the author, March 1, 2018.

38. Tara Dennehy and Nilanjana Dasgupta, "Female Peer Mentors Early in College Increase Women's Positive Academic Experiences and Retention in Engineering," *Proceedings of the National Academy of Sciences* 114, no. 23 (2017): 5964–69; Nilanjana Dasgupta, "Ingroup Experts and Peers as Social Vaccines Who Inoculate the Self-Concept: The Stereotype Inoculation Model," *Psychological Inquiry* 22, no. 4 (2011): 231–46.

39. India Johnson et al., "Exploring Identity-Safety Cues and Allyship Among Black Women Students in STEM Environments," *Psychology of Women Quarterly* 43, no. 2 (2019): 131–50.

40. Asegun Henry, interview with the author, September 24, 2020.

41. "Top 100 Women," International Chess Federation, October 2020, https://ratings.fide.com/top_lists.phtml?list=women.

42. Lori Beaman, Esther Duflo, Rohini Pande, and Petia Topalova, "Female Leadership Raises Aspirations and Educational Attainment in Girls: A Policy Experiment in India," *Science* 335, no. 6068 (2012): 582–86.

43. Anette Hosoi, interview with the author, March 7, 2018.

44. MIT students, interviews with the author, March 13, 2018.

45. Betar Gallant, interview with the author, March 1, 2018.

46. Alice Nasto, interview with the author, March 13, 2018.

47. Djuna-Copley Woods, interview with the author, March 18, 2018.

48. Raj Chetty, Nathaniel Hendren, and Lawrence Katz, "The Effects of Exposure to Better Neighborhoods on Children: New Evidence from the Moving to Opportunity Project," *American Economic Review* 106, no. 4 (2016): 855–902; and Lori Beaman, Raghabendra Chattopadhyay, Esther Duflo, Rohini Pande, and Petia Topalova, "Powerful Women: Does Exposure Reduce Bias?," *Quarterly Journal of Economics* 124, no. 4 (2009): 1497–540.

49. Beaman et al., "Female Leadership Raises Aspirations and Educational Attainment for Girls"; Lori Beaman et al. "Political Reservation and Substantive Representation: Evidence from Indian Village Councils," *India Policy Forum* (2010–11): 151–91.

50. Asegun Henry, interview with the author, September 24, 2020; Wei Lv and Asegun Henry, "Examining the Validity of the Phonon Gas Model in Amorphous Materials," *Scientific Reports* 6 (2016): 37675.

51. Diana Rhoten and Stephanie Pfirman, "Women in Interdisciplinary Science: Exploring Preferences and Consequences," *Research Policy* 36, no. 1 (2007): 56–75.

52. Nicholas R. Jones, Zeshan Qureshi, Robert Temple, Jessica Larwood, Trisha Greenhalgh, and Lydia Bourouiba, "Two Metres or 1? What Is the Evidence Base for Physical Distancing in the Context of COVID-19?," *British Medical Journal* 370, no. 8259 (2020); and Lydia Bourouiba, "Turbulent Gas Clouds and Respiratory Pathogen Emissions: Potential Implications for Reducing Transmission of COVID-19," *JAMA* 323, no. 18 (2020): 1837–38.

53. Christopher Begeny et al., "In Some Professions, Women Have Become Well Represented, Yet Gender Bias Persists—Perpetuated by Those Who Think It Is Not Happening," *Science Advances* 6, no. 26 (2020): EABA7814.

54. Doug Hart, interview with the author, February 28, 2018

55. Doug Hart, interview with the author, February 28, 2018.

56. Yang Shao-Horn, interview with the author, March 7, 2018.

CHAPTER 9. THE ARCHITECTURE OF INCLUSION

1. Uché Blackstock, interviews with the author, October 3 and 30, 2019.
2. Linda Pololi, Lisa Cooper, and Phyllis Carr, "Race, Disadvantage, and Faculty Experiences in Academic Medicine," *Journal of General Internal Medicine* 25 (2010): 1363–69; Herschel Alexander and Jonathan Lang, "The Long-Term Retention and Attrition of U.S. Medical School Faculty,"*Association of American Medical Colleges* 8, no. 4, (2008); Phyllis Carr et al., "Inadequate Progress for Women in Academic Medicine: Findings from the National Faculty Study," *Journal of Women's Health* 24, no. 3 (2015): 190–99; Destiny Peery, Paulette Brown, and Eileen Letts, "Left Out and Left Behind: The Hurdles, Hassles, and Heartaches of Achieving Long-Term Legal Careers for Women of Color," American Bar Association, Commission on Women in the Profession, 2020; "Journalism's Bad Reflection," *Columbia Journalism Review* (Fall 2018), https://www.cjr.org/special_report/10-newsrooms-racial-disparity.php; Richard Prince, "7 Journalists of Color Leaving Houston Chronicle," *Root*, August 6, 2017, https://journalisms.theroot.com/7-journalists-of-color-leaving-houston-chronicle-1797574585; Nancy Cassutt, "Racial Bias in MPR's Work? We Want to Know," *MPR News*, January 22, 2019, https://www.mprnews.org/story/2019/01/18/mpr-news-changing-racial-narratives; Catherine Hill et al., *Why So Few? Women in Science, Technology, Engineering, and Mathematics* (Washington, DC: American Association of University Women, 2010); Susan S. Silbey, "Why Do So Many Women Who Study Engineering Leave the Field?," *Harvard Business Review*, August 23, 2016; and Michàlle Mor Barak et al., "Organizational and Personal Dimensions in Diversity Climate: Ethnic and Gender Differences in Employee Perceptions," *Journal of Applied Behavioral Science* 34, no. 1 (1998): 82–104. See also Samantha Kaplan et al., "Race/Ethnicity and Success in Academic Medicine: Findings from a Longitudinal Multi-institutional Study," *Academic Medicine* 93, no. 4 (2018): 616-22.
3. Ajilli Hardy, interview with the author, April 21, 2019.
4. Lisa Nishii, "The Benefits of Climate for Inclusion for Gender-Diverse Groups," *Academy of Management Journal* 56, no. 6 (2013): 1754–74.
5. Robin Ely, interview with the author, October 10, 2019.
6. Yang Shao-Horn, interview with the author, March 7, 2018; and Doualy Xaykaothao, interview with the author, August 30, 2019.
7. Doualy Xaykaothao, interview with the author, August 30, 2019; and Robin Ely, interview with the author, October 10, 2019.
8. Robert Dailey and Delaney Kirk, "Distributive and Procedural Justice as Antecedents of Job Dissatisfaction and Intent to Turnover," *Human Relations* 45, no. 3 (1992): 305–17.
9. Taj history from Gianmarco Monsellato, interview with the author, March 30, 2018; Linda Buisson, interview with the author, March 27, 2018; Sophie Blégent-Delapille, interview with the author, April 20, 2018; and Deloitte, "Seven Lessons

in Gender Diversity: How Values-Driven Leadership Leads to the Advancement of Women," March 1, 2013, https://www2.deloitte.com/content/dam/Deloitte /global/Documents/dttl_Diversity_lessons_from%20Taj_March2013.pdf.

10. Gianmarco Monsellato, personal correspondence with the author, February 22, 2021; and Gianmarco Monsellato, interview with the author, March 30, 2018. Revenue from *Juristes Associes* annual reports, 2004 to 2016, provided to author March 26, 2021.

11. Malik Douaoui, interview with the author, May 13, 2019.

12. Frank Dobbin, Daniel Schrage, and Alexandra Kalev, "Rage Against the Iron Cage: The Varied Effects of Bureaucratic Personnel Reforms on Diversity," *American Sociological Review* 80, no. 5 (2015): 1014–44; Frank Dobbin and Alexandra Kalev, "Why Diversity Programs Fail," *Harvard Business Review*, July 1, 2016, https://hbr.org/2016/07/why-diversity-programs-fail; Alexandra Kalev, Frank Dobbin, Erin Kelly, "Best Practices or Best Guesses? Assessing the Efficacy of Corporate Affirmative Action and Diversity Policies," *American Sociological Review* 71, no. 4 (2006): 589–617; and Frank Dobbin, personal correspondence with the author, October 30, 2019.

13. Robin Ely and David Thomas, "Cultural Diversity at Work: The Effects of Diversity Perspectives on Work Group Processes and Outcomes," *Administrative Science Quarterly* 46, no. 2 (2001): 229–73.

14. Martin Davidson, interview with the author, November 25, 2019.

15. Martin Davidson, interview with the author, November 25, 2019; and Evelyn Carter, interview with the author, February 21, 2019.

16. Sara Ahmed, *On Being Included: Racism and Diversity in Institutional Life* (Durham, NC: Duke University Press, 2012), 102.

17. Gianmarco Monsellato, interview with the author, March 30, 2018.

18. Sophie Blégent-Delapille, interview with the author, April 20, 2018.

19. Katherine Williams Phillips and Charles O'Reilly III, "Demography and Diversity in Organizations: A Review of 40 Years of Research," *Research in Organizational Behavior* 20 (January 1998): 77–140; and Wendy DuBow and J. J. Gonzalez, *NCWIT Scorecard: The Status of Women in Technology* (Boulder, CO: NCWIT, 2020).

20. Martha L. Maznevski, "Understanding Our Differences: Performance in Decision-Making Groups with Diverse Members," *Human Relations* 47, no. 5 (1994): 531–52, https://doi.org/10.1177/001872679404700504; Robin Ely and David Thomas, "Getting Serious About Diversity: Enough Already with the Business Case," *Harvard Business Review* 98, no. 6 (November 2020); and Robin Ely, Irene Padavic, and David Thomas, "Racial Diversity, Racial Asymmetries, and Team Learning Environment: Effects on Performance," *Organization Studies* 33, no. 3 (2012): 341–62.

21. Robin Ely, interview with the author, October 10, 2019.

22. Terrell Strayhorn, "When Race and Gender Collide: Social and Cultural Capital's Influence on the Academic Achievement of African American and Latino

Males," *Review of Higher Education* 33, no. 3 (2010): 307–22; Anthony Patterson, "'It Was Really Tough': Exploring the Feelings of Isolation and Cultural Dissonance with Black American Males at a Predominantly White Institution," *Journal of College Student Retention: Research, Theory & Practice*, September 2018; Amanda Glazer, "National Mathematics Survey," MIT Women in Mathematics, https://math.mit.edu/wim/2019/03/10/national-mathematics-survey/; Daniel Grunspan et al., "Males Under-Estimate Academic Performance of Their Female Peers in Undergraduate Biology Classrooms," *PLoS One* 11, no. 2 (2016): e0148405; and Brittany Bloodhart et al., "Outperforming Yet Undervalued: Undergraduate Women in STEM," PLoS One 15, no. 6 (2020): e0234685.

23. Federico Ardila-Mantilla, interview with the author, April 15, 2019; Andrés Vindas-Meléndez, interview with the author, May 16, 2019; Federica Ardila-Mantilla, "Todos Cuentan: Cultivating Diversity in Combinatorics," *Notices of the American Mathematical Society* 63, no. 10 (2016): 1164–70; and Aram Dermenjian, interview with the author, May 16, 2019.

24. Federico Ardila-Mantilla, interview with the author, April 15, 2019; Federico Ardila-Mantilla, "encuentro colombiano de combinatoria: community agreement," math.sfsu.edu/federico/SFSUColombia/eccoagreement.pdf; Federico Ardila-Mantilla and Carolina Bendetti, "Todxs Cuentan in ECCO: Community and Belonging in Mathematics," 2020, https://arxiv.org/abs/2008.02877; and "Todxs Cuentan: Building Community and Welcoming Humanity from the First Day of Class," math.sfsu.edu/federico/society.html, accessed April 2, 2021.

25. Aram Dermenjian, interview with the author, May 16, 2019.

26. Nolan Cabrera et al., "Missing the (Student Achievement) Forest for All the (Political) Trees: Empiricism and the Mexican American Studies Controversy in Tucson," *American Educational Research Journal*, 51, no. 6 (2014), 1084–1118. See also Thomas Dee and Emily Penner, "The Causal Effects of Cultural Relevance: Evidence from an Ethnic Studies Curriculum," National Bureau of Economic Research, 2016, https://doi.org/10.3386/w21865.

27. Erica Klarreich, "A Mathematician Who Dances to the Joys and Sorrows of Discovery," *Quanta Magazine*, November 20, 2017, https://www.quantamagazine.org/mathematician-federico-ardila-dances-to-the-joys-and-sorrows-of-discovery-20171120/.

28. Tara Dennehy and Nilanjana Dasgupta, "Female Peer Mentors Early in College Increase Women's Positive Academic Experiences and Retention in Engineering," *Proceedings of the National Academy of Sciences* 114, no. 23 (2017): 5964–69; Nilanjana Dasgupta, "Ingroup Experts and Peers as Social Vaccines Who Inoculate the Self-Concept: The Stereotype Inoculation Model," *Psychological Inquiry* 22, no. 4 (2011): 231–46; Catherine Good, Aneeta Rattan, and Carol Dweck, "Why Do Women Opt Out? Sense of Belonging and Women's Representation in Mathematics," *Journal of Personality and Social Psychology* 102, no. 4 (2012): 700–17; Gregory Walton and Geoffrey Cohen, "A Question of Belonging: Race, Social Fit, and Achievement," *Journal of Personality*

and Social Psychology 92, no. 1 (2007): 82–96; and Sophie Kuchynka, Danielle Findley-Van Nostrand, and Richard Pollenz, "Evaluating Psychosocial Mechanisms Underlying STEM Persistence in Undergraduates: Scalability and Longitudinal Analysis of Three Cohorts from a Six-Day Pre–College Engagement STEM Academy Program," *CBE—Life Sciences Education* 18, no. 3 (2019): ar41, https://doi.org/10.1187/cbe.19-01-0028. For a review, see Gloriana Trujillo and Kimberly Tanner, "Considering the Role of Affect in Learning: Monitoring Students' Self-Efficacy, Sense of Belonging, and Science Identity," *CBE Life Sciences Education* 13, no. 1 (2014): 6–15.

29. Jessica Good, Kimberley Bourne, and Grace Drake, "The Impact of Classroom Diversity Philosophies on the STEM Performance of Undergraduate Students of Color," *Journal of Experimental Social Psychology* 91 (2020): 104026.

30. Dom deGuzman, interview with the author, April 10, 2019.

31. Leslie Miley, interview with the author, September 20, 2016.

32. Mekka Okereke, "Building Inclusive Engineering Teams," Calibrate 2018, May 9, 2018, https://www.youtube.com/watch?v=SYsI-6_csMY&feature=emb_logo; and Mekka Okereke, personal correspondence with the author, March 2021.

33. Yolanda Davis, interview with the author, April 4, 2019.

CHAPTER 10. UNBREAKING CULTURE

1. Glenn Adams et al., "Beyond Prejudice: Toward a Sociocultural Psychology of Racism and Oppression," in *Commemorating Brown: The Social Psychology of Racism and Discrimination*, ed. G. Adams et al. (Washington, DC: American Psychological Association, 2008), 236.

2. Heba Y. Amin, "'Arabian Street Artists' Bomb Homeland: Why We Hacked an Award-Winning Series," http://www.hebaamin.com/arabian-street-artists -bomb-homeland-why-we-hacked-an-award-winning-series/.

3. Markus Brauer and Abdelatif Er-rafiy, "Increasing Perceived Variability Reduces Prejudice and Discrimination," *Journal of Experimental Social Psychology* 47, no. 5 (2011): 871–81; Abdelatif Er-rafiy, Markus Brauer, and Serban Musca, "Effective Reduction of Prejudice and Discrimination: Methodological Considerations and Three Field Experiments," *Revue Internationale de Psychologie Sociale/International Review of Social Psychology* 23, no. 2 (2010): 57–95; Abdelatif Er-rafiy and Markus Brauer, "Modifying Perceived Variability: Four Laboratory and Field Experiments Show the Effectiveness of a Ready-to-Be-Used Prejudice Intervention," *Journal of Applied Social Psychology* 43, no. 4 (2013): 840–53; and Markus Brauer et al., "Describing a Group in Positive Terms Reduces Prejudice Less Effectively Than Describing It in Positive and Negative Terms," *Journal of Experimental Social Psychology* 48, no. 3 (2012): 757–61.

4. Brauer and Er-rafiy, "Increasing Perceived Variability Reduces Prejudice and Discrimination."

5. Charlotte Cole, interview with the author, July 12, 2019; Jordan Bliss et al.,

"Sesame Workshop: Going Global with Muppets," Columbia Business School Student Research Paper, 2006, https://www0.gsb.columbia.edu/mygsb/faculty /research/pubfiles/2355/Sesame%20Workshop.pdf; and Shalom Fisch, *Children's Learning from Educational Television: Sesame Street and Beyond* (Abingdon, U.K.: Routledge, 2014), 105–21.

6. "Sesame Street: The World According to Sesame Street," *The Orchard Entertainment*, PBS, October 24, 2006.

7. "Sesame Street: The World According to Sesame Street."

8. Charlotte Cole, interview with the author, July 12, 2019.

9. Charlotte Cole, interview with the author, July 12, 2019; and "Same Different Song," *Rruga Sesam/Ulica Sezam*, Sesame Street International Social Impact, October 22, 2015, https://www.youtube.com/watch?v=_nv_NKgijV8.

10. "Assessment of Educational Impact of *Rruga Sesam* and *Ulica Sezam* in Kosovo, Report of Findings: Prepared for Sesame Workshop," Fluent, 2008, http://downloads.cdn.sesame.org/sw/SWorg/documents/FullKosovoReport _Jan+2008.pdf; and "Ripple Effects: Using Sesame Street to Bridge Group Divides in the Middle East, Kosovo, Northern Ireland, and Elsewhere," in *The Sesame Effect: The Global Impact of the Longest Street in the World*, ed. Charlotte Cole and June Lee (New York: Routledge, 2016), 162–69.

11. Cole and Lee, *The Sesame Effect*, xxiii.

12. Ervin Staub, "Reconciliation After Genocide, Mass Killing, or Intractable Conflict: Understanding the Roots of Violence, Psychological Recovery, and Steps Toward a General Theory," *Political Psychology* 27, no. 6 (2006): 867–94; Mahmood Mamdani, *When Victims Become Killers: Colonialism, Nativism, and the Genocide in Rwanda* (Princeton, NJ: Princeton University Press, 2001), 98–102; Susan Pederson, *The Guardians: The League of Nations and the Crisis of Empire* (New York: Oxford University Press, 2015), 241; Timothy Longman. "Identity Cards, Self-Perception, and Genocide in Rwanda," in *Documenting Individual Identity: The Development of State Practices in the Modern World*, ed. Jane Caplan and John Torpey (Princeton, NJ: Princeton University Press, 2001), 345–57; Felix Mukwiza Ndahinda, "Ethnicities in Rwanda: The Mythical Foundations of a Contemporary Reality," *Love Radio Rwanda*, April 2014, http://www.loveradio-rwanda.org/episode/2/onair/ essay; Timothy Longman, personal correspondence with the author, February 8, 2021; and Filip Reyntjens, personal correspondence with the author, February 26, 2021.

13. RTLM radio transcripts, March 23, 1994, http://migs.concordia.ca/links /RwandanRadioTrascripts_RTLM.htm, accessed January 20, 2021.

14. Darryl Li, "Echoes of Violence," *Dissent* (Winter 2002); David Yanagizawa-Drott, "Propaganda and Conflict: Evidence from the Rwandan Genocide," *Quarterly Journal of Economics* 129, no. 4 (2014): 1947–94; and *Prosecutor v. Ferdinand Nahimana, Jean-Bosco Barayagwiza, and Hassan Ngeze*, ICTR-99-52-T, 34936 (International Criminal Tribunal for Rwanda).

15. "La Benevolencija in Rwanda," La Benevolencija Humanitarian Tools Foundation, 2010–2013, http://www.labenevolencija.org/rwanda/la-benevolencija-in-rwanda/; Ndahinda, "Ethnicities in Rwanda." For a retelling of New Dawn, see the Web documentary Love Radio Rwanda, http://www.loveradio-rwanda.org/episode/2/onair/essay.

16. Elizabeth Levy Paluck, "Reducing Intergroup Prejudice and Conflict Using the Media: A Field Experiment in Rwanda," Journal of Personality and Social Psychology 96, no. 3 (2009): 57–87; and Elizabeth Levy Paluck, interview with the author, July 7, 2016.

17. Elizabeth Levy Paluck, interview with the author, July 7, 2016.

18. Noah Goldstein, Robert Cialdini, and Vladas Griskevicius, "A Room with a Viewpoint: Using Social Norms to Motivate Environmental Conservation in Hotels," Journal of Consumer Research 35, no. 3 (2008): 472–82.

19. Jessica Nolan, Paul Schultz, and Robert Cialdini, "Normative Social Influence Is Underdetected," Personality and Social Psychology Bulletin 34, no. 7 (2008): 913–23.

20. Robert Cialdini et al., "Managing Social Norms for Persuasive Impact," Social Influence 1, no. 1 (2006): 3–15; and Robert Cialdini, "Crafting Normative Messages to Protect the Environment," Current Directions in Psychological Science 12, no. 4 (2003): 105–9.

21. Elizabeth Levy Paluck, interview with the author, July 7, 2016.

22. Gretchen B. Sechrist and Lisa R. Milford, "The Influence of Social Consensus Information on Intergroup Helping Behavior," Basic and Applied Social Psychology 29, no. 4 (2007): 365–74, https://doi.org/10.1080/01973530701665199.

23. Stacey Sinclair et al., "Social Tuning of Automatic Racial Attitudes: The Role of Affiliative Motivation," Journal of Personality and Social Psychology 89, no. 4 (2005): 583–92.

24. Charlotte Cole, interview with the author, July 12, 2019.

25. History and overview of the program comes from interviews with Lotte Rajalin and teachers at Egalia and Nikolaigarden, October 13, 2017; Lotte Rajalin, interviews with the author, July 10 and October 13, 2017; Sverige Delegationen för jämställdhet i förskolan, "Jämställdhet i förskolan: om betydelsen av jämställdhet och genus i förskolans pedagogiska arbete: slutbetänkande" (Stockholm: Tryckt av Edita Sverige, 2006); and Tuba Acar Erdol, "Practicing Gender Pedagogy: The Case of Egalia," Egitimde Nitel Arastirmalar Dergisi—Journal of Qualitative Research in Education 7, no. 4 (2019): 1365–85.

26. Mary Leinbach and Beverly Fagot, "Categorical Habituation to Male and Female Faces: Gender Schematic Processing in Infancy," Infant Behavior and Development 16, no. 3 (1993): 317–32; Peter LaFreniere, Fred Strayer, and Roger Gauthier, "The Emergence of Same-Sex Affiliative Preferences in Children's Play Groups: A Developmental/Ethological Perspective," Child Development 55 (1984): 1958–65.

27. Kristin Shutts et al., "Early Preschool Environments and Gender: Effects of

Gender Pedagogy in Sweden," *Journal of Experimental Child Psychology* 162 (2017): 1–17.

28. Lotta Rajalin, interview with the author, October 13, 2017.

29. Ana Garcia Rodriguez, interview with the author October 13, 2017.

30. Garcia Rodriguez, interview with the author, October 13, 2017.

31. Michael Wells and Disa Bergnehr, "Families and Family Policies in Sweden," in *Handbook of Family Policies Across the Globe*, ed. Mihaela Robila (New York: Springer, 2014), 91–108; Celia Modig, "Never Violence: Thirty Years On from Sweden's Abolition of Corporal Punishment," publication of Ministry of Health and Social Affairs, Sweden, and Save the Children, Sweden, trans. Greg McIvor (Stockholm, 2009).

32. Chester Pierce and Gail Allen, "Childism," *Psychiatric Annals* 5, no. 7 (1975): 15–24; and Ezra Griffith and Chester Pierce, *Race and Excellence: My Dialogue with Chester Pierce* (Iowa City: University of Iowa Press, 1998), 140–41.

CONCLUSION

1. Quoted in Alison Bailey, "On White Shame and Vulnerability," *South African Journal of Philosophy* 30 (no. 4), 2011: 472–83.

2. Evelyn Carter, personal correspondence with the author, December 20, 2019.

3. Mara Lynn Keller, The Eleusian Mysteries of Demeter and Persephone," *Journal of Feminist Studies of Religion* 4, no. 1 (1988); and Steven Roberts et al., "God as a White man: A Psychological Barrier to Conceptualizing Black People and Women as Leadership Worthy," *Journal of Personality and Social Psychology* (January 30, 2020).

4. Susan Faludi, "Death of a Revolutionary," *New Yorker*, April 8, 2013.

5. James Baldwin, "A Letter to My Nephew," *Progressive*, December 1, 1961; John Haller, *Outcasts from Evolution: Scientific Attitudes of Racial Inferiority, 1859–1900* (Urbana: University of Illinois Press, 1971), 41–61; Charles Bacon, "The Race Problem," *Medicine* 10 (May, 1903): 338–43; and Audre Lorde, "A Litany for Survival," from *The Collected Poems of Audre Lorde* (New York: Norton, 1978).

6. Lisa Tessman, *Burdened Virtues: Virtue Ethics for Liberatory Struggles* (New York: Oxford University Press, 2005), 327; and Samantha Vice, "'How Do I Live in This Strange Place?,'" *Journal of Social Philosophy* 41, no. 3 (2010): 323–42.

7. Nancy Sherman, *Afterwar: Healing the Moral Wounds of Our Soldiers* (New York: Oxford University Press, 2015)

8. Audre Lorde, "Age, Race, Class, and Sex: Women Redefining Difference," paper delivered at the Copeland Colloquium, Amherst College, April 1980, reprinted in Audre Lorde, *Sister Outsider* (Trumansburg, NY: Crossing Press, 1984).

9. Pumla Gobodo-Madikizela, *What Does It Mean to Be Human in the Aftermath of Historical Trauma?* (Uppsala, Sweden: Nordic Africa Institute, 2016); and Pumla Gobodo-Madikizela, "Forgiveness and the Maternal Body: An African Ethics of Interconnectedness," Essays on Exploring a Global Dream, Fetzer Institute, Spring 2011.

ACKNOWLEDGMENTS

This project would not exist without the generosity, wisdom, love, and skill of countless people. I want to thank the researchers who gave so generously of their time and expertise, spending hours with me on the issues they study, sometimes fielding countless follow-up phone calls and emails. Many I met in person; others I got to know through their work; in all cases, their passion and care astound. The same goes for dozens of experts across numerous fields, from Assyria to robotics, as well as employees at myriad organizations and countless citizens whom I thank for many incisive, honest, and vulnerable conversations.

For research assistance I thank Mackenzie McDonald, Sandesh Ghimire, and Alex Yablon and Abby Sanders, who did extensive and yeoman's work. My ongoing conversations with Noah Kim and Ruby Bilger were a weekly source of inspiration and made the monastic process of writing more joyful. Peregrine Stevens stepped in with her raptor-like vision at several key moments.

My utmost gratitude to the many friends, colleagues, and experts who read this book, in part or in whole, and offered astute comments and criticism: David Otero, Ajilli Hardy, Kao Kalia Yang, Madeleine Baran,

Nick Adams, Scott Schofield, Lindsay Nordell, and V. V. Ganeshananthan. Thanks to Jack Dovidio, David Amodio, Rhonda Magee, Patricia Devine, Will Cox, Patrick Forscher, Uché Blackstock, Becky Bigler, Robin Ely, David Redish, Kristen Schilt, Yoona Kang, Daniel Kahneman, Federico Ardila, Mekke Okereke, Travis Dixon, Filip Reyntjens, Timothy Longman, and Nilanjana Dasgupta for extra review and clarifications. Evelyn Carter's insight, expertise, and sensitive attention to the science improved every chapter. The heroines at Silver Street Strategies, Kristen Bartoloni, Alex Platkin, and Allison Kelly, checked thousands of facts and somehow came out still standing. Any remaining errors are mine. Special thanks to Kim Todd and Mara Hvistendahl whose ongoing support, criticism, and conversation vastly improved this book.

Thanks to Kenny Joseph, who embarked on a zany and never-ending agent-based model with me, and to Yuhao Du for incredible help and for solving problems that stumped Kenny and me.

I'm thankful for the early support of Dave Grossman and Kate Tomford and for the advice and input of Annalee Newitz, Rebecca Traister, and Adam Grant. Thanks to Ellen Guettler and Ben Pofahl, Susan Pagani, and Siri and Mike Myrholm for providing places to concentrate and write and to Rob White for essential mentoring.

For conversation, friendship, and general wonderfulness, I couldn't be more grateful to Ellen Guettler, Julia Lipman, Sofia Krans, Emily Lichtenheld, Jess Shryack, and Susan Pagani. For his early belief, and for keeping the faith, I thank Amaud Jamaul Johnson. Elif Batuman's counsel and kindness were a necessary balm and buoy. For encouragement and illuminating conversations, I'm grateful to Dahni-El Giles. Many people jumped in to help during the difficulties that come with being a person: Brenda Hartman, Julie Siple, Lucy Lyon, Rachel Hutton, Madhuri Kasat Shors and Luke Shors, Kyrra and Jerome Rankine. For a twenty-year conversation, thanks to Rich Lachman; for beginning one, Jet Lachman. Brian Heller, Eric Plosky, Arjan Schutte, Karin Fong, Rachel Widome, and Gabe Cheifetz have been true and steadfast friends. Special thanks to Madeleine Baran for laughter and companionship during a hard time and for editorial acumen.

Thanks to Vauhini Vara, Paul Bisceglio, and David Bornstein, who commissioned pieces in which some of these ideas were first developed. It was a pleasure to work with such skilled and thoughtful editors.

My agent, Adam Eaglin, has been a brilliant guide through the thicket of publishing, and I treasure his wisdom and friendship. Thanks to everyone at the Elyse Cheney agency, especially Elyse, who responded to a cold email some six years ago. Thanks to Isabel Mendía, Claire Gillespie, and Alex Jacob.

Gratitude to everyone at Metropolitan Books. First, my editor, Riva Hocherman, who championed this book from the very beginning, whose incisive pencil shaped every chapter, and whose dedication to this book and its author are unparalleled. Thank you, Riva: every page is better because of you. Thanks to Sara Bershtel, Brian Lax, Carolyn O'Keefe and Patricia Eisemann, Maggie Richards and Caitlin O'Shaughnessy, and Maia Sacca-Schaeffer. Grigory Tovbis contributed many sharp and essential insights, and Chris O'Connell shepherded countless revisions. Thanks to Chris Sergio for his vision and to Grace Han for her beautiful book design. I am grateful for the astute and generous input of Anne Meadows; working with Rowan Cope and the entire Granta team has been nothing but delightful.

To the countless writers whose words expanded my understanding of what it means to be human: Adrienne Rich, James Baldwin, Audre Lorde, Pumla Gobodo-Madikizela, Stanley Kunitz, George Lamming, and so many others: your spirits guide me. To Rhonda Magee for her wonderful book *The Inner Work of Racial Justice*, and to Suzy Hansen for *Notes on a Foreign Country*, my literary companion on this journey, I offer deep thanks.

To my family: I love you. Thanks to Diane Nordell for teaching fearlessness, to Charles Nordell for teaching kindness, and to Lindsay Nordell for being a jewel in my life and a soul companion. Finally, this book would likely have never been written if not for my husband, Andrew, who generously responded to every point in this book, and whose encouragement, humor, and fortitude make everything possible. His friendship is my rock; his love is my home.

And to you, reader: thank you for the gift of your attention. May we create something new together.

INDEX

Entries in italics refer to illustrations.

ABOUT THE AUTHOR

JESSICA NORDELL is a science and culture journalist whose writing has appeared in the *Atlantic*, the *New York Times*, the *New Republic*, and many other publications. A former writer for public radio and producer for American Public Media, she graduated from Harvard University and the University of Wisconsin, Madison. She lives in Minneapolis, Minnesota. *The End of Bias: A Beginning* is her first book.